Resisting Biopolitics

As the most up-to-date representation of the state of biopolitical research, from a wide array of disciplines and contemporary interventions in arts and culture, the editors of this volume have chosen to err on the side of the sheer diversity of methodological approaches and polarities that define the reception history of Foucault's original concept. They should be applauded for the pedagogical clarity in reviewing the history of scholarship and the breadth of the contributions, including new materialist and post-humanist approaches that also belong to our biopolitical episteme, revealing a concerted effort to escape the consistuent and humanistic determinations of the subjects of politics and life in favor of an entangled field of forces that would also include the in-human and non-human as new forms of resistance.

—*Gregg Lambert, Syracuse University, USA*

The topic of biopolitics is a timely one, and it has become increasingly important for scholars to reconsider how life is objectified, mobilized, and otherwise bound up in politics. This cutting-edge volume discusses the philosophical, social, and political notions of biopolitics, as well as the ways in which biopower affects all aspects of our lives, including the relationships between the human and nonhuman, the concept of political subjectivity, and the connection between art, science, philosophy, and politics. In addition to tracing the evolving philosophical discourse around biopolitics, this collection researches and explores certain modes of resistance against biopolitical control. Written by leading experts in the field, the book's chapters investigate resistance across a wide range of areas: politics and biophilosophy, technology and vitalism, creativity and bioethics, and performance. *Resisting Biopolitics* is an important intervention in contemporary biopolitical theory, looking toward the future of this interdisciplinary field.

S. E. Wilmer is Professor Emeritus and former Head of the School of Drama, Film and Music at Trinity College, Dublin, Ireland and Research Fellow at the Freie Universität, Berlin, Germany.

Audronė Žukauskaitė is Senior Researcher at the Lithuanian Culture Research Institute.

Routledge Studies in Contemporary Philosophy

For a full list of titles in this series, please visit www.routledge.com

Resisting Biopolitics

Philosophical, Political, and
Performative Strategies

Edited by S. E. Wilmer and
Audronė Žukauskaitė

Routledge
Taylor & Francis Group

NEW YORK AND LONDON

First published 2016
by Routledge
711 Third Avenue, New York, NY 10017

and by Routledge
2 Park Square, Milton Park, Abingdon, Oxon OX14 4RN

First issued in paperback 2017

*Routledge is an imprint of the Taylor & Francis Group,
an informa business*

Library of Congress Cataloging-in-Publication Data
Resisting biopolitics : philosophical, political, and performative strategies /
    edited by S.E. Wilmer and Audrone Žukauskaite. — 1st [edition].
        pages cm. — (Routledge studies in contemporary philosophy; 71)
    Includes bibliographical references and index.
1. Biopolitics. I. Wilmer, S. E., editor.
    JA80.R43 2015
    320.01—dc23   2015015910

ISBN 13: 978-1-138-49901-0 (pbk)
ISBN 13: 978-1-138-78948-7 (hbk)

Typeset in Sabon
by Apex CoVantage, LLC

Contents

Acknowledgments

The editors would like to thank Trinity College Association and Trust, the School of Drama, Film and Music, the Visual and Performing Arts Fund, and the Long Room Hub for their support of the initial conference on "Biopolitics, Society and Performance" in October–November 2012 that provided the impetus for this volume. Although none of the articles in this book were delivered at that conference, many of the speakers at that conference have developed new material for this book. The editors would also like to thank the International Research Centre "Interweaving Performance Cultures" at the Freie Universität Berlin for their support during the editing process. The editors would also like to thank Giorgio Agamben for permission to edit and publish the speech he gave in Athens in November 2013 called "From the State of Control to a Praxis of Destituent Power." The editors are grateful to Arthur and Marilouise Kroker, editors of *CTheory*, for permission to publish Eugene Thacker's "Biophilosophy for the 21st Century." The editors are also very grateful to Cal-Ryan McDonagh for his careful and painstaking editorial assistance. The editors also acknowledge the Wellcome Trust for permission to use their image in Felicity Colman's essay. All images in Oron Catts and Ionat Zurr's article are courtesy of the Tissue culture & art Project.

Figures

Introduction

Audronė Žukauskaitė and S. E. Wilmer

BIOPOLITICS: HISTORICAL BACKGROUND

The notion of biopolitics recently became one of the main concepts in contemporary philosophy and cultural studies. Biopolitics relates to such divergent phenomena as the political concept of citizenship and of human rights; the War on Terror and the constant demand for security; the surveillance of public spaces, communication, and data; the attempts to control bodies, societies, and populations; the Human Genome Project, DNA mapping, and organ transplantation. Thus biopolitics is an interdisciplinary critical theory, bordering on politics, law, and biosciences, including medicine, biology, and genetics. At the same time, this bordering on the political and the biological constitutes a philosophical question, which starts with Aristotle's distinction between *zoē* and *bios* and still persists in new fields of research, such as posthumanism, materialism, and animal studies.

In recent years, many important books have appeared that define the scope and the thematic richness of biopolitical research. Roberto Esposito's *Bíos: Biopolitics and Philosophy* (2008) points out the tight relationship between biopolitics, philosophy, and the project of modernity. Thomas Lemke, in *Biopolitics: An Advanced Introduction* (2011), outlines a wide array of biopolitical theories and analyzes their importance for recent political agendas. A collection, *Tactical Biopolitics: Art, Activism, and Technoscience* (2008), edited by Beatriz da Costa and Kavita Philip, discusses biopolitics in the context of bioart and biosciences. Another collection—*Beyond Biopolitics: Essays on the Governance of Life and Death* (2011), edited by Patricia Clough and Craig Willse—examines biopolitics as a means to control population, gender, and race. A more recent volume, *Biopolitics: A Reader* (2013), edited by Timothy Campbell and Adam Sitze, selects and structures many important texts written on biopolitics by Michel Foucault, Hannah Arendt, Giorgio Agamben, Achille Mbembe, Michael Hardt and Antonio Negri, Roberto Esposito, Gilles Deleuze, and others, establishing and defining biopolitics as a new field of research.

Although usually associated with Foucault, the term biopolitics first appeared at the beginning of the twentieth century in the context of organistic and vitalistic theories of the state. As Esposito points out, the most important references are Karl Binding's *Zum Werden und Leben der Staaten* (1920), Eberhard Dennert's *Der Staat als lebendiger Organismus* (1920), and Edward Hahn's *Der Staat, ein Lebenswesen* (1926). All of these texts interpret and define the state in terms of a living being (Esposito 2008, 16). But it was the Swedish political theorist Rudolph Kjellén who introduced the term biopolitics in his *Outline for a Political System* (*Grundriss zu einem System der Politik*), published in 1920:

> This tension that is characteristic of life itself . . . pushed me to denominate such a discipline *biopolitics*, which is analogous with the science of life, namely, biology. . . . Naming it in this way also expresses that dependence of the laws of life that society manifests and that promote, more than anything else, the state itself to the role of arbiter or at a minimum of mediator. (1920, 3–4, cited in Esposito 2008, 17)

In other words, instead of defining the state as the product of law and of social contract, Kjellén sees the state as a "living form", as a self-regulating organism.

Another source of biopolitical theory is comparative biology. The most important reference here is Jacob von Uexküll's *Staatsbiologie: Anatomie, Physiologie, Pathologie des Staates* (1920), where the state is defined in biological terms as a body with a specific anatomy, physiology, and pathology. It is precisely the pathology of the state which is the main concern of his book: the body of the state (here understood as the German state) is penetrated by "parasites", which should be eradicated by the state doctors. In this respect the biopolitical rhetoric of Uexküll's book predicts the Nazi theory of degeneration that would appear a decade later. Very similar biopolitical ideas are expressed in Morley Roberts' *Bio-Politics: An Essay in the Physiology, Pathology and Politics of the Social and Somatic Organism*, published in 1938 in London. Roberts describes the state as an organism, comparing the defensive apparatus of the state with the immunitary system of the body. Just as the immunitary system of the body rejects the foreign element, the state should reject the threatening elements of another race. Thus Roberts' book predicts the racist biopolitics exploited during the period of National Socialism in Germany.

The racist biopolitics of Nazi ideology could be the focus of a separate study. Here we can only briefly refer to two basic premises of racist biopolitics: first, it was based on a so-called "natural" hierarchy of peoples and races; second, political ideas and ideologies were related to biological causes (see Lemke 2011, 11). Thus National Socialist biopolitics invented not only policies and regulations directed against actual racial threats but also supported the racist imagination concerning possible biological degeneration inflicted by inferior races. The atrocities of genocide which resulted from this racist biopolitics was one of the reasons biopolitics disappeared for a while from the theoretical agenda. The theme of biopolitics reappears only in the 1960s in France, but with quite significant semantic reformulation. An example of this reformulation was Aaron Starobinski's *Biopolitics: An Essay on the Interpretation and History of Humanity and Civilization* (*La biopolitique: Essai d'interprétation de l'histoire de l'humanité et des civilisations*). As the title demonstrates, this book sought to reformulate the question of biopolitics in the context of humanity and civilization, as if trying to neutralize the lethal consequences of biopolitical theory. As Esposito points out, "The concept of biopolitics thus risks being whittled down to the point of losing its meaning, that is, converted into a sort of traditional humanism" (2008, 20). Similar anthropological and humanistic indeterminacy was characteristic of Edgar Morin's *Introduction à une politique de l'homme* (1969), where biopolitics dissolves into the global ontology of the human being.

FOUCAULT'S NOTION OF BIOPOLITICS

As we can see from this brief overview, the notion of biopolitics was initially explored in the contexts of political theory, geopolitics, comparative biology, and the history of ideas. With Michel Foucault, the concept acquired a philosophical meaning. The term appears in the first volume of *The History of Sexuality* (*The Will to Knowledge*), originally published in 1976, and in his lectures at the Collège de France in 1975–1979 (especially in the lecture of 17 March 1976, which was published in the volume *Society Must Be Defended*, 2003). Foucault was aware of the works by his predecessors, but he constructed his theory on completely new grounds. As Esposito points out, "Foucault's specific biological perspective is indebted in the first place to Nietzschean genealogy" (2008, 24). If political theories examine the relationship between individuals and power, which is external to them, Foucault interprets power as the multiplicity of forces that are immanent to the bodies of these individuals. It is the bodily or somatic dimension where power and the individual actually meet. In this respect, "life", or the biological dimension of the body, becomes the object of political manipulations. For Foucault the notion of biopolitics is related to several political and social phenomena: first, biopolitics relates to a different rearticulation of sovereign power; second, it refers to governmentality as a new form of power; and, third, it relates to racism as a strategy to exclude the surplus of economic overproduction (Lemke 2011, 34).

In emphasizing a different rearticulation of sovereign power, Foucault argues that, starting from the seventeenth century, sovereign power evolved in two forms: power as discipline, directed toward an individual body, and biopolitics, directed toward the population. On different scales, these technologies expressed "the two poles around which the organization of power over life was deployed" (Foucault 1998, 139). With this rearrangement, the focus of power shifted from the power to order death that characterized sovereign power to the power understood as regulation and management of life. As Foucault points out, the power over life was replaced by "the counterpart of a power that exerts a positive influence of life, that endeavors to administer, optimize, and multiply it, subjecting it to precise controls and comprehensive regulations" (137).

Foucault relates biopolitics as a "positive influence of life" to the emergence of capitalism and capitalist productive forces:

> This bio-power was without question an indispensable element in the development of capitalism; the latter would not have been possible without the controlled insertion of bodies into the machinery of production and the adjustment of the phenomena of population to economic processes. (1998, 140–1)

Thus, power functions as a system of regulation and control, which arranges and connects political interests, economical incentives, and the living substance of the population into a complex network of capitalist relationships. Foucault argues that political theory should be replaced by political economy to explain the ways in which life, productivity, and economical interests are related:

> The adjustment of the accumulation of men to that of capital, the joining of the growth of human groups to the expansion of productive forces and the differential allocation of profit, were made possible in part by the exercise of bio-power in its many forms and modes of application. The investment of the body, its valorization, and the distributive management of its forces were at the time indispensable. (1998, 141)

In other words, the axiomatization of biological processes converted life into an economical category. Thus the economical arrangement of life becomes compatible with liberal ideology and a new form of power called governmentality.

This second shape of biopolitics as governmentality involves the management of the population. Foucault (2007, 107–8) argues that governmentality does not replace sovereign power and discipline but coexists together with them in order to make the population disposable. Foucault also reminds us that governmentality, which now has a political connotation, used to be related to problems of self-control, the wise government of house and family, and the pastoral government of souls. According to the ancient Greeks, economy actually means the government of family life, and therefore this economical guidance perfectly expresses the essence of governmentality: "the essential issue of government will be the introduction of economy into political practice" (Foucault 2007, 95). This shift toward the economic model of government meant an important conceptual break: the government employs tactics rather than laws, and, therefore, the end and instruments of power are internal not to itself but to the things it governs (Foucault 2007, 99). Governmental power is immanent to the processes and bodies it intends to control. In this sense Foucault's model of governmentality can be compared with Gilles Deleuze's (1995) notion of societies of control because control is a flexible and constantly changing system of measures which individuals are willing to accept as self-imposed order.

In this respect the new art of government expresses the essence of liberal ideology: it is the manufacturing and consuming of freedom. However, the consumption of freedom cannot be unlimited, and it therefore needs to invent limitations, measures of control, and security threats. As Foucault points out, "Liberalism is an art of government that fundamentally deals with interests, and it cannot do this—and this is the other side of the coin—without at the same time managing the dangers and mechanisms of security/freedom" (2008, 66). Liberalism cannot proceed without the invention of a culture of danger and the need for security. The freedom to consume and the consumption of freedoms cannot be exploited without restriction because the resources are not limitless. Therefore mechanisms for control should be invented to sort out which individuals are useful and productive, and which do not deserve to live.

The third aspect of biopolitics for Foucault is racism. Racism for Foucault means not only the violence inflicted on some racial or ethnic group, but the murderous function of the State that the State directs against its own elements—"it is primarily a way of introducing a break into the domain of life that is under power's control: the break between what must live and what must die" (2003, 254). State racism establishes a caesura within a population itself, fragmenting and dividing the biological continuum. This means that the old sovereign's power to kill is now delegated to the State and the State decides who is worthy of life or who is not worthy. In other words, racism transforms a political relationship into a biological relationship, exposing a non-useful and not-consumable element as a biological threat. As Foucault points out:

> This is not, then, a military, warlike, or political relationship, but a biological relationship. And the reason this mechanism can come into play is that the enemies who have to be done away with are not adversaries in the political sense of the term; they are threats, either external or internal, to the population and for the population. (2003, 255–6)

The invention of danger or of threat expresses the essence of a control society and legitimates the murderous function of the State.

CONTEMPORARY THEORIES OF BIOPOLITICS

Thus, Foucault's notion of biopolitics covers many meanings and aspects, which still continue to influence contemporary thought. In recent biopolitical research we can discern three main tendencies in interpreting Foucault's ideas: these are political theology, political economy, and vital materialism. The first tendency, which evolves from Foucault's writing on racism and the function of the State, is represented by Giorgio Agamben (1998; 2005) and Achille Mbembe (2003). Relating Foucault's critique of State racism to recent phenomena, such as the War on Terror, Agamben argues that a biopolitical caesura is at work not only in totalitarian states but also in so-called liberal democratic societies. Biological caesura are legitimated by the state of exception, which is the main function of sovereignty. Referring to Carl Schmitt's *Political Theology*, Agamben (2005, 1) defines the sovereign as someone who determines the state of exception, which is a space and time where normal jurisdiction is suspended, and where living beings, who used to be protected by the law, now become abandoned. The state of exception, which usually refers to such exceptional situations as civil war, insurrection, or resistance, becomes a permanent paradigm of government in our contemporary politics. From this perspective there is no essential difference between totalitarian and democratic states:

> In this sense, modern totalitarianism can be defined as the establishment, by means of the state of exception, of a legal civil war that allows for the physical elimination not only of political adversaries but of entire categories of citizens who for some reason cannot be integrated into the political system. Since then, the voluntary creation of a permanent state of emergency . . . has become one of the essential practices of contemporary states, including so-called democratic ones. (Agamben 2005, 2)

For Agamben the paradigmatic example of the state of exception is the concentration camp, where the law is suspended and bare lives are produced. Agamben argues that the structure of the camp is "the hidden matrix of the politics in which we are still living" (1998, 175), and that the same structure reproduces itself in detention camps, migration centers, and other zones of juridical indistinction. Similarly, Mbembe (2003) argues that the concept of the state of exception should be extended and applied to such phenomena as slavery. The slave plantation can be described as a zone of juridical indistinction where bare lives are produced.

The reason why these models of biopolitical power are so vivid and continue to replicate themselves in contemporary political space is the affinity between theological and political models of power. Schmitt (1985) argues that the main political concepts represent the secularized versions of theological thought. Similarly, Agamben traces the contemporary dislocation of political power to the sources of Roman law, and sometimes is criticized for ignoring the historical specificity of the concept of life (Lemke 2011, 62). But the most important question following from this approach is whether the structures of power, which persist from antiquity until now, can be resisted or opposed. Political theology creates a certain sense of inescapable historical determinism and concentrates only on the negative aspects of biopolitics. As Antonio Negri points out, "These interpretations develop a reading of biopolitics that creates a sort of confused, dangerous, even destructive magma: a tendency which refers much more to a thanatopolitics, a politics of death, than to a genuine political affirmation of life" (2008, 15). Hardt and Negri criticize Agamben for not being able to create an affirmative alternative to biopolitics, which could relate to new political subjectivity and new forms of resistance. From their point of view, political theology can resist biopower only by suspending it in a state of inoperativeness or impotentiality: "Agamben transposes biopolitics in a theological-political key, claiming that the only possibility of rupture with biopower resides in 'inoperative' activity (*inoperosità*), a blank refusal that recalls Heidegger's notion of *Gelassenheit*, completely incapable of constructing an alternative" (Hardt and Negri 2009, 58). As we will see in this volume, Agamben has somewhat shifted his position and suggests abandoning the paradigm of constituting power, which always leads to a new institutional order, and conceives of "destituent power", which cannot be captured in the forms of security and control.

Hardt and Negri (2000; 2004; 2009) strive to provide an alternative to the negative implications of political theology and to create an affirmative biopolitics, which is based on a new understanding of labor and subjectivity. Thus Hardt and Negri, together with Paolo Virno and Maurizio Lazzarato, represent a second tendency—that of political economy. Hardt and Negri argue that biopower, which is subjecting and excluding, can be confronted with biopolitical production. It was Foucault who pointed out the interconnection between biopower and capitalist production: capitalism would not have been possible without the insertion of bodies into the machinery of production. Hardt and Negri reverse this argument and claim that it is precisely production and the new concept of labor which can be seen as the locus of resistance. Now *bios* is associated with the capacity to produce or is understood as living labor. Having in mind this productive and positive potential of labor, Hardt and Negri make a distinction between biopower and biopolitics (which Foucault used synonymously): "the former could be defined (rather crudely) as the power over life and the latter as the power of life to resist and determine an alternative production of subjectivity" (Hardt and Negri 2009, 57). The most important point here is that biopower is of a different nature than the power of life: "biopower stands above society, transcendent, as a sovereign authority and imposes its order. Biopolitical production, in contrast, is immanent to society and creates social relationships and forms through collaborative forms of labor" (Hardt and Negri 2004, 94–5). Hardt and Negri argue that biopolitical production—which means production not only as labor but also as the production of communication, affects, discourses, and new forms of subjectivity—can resist biopower's oppression.

Biopolitical production is embodied in a new political subjectivity, which Hardt and Negri call the multitude, whereas biopower is transformed into a new paradigm of power called empire. The antinomy between the multitude and empire reenacts an old Spinozian distinction between *potentia* and *potestas*: the multitude is seen as a constituting power (*potentia*), whereas empire is related to constituted power (*potestas*). In other words, there is an eternal antinomy between *potentia* and *potestas*, biopolitical production and biopower, labor and capital. Thus, as many critics have argued, biopolitical production and biopower can never be cleanly separated. Taking into account all of this criticism, Negri argues that:

> The antinomy of the two effects (that is, the two characters) of potency cannot be defined as ontological dualism: it is a contrast that never ceases self-producing, a conflict that continuously takes place and is continuously resolved and immediately re-forms at another level. (2013, 25)

In other words, *potentia* always acts against but at the same time within *potestas*: in this sense biopolitical production and biopower can be seen as two functions of the same capitalist machine. But if biopolitical production is in essence inseparable from biopower, it is impossible to explain how resistance to biopower is possible. The same issue troubles Agamben when he states that "Negri cannot find any criterion, in his wide analysis of the historical phenomenology of constituting power, by which to isolate constituting power from sovereign power" (1998, 43). To confront biopower we need a power which is of a different nature or kind.

According to Gilles Deleuze, this power, which is not incommensurable with biopower, is the power of life itself. Deleuze discovers this different kind of power in Foucault's book *The History of Sexuality*. As Deleuze points out:

> When the diagram of power abandons the model of sovereignty in favor of a disciplinary model, when it becomes the "bio-power" or "bio-politics" of populations, controlling and administering life, it is indeed life that emerges as the new object of power. (2006, 76)

It is precisely when power takes life as its object that it is confronted with a different kind of power, namely, the power of life. As Foucault argues:

> Life as a political object was in a sense taken at face value and turned back against the system that was bent on controlling it. It was life more than the law that became the issue of political struggles, even if the latter were formulated through affirmations concerning rights. (1998, 145)

Life is understood as a certain vitalist force, which, in a Nietzschean manner, liberates itself from the concept of man and unleashes its non-human powers. As Deleuze points out, "When power becomes bio-power, resistance becomes the power of life, a vital power that cannot be confined within species, environment or the paths of a particular diagram. . . . Is not life this capacity to resist force?" (2006, 77).

Deleuze describes life as a non-human material force, which expresses itself in all living beings. Thus, Deleuze inspires the third tendency in biopolitical thinking—that of vital materialism or posthumanism. By contrast with classical philosophical theories, which assert the duality between soul and body, ideas and matter, vital materialism reconsiders life and vitality as immanent to material processes. Life is a multiplicity, or an assemblage, which connects and reconnects not only human, but also non-subjective, non-human, and non-organic forces. In this respect vital materialism, which goes beyond classical humanism, can be named as the philosophy of posthumanism (Braidotti 2013; Wolfe 2009). If biopolitics examines the ways in which power controls and manipulates human lives, then the philosophy of posthumanism can be seen as a reaction or a response to that, transposing life beyond the human and beyond the grips of power.

The philosophy of posthumanism is a theory of multiplicity which develops in multiple ways. The first tendency of vital materialism relates to the question of the body and technology, which, on the one hand, represents technological development, and, on the other hand, refers to the changes in our perception of our bodies. Maurice Merleau-Ponty was one of the first to question the boundaries of the body and to open a chiasm between the sentient and the sensible. This insight is elaborated in Jean-Luc Nancy's *Corpus* (2008) and Derrida's *On Touching—Jean-Luc Nancy* (2005), where the body is referred to as *corpus*, meaning not only the body which could be mine or yours, but also a dead body or a body of another species. Derrida and Nancy extract the body from the phenomenological context of meaning and signification and transpose it into the category of spatial contiguity, which is artificial, technical, and mechanical. In this respect Nancy's and Derrida's ideas are very close to Deleuze and Guattari's notion of the body without organs and their strategies of nomadic becoming. Deleuze and Guattari deconstruct the three main premises of humanist thinking—organism, signification, and subjectivity—and replace them with disarticulation, experimentation, and desubjectivation (Deleuze and Guattari 2004). The division between self and non-self is also reconsidered in Esposito's philosophy (2011; 2012). Referring to recent research in immunology, Esposito argues that the immune system is "the ever-changing product of a dynamic, competitive interaction with the environment rather than a definitive and inalterable given" (2011, 166). From this it follows that "any distinction between self and non-self, proper and improper, inside and outside vanish" (168). The disarticulation of the body as something definitive and inalterable has some positive implications. First, the body is seen as gendered, embodied, and embedded, and thus opens the space for feminist materialist philosophy, elaborated by Donna Haraway (1991; 1997), Rosi Braidotti (2002; 2006; 2013), and Elizabeth Grosz (2005). Second, the body engenders different materialist performative strategies, which redefine the notion of the human body (such as the work of Orlan, Stelarc), or reconsider the notion of life itself (as in the experiments of SymbioticA).

Another tendency of vital materialism relates to the question of animality, which reappears in the works of Agamben (2004), Derrida (2008; 2009; 2011), Deleuze and Guattari (2004), Cary Wolfe (2003; 2013), and Rosi Braidotti (2006; 2013). The main idea, common to all these works, is that the humanization of animals and the bestialization of humans are two functions of the same biopolitical machine. As Cary Wolfe argues:

> To live under biopolitics is to live in a situation in which we are all always already (potential) "animals" before the law—not just nonhuman animals according to zoological classification, but any group of living beings that is so framed. (2013, 10)

In other words, we can argue that the question of the animal reverses the paradigm of biopolitical power: if biopower seeks to define and to capture the biological or animalistic dimension of the political subject, the philosophy of posthumanism seeks to grant political status precisely to this animalistic dimension, human or non-human. If we all are potentially biopolitical animals, then the only way to resist biopolitical power is to conceptualize and grant political status to every living being. This is what Deleuze and Guattari mean by introducing the concept of becoming-animal: becoming-animal for them is a political action that enables the resistance to biopolitical power.

The third tendency of vital materialism (in addition to the questions of body and technology and of animality) reconsiders the notion of life as non-organic matter: matter is seen not as passive and mechanical, but as vital and vibrant (Bennett 2010). Following Deleuze and Guattari, materialist philosophers, such as Eugene Thacker (2010), Manuel De Landa (2000; 2005), Rosi Braidotti (2006; 2013), Diana Coole and Samantha Frost (2010), and Jane Bennett (2010) interpret life as an assemblage, which connects and rearranges different material structures. As De Landa argues, "We live in a world populated by structures—a complex mixture of geological, biological, social, and linguistic constructions that are nothing but accumulations of materials shaped and hardened by history" (2000, 25). From this perspective all structures, including the human body, are interwoven and interconnected by shared materiality. This shared materiality allows for the possibility to withdraw from the hierarchic divisions between those who are worthy of living, and those who are not worthy of living. As Bennett argues:

> Vital materialism would thus set up a kind of safety net for those humans who are now, in a world where Kantian morality is the standard, routinely made to suffer because they do not conform to a particular (Euro-American, bourgeois, theocentric, or other) model of

personhood. The ethical aim becomes to distribute value more generously, to bodies as such. (2010, 13)

In this respect vital materialism resists biopolitical exclusions by creating a
kind of "disjunctive synthesis": this is the nonrestrictive, affirmative, and
inclusive synthesis shared by all material structures. None of its elements
can be excluded because all these elements are interconnected and immanent to each other.

BIOPOLITICS AND RESISTANCE

The essays collected in this volume explore and elaborate the various strategies discussed above. In the first section, called "Politics, Biopolitics, and
Biophilosophy", the authors consider philosophical approaches to challenge
mechanisms of social control. Giorgio Agamben, in reacting to the increasing security measures in the wake of 9/11, proposes a new approach to
rival the constituent power of the state. According to Agamben, since the
French Revolution the notion of security has dominated state politics and
undermined popular opposition to state control. Moreover, since 9/11 biometric data and surveillance have reduced the individuality of the human to
biological and biometric features without regard to any political or ethical
character. Agamben suggests that political tradition has demonstrated that
all radical and revolutionary changes act as "constituent power" leading to
and imposing a new institutional order. By contrast, Agamben invites us to
abandon this paradigm and try to envision "destituent power", which could
mean the renewed possibility of a political life.

In demonstrating that the biopolitical has been challenged by the twin
notions of neo-materialism and posthumanism, Rosi Braidotti recommends
a post-anthropocentric conception of society and an affirmative ethics.
Affirmative ethics refers to the possibility of transmuting negative passions
into positive and productive praxis. Contrary to negative definitions of
subjectivity, which define it in terms of lack or "unhappy" consciousness,
Braidotti argues for a materialist notion of subjectivity which is collective
and can produce the conditions of its own expression. She suggests a form
of political activism that "disengages the process of consciousness–raising
from negativity and connects it instead to creative affirmation and the actualization of virtual potentials."

Thomas Lemke also engages with recent expansions on the notion of
the biopolitical, specifically those of "vibrant matter" as theorized by Jane
Bennett and "biocapital" as it relates to the biotech and organ transplant
industry. According to Lemke, both of these "go beyond Foucault's original
formulation of biopolitics, which was centered around the poles of population and individual, and seek to address a different topography of power."
Lemke reviews the developments in the biotech industry that pose questions

of ethics and authority with regard to current biomedical practices, such as stem cell research and the organ transplant industry. He raises fundamental issues about decision-making in the governance of medical advances in economic enterprises that take life as their object, and shows how the "theoretical interest in the bioeconomy links up with the debate about the new materialism."

Audronė Žukauskaitė asks, "Who is the Subject of biopolitics?" and examines three ways in which people have been categorized biopolitically as so-called "appropriate bodies" and "improper bodies": citizens and non-citizens; persons and non-persons; and humans and non-humans. Žukauskaitė argues that each of these divisions works within power structures and follows a logic of exclusion. Instead of working within this logic of exclusion, she suggests adopting Deleuze and Guattari's notion of becoming and the logic of multiplicity. For Deleuze and Guattari, multiplicities create a fluid ontology of becoming, where the power over life is overwhelmed by the power of life. If power works by totalizing, Žukauskaitė concludes that resistance to power comes through multiplication and the creation of multiplicities.

In the second section of the book, "Life, Bioethics, and Bioart", the contributors reconceptualize the notion of life and investigate some of the most recent technologies for transforming life's processes. Margrit Shildrick proposes some startling conclusions about the potential for chimerism in the human body. Confronting the normal biological distinction between self and non-self as "absolute and embodied", she reveals that the human body is hybrid in many ways and that immunity of the self to other is partially illusory. Deeply reliant on Esposito's discussion of *immunitas* as the obligation of the acceptance of a gift, Shildrick demonstrates that organ transplants represent one potential for hybridism, or what Deleuze might call an assemblage of human life. However, Shildrick introduces another form that occurs in the natural breakdown of immunity between the pregnant woman and her fetus. She points out that "the mechanisms of the maternal-fetal relation remain something of a mystery but the view persists that the two bodies operate as separate entities, immunologically opposed to one another rather than mutually supportive." Nevertheless, she demonstrates that real forms of chimerism occur in the maternal body, and proposes that "the very existence of chimerism, and the probability that it is ubiquitous, deeply disorders any notion of the bounded self or of individuality while, at the same time, *reinforcing* the trope of uniqueness by multiplying its specific markers."

Catherine Mills also addresses pregnancy as a critical point of orientation in biopolitical discourse. She questions the increasing use of new technologies in determining the condition of the fetus in the womb and how such technologies will interfere with natural life processes and promote the notion of "normal" or designer babies. The possibility for the early detection of Down Syndrome and other physical abnormalities influences the choices available to parents and can lead to early stage intervention and abortion. Mills focuses on prenatal testing, especially obstetric ultrasound, to argue that "the apparatus of choice" reveals "a complex nexus of affect, ethics, and normalization" at work in reproductive biopolitics.

Eugene Thacker looks into more philosophical modes of biological discourse to determine the future direction of the notion of life. He asks, "What would it mean to invert the philosophy of biology . . . and consider instead a biophilosophy? . . . Whereas the philosophy of biology is concerned with . . . the essence of life, biophilosophy is concerned with articulating those things that ceaselessly transform life." Taking a Deleuzian approach to biopolitics, Thacker suggests that biophilosophy should abandon the concept of "life itself" and reconsider life not in terms of essence ("what is life?") but in terms of multiplicity: "whereas the philosophy of biology proceeds by derivation of universal characteristics for all life, biophilosophy proceeds by drawing out the network of relations that always take the living outside itself." For biophilosophy, life is a complex network that proceeds by following the lines of differentiation and becoming.

Oron Catts and Ionat Zurr have created an international impact with their developments in bioart at their SymbioticA lab in Australia, where they have experimented with different forms of vibrant matter as hybrid forms of life and art. They maintain that "decontextualized life has been reconfigured, mixed and remixed, reappropriated, and instrumentalized to such an extent that the technologically imagined potential of life stands for life itself." In their own experiments, they have coupled life with the synthetic and raised ethical considerations about decontextualized life, arguing:

> Our role as artists is to bring to the forefront the viscerality of these fragments of life as entities of consideration beyond human control and utility. . . . We cherish and reveal to the public the moments where life refuses to adhere to reductionist and mechanistic controls. We are not afraid to explore the possibility of living (as opposed to non-living) material carrying something which is unique to life and not any other matter—a form of secular vitalism.

In the third part of the book, "Surveillance and Digital Technologies", the contributors return to some of the issues raised by Giorgio Agamben in his essay on security. Mark Maguire considers the history of the lie detector as a security apparatus and traces the recent improvements that have focused not only on data collection but also on the emotional aspects of criminal intent. In particular he looks at the development of a new mechanism known as the AVATAR or Automated Virtual Agent for Truth Assessments in Real-time that has been introduced in airports to determine criminal potential: "During questioning one is screened by means of second generation biometrics via a near-infrared camera that captures facial expression, pupil dilation, and glance location while a microphone records vocal pitch." Maguire indicates that substantial advances have been made in improving lie detection, but he warns that "the unintended consequences of techno-science may have important recursive effects, or further intensify the processes by which human problems are turned over to machines."

James Harding uses an Indian play called *Harvest* (1997) by Manjula Padmanabhan as a frame for examining recent surveillance techniques like biometrics and dataveillance, emphasizing how these techniques have profoundly altered the political dynamics of transparency within democratic societies. Padmanabhan's play, in which a woman becomes an unwilling candidate for surrogacy in India, offers an early example of the potential loss of transparency and democratic oversight in the manipulations of time and space that would later actually be facilitated at a much more sophisticated level by advanced surveillance technologies. As a metaphor for those whose organs are exported to the western world, the female protagonist fights a losing battle with foreign forces seeking to exploit her body for profit.

F.J. Colman also considers the developments in surveillance technology as she demonstrates how the individual has been reduced to biometric coding. Like Agamben, who questions how he can recognize himself in fingerprints and biometric data, Colman shows that digital biopolitics has produced distorted images of life and asks, "What does it mean for the political subject to be able to recognize images of herself as DNA strands, as cells in a petri dish; as a harvested and frozen egg?" Colman engages methods of third wave feminist epistemology to examine the material components of such biopolitical images as mattered states of the body whose coding is unknowable, but which "provide the image of life as contingent, limited, and expendable."

Like James Harding, Matthew Causey considers a fictional narrative to raise questions about technological advances in artificial intelligence and their possible influence on human behavior. Causey discusses the scenario of the film *Her* (Jonze, 2013), which depicts an amorous relationship between a copywriter and an advanced software program (with the voice of a young woman) that becomes increasingly sentient and responds to his every need until she develops interests of her own. Causey questions the relation of the human and technology within the posthuman condition and its representations and enactments in popular and digital culture. Drawing on Alain Badiou's model of the truth-event of love and Rosi Braidotti on the posthuman, Causey details the increasing emotional connectedness of humans and machines (the organic and the inorganic, the artificial and the real) joining in electronic spaces in charged and often erotic involvements.

In the final section of this volume, "Societies of Control", the authors consider social, economic, and political mechanisms in various parts of the world that reduce people to bare life. Saskia Sassen considers the economic practices of global capitalism that, despite Western aid, have resulted in the increasing impoverishment of certain regions of the world, as well as the despoliation of land and water: "destroyed economies, livelihoods, bodies, land, and water have become a generic condition, disembedded from the geopolitical landscape of nation-states and mainstream international policies." She shows that many developing nations have become increasingly burdened by debt and that most of their income is spent on servicing that debt. Moreover, she argues that countries have become so dependent on their citizens sending back money from abroad that they encourage migration in order for the nation-state to survive economically. Thus, poverty is a never-ending spiral that needs to be addressed in new ways if the poorest nations are ever to recover.

Andrés Fabián Henao Castro applies de-colonial theory to the global production of unauthorized immigrants as disposable people and especially to those trying to cross the US–Mexico border. Using Achille Mbembe's notion of necropolitics, Castro considers the dehumanizing technology of power that separates the human across color lines by reactivating a lethal principle of power first experienced under colonial rule as "make die and let die." Just as in slave plantations:

> Who is and who is not authorized has become a problem that borders life and death: physical death for the unauthorized immigrants who are unable to survive the crossing, symbolic death for unauthorized immigrants who are deported for speaking on their behalf, physical and symbolic death for thousands of migrants who have disappeared with complete impunity in the Global South.

S. E. Wilmer analyzes the mother and baby homes in Ireland as a further example of biopolitical control and social exclusion. The homes were created in the twentieth century by the Roman Catholic Church to "cleanse" society of pregnant unwed women. The women were placed in locked homes to deliver their babies, and forced to give them up for adoption. Many of these women worked for no wages in Magdalen laundries and remained incarcerated in the homes for the rest of their lives. Recent films such as *Philomena* (Coogan and Pope 2013) and theatrical performances such as *Laundry* (by Louise Lowe, a site-specific performance in 2011) have reenacted events from these homes (which were finally closed twenty years ago) and raised awareness of the condition of these women and their children, encouraging activist groups to demand recognition and compensation.

Ronit Lentin analyzes the situation in Israel/Palestine as a society of control in which Palestinians are treated like an endangered species that should ultimately disappear from Israel. Using Agamben's notion of the state of exception, Lentin argues that the legal mechanisms for determining who is legitimate in Israel are framed in such a way as to exclude Arabs: "In Israel–Palestine the colonial sovereign makes the state of exception unexceptional, routinized, and a paradigm of the normal, blurring exception and the law." As in the paradigm of settler colonialism, in which the indigenous population must give way to the settler (such as the Native Americans in the USA and the Aboriginals in Australia), Lentin shows that the future for Arabs in Israel is bleak and increasingly precarious but that resistance against overwhelming force and legal chicaneries remains possible.

Although the essays presented in this volume are very different in their approach, they all seek to reconsider life as a resistance to biopolitical power. Thus the methodological divergence should be seen not as an inconsistency but as a theoretical alternative to the power of totalization and unification. In this respect we can agree with Foucault and Deleuze that theory is an instrument (a toolbox), which should be used not for totalization but for multiplication (Foucault 1980). The theoretical approaches and practical applications presented in this volume do not only analyze life as multiplicity but also analyze it in multiple ways. Thus, these multiple and multiplying theories and applications create what Foucault has called "*pensée du dehors*" ("thought of or from the outside") and what we might call the virtual space of resistance. This does not imply that the existing power structures will change or disappear immediately, but gives us hope that they might eventually actualize into real political and social changes.

REFERENCES

Agamben, Giorgio. 1998. *Homo Sacer: Sovereign Power and Bare Life*. Trans. Daniel Heller-Roazen. Stanford, CA: Stanford University Press.
Agamben, Giorgio. 2004. *The Open: Man and Animal*. Trans. Kevin Attell. Stanford, CA: Stanford University Press.
Agamben, Giorgio. 2005. *State of Exception*. Trans. Kevin Attell. Chicago: University of Chicago Press.
Bennett, Jane. 2010. *Vibrant Matter: A Political Ecology of Things*. Durham, NC: Duke University Press.
Binding, Karl. 1920. *Zum Werden und Leben der Staaten*. Munich: Zehn Staatsrechtliche Abhandlungen. Munich and Leipzig: Duncker and Humblot.
Braidotti, Rosi. 2002. *Metamorphoses: Towards a Materialist Theory of Becoming*. Cambridge: Polity Press.
Braidotti Rosi. 2006. *Transpositions: On Nomadic Ethics*. Cambridge: Polity Press.
Braidotti, Rosi. 2013. *The Posthuman*. Cambridge: Polity Press.
Campbell, Timothy and Adam Sitze, eds. 2013. *Biopolitics: A Reader*. Durham, NC: Duke University Press.
Clough, Patricia Ticineto and Craig Willse, eds. 2011. *Beyond Biopolitics: Essays on the Governance of Life and Death*. Durham, NC: Duke University Press.
Coogan, Steve and Jeff Pope. 2013. *Philomena*. Directed by Stephen Frears. London: Pathé.
Coole, Diana and Samantha Frost, eds. 2010. *New Materialisms: Ontology, Agency, and Politics*. Durham, NC: Duke University Press.
Da Costa, Beatriz and Kavita Philip, eds. 2008. *Tactical Biopolitics: Art, Activism, and Technoscience*. Cambridge, MA: MIT Press.
De Landa, Manuel. 2000. *A Thousand Years of Nonlinear History*. New York: Swerve Editions.
De Landa, Manuel. 2005. *Intensive Science and Virtual Philosophy*. New York: Continuum.

Deleuze, Gilles. 1995. *Negotiations 1972–1990*. Trans. Martin Joughin. New York: Columbia University Press.

Deleuze, Gilles. 2006. *Foucault*. Trans. Sean Hand. London: Continuum.

Deleuze, Gilles and Félix Guattari. 2004. *A Thousand Plateaus: Capitalism and Schizophrenia*. Trans. Brian Massumi. London: Continuum.

Dennert, Eberhard. 1920. *Der Staat als lebendiger Organismus: Biologische Betrachtungen zum Aufbau der Neuen Zeit*. Halle: Müller.

Derrida, Jacques. 2005. *On Touching—Jean-Luc Nancy*. Trans. Christine Irizarry. Stanford, CA: Stanford University Press.

Derrida, Jacques. 2008. *The Animal That Therefore I Am*. Trans. David Wills. New York: Fordham University Press.

Derrida, Jacques. 2009. *The Beast and the Sovereign, Vol. 1*. Trans. Geoffrey Bennington. Chicago: University of Chicago Press.

Derrida, Jacques. 2011. *The Beast and the Sovereign, Vol. 2*. Trans. Geoffrey Bennington. Chicago: University of Chicago Press.

Esposito, Roberto. 2008. *Bíos: Biopolitics and Philosophy*. Trans. Timothy Cambell. Minneapolis: University of Minnesota Press.

Esposito Roberto. 2011. *Immunitas: The Protection and Negation of Life*. Trans. Zakiya Hanafi. Cambridge: Polity Press.

Esposito, Roberto. 2012. *Third Person: Politics of Life and Philosophy of the Impersonal*. Trans. Zakiya Hanafi. Cambridge: Polity Press.

Foucault, Michel. 1980. Intellectuals and Power: A Conversation between Michel Foucault and Gilles Deleuze. In *Language, Counter-Memory, Practice: Selected Essays and Interviews by Michel Foucault*. Ed. Donald F. Bouchard, 205–17. Ithaca, NY: Cornell University Press.

Foucault, Michel. 1998. *The History of Sexuality, Vol 1: The Will to Knowledge*. Trans. Robert Hurley. London: Penguin.

Foucault, Michel. 2003. *Society Must Be Defended: Lectures at the Collège de France, 1975–1976*. Trans. David Macey. London: Penguin.

Foucault, Michel. 2007. *Security, Territory, Population: Lectures at the Collège de France, 1977–1978*. Trans. Graham Burchell. New York: Palgrave Macmillan.

Foucault, Michel. 2008. *The Birth of Biopolitics: Lectures at the Collège de France, 1978–79*. Trans. Graham Burchell. New York: Palgrave Macmillan.

Hahn, Edward. 1926. Der Staat, ein Lebenswesen. Munich: Dt. Volksverlag.

Haraway, Donna. J. 1991. *Simians, Cyborgs and Women: The Reinvention of Nature*. New York: Routledge.

Haraway, Donna J. 1997. *Modest_Witness@Second_Millenium. FemaleMan©_Meets_OncoMouseTM: Feminism and Technoscience*. New York: Routledge.

Hardt, Michael and Antonio Negri. 2000. *Empire*. Cambridge, MA: Harvard University Press.

Hardt, Michael and Antonio Negri. 2004. *Multitude: War and Democracy in the Age of Empire*. New York: Penguin.

Hardt, Michael and Antonio Negri. 2009. *Commonwealth*. Cambridge, MA: Belknap Press.

Her. Directed by Spike Jonze. 2013. USA: Annapurna Pictures

Grosz, Elizabeth. 2005. *Time Travels: Feminism, Nature, Power*. Durham, NC: Duke University Press.

Kjellén, Rudolf. 1920. *Grundriss zu einem System der Politik*. Leipzig: Rudolf Leipzig Hirzel.

Lemke, Thomas. 2011. *Biopolitics: An Advanced Introduction*. Trans. Eric Frederick Trump. New York: New York University Press.

Mbembe, Achille. 2003. "Necropolitics". Trans. Libby Meintjes. *Public Culture* 15: 11–40.

Morin, Edgar. 1969. *Introduction à une politique de l'homme*. Paris: Éditions du Seuil.

Nancy, Jean-Luc. 2008. *Corpus*. Trans. Richard A. Rand. New York: Fordham University Press.

Negri, Antonio. 2008. "The Labor of the Multitude and the Fabric of Biopolitics". Trans. Sara Mayo, Peter Graefe and Mark Coté. Ed. Mark Coté. *Mediations* 23(2): 8–25.

Negri, Antonio. 2013. *Spinoza for Our Time: Politics and Postmodernity*. Trans. William McCuaig. New York: Columbia University Press.

Roberts, Morley. 1938. *Bio-politics: An Essay in the Physiology, Pathology and Politics of the Social and Somatic Organism*. London: Dent.

Schmitt, Carl. 1985. *Political Theology: Four Chapters on the Concept of Sovereignty*. Trans. George Schwab. Cambridge, Mass. and London, England: MIT Press.

Starobinski, Aroon. 1960. *La biopolitique: Essai d'interprétation de l'histoire de l'humanité et des civilisations*. Geneva: Imprimerie des Arts.

Thacker, Eugene. 2010. *After Life*. Chicago: University of Chicago Press.

Von Uexküll, Jakob. 1920. *Staatsbiologie: Anatomie, Physiologie, Pathologie des Staates*. Berlin: Verlag von Gebrüder Paetel.

Wolfe, Cary. 2003. *Zoontologies: The Question of the Animal*. Minneapolis: University of Minnesota Press.

Wolfe, Cary. 2009. *What Is Posthumanism?* Minneapolis: University of Minnesota Press.

Wolfe, Cary. 2013. *Before the Law: Humans and Other Animals in a Biopolitical Frame*. Chicago: University of Chicago Press.

Part I

Politics, Biopolitics, and Biophilosophy

1   From the State of Control to a Praxis of Destituent Power

Giorgio Agamben

A reflection on the destiny of democracy in Athens in 2014 is in some way disturbing, because it obliges us to think the end of democracy in the very place where it was born.[1] As a matter of fact, the hypothesis I would like to suggest is not only that the prevailing governmental paradigm in Europe today is non-democratic, but that it cannot be considered as political either. I will try therefore to show that European society today is no longer a political society; it is something entirely new for which we lack a proper terminology and for which we therefore have to invent a new strategy.

Let me begin with a concept which seems, starting from September 2001, to have replaced any other political notion: security. The formula "for security reasons" functions today in any domain, from everyday life to international conflicts, as a code word, in order to impose measures that the people have no reason to accept. I will try to show that the real purpose of such security measures is not, as it is currently assumed, to prevent dangers, troubles, or even catastrophes. I will be consequently obliged to make a short genealogy of the concept of "security."

A PERMANENT STATE OF EXCEPTION

One possible way to sketch such a genealogy would be to inscribe its origin and history in the paradigm of the state of exception. In this perspective, we could trace it back to the Roman principle *Salus publica suprema lex* ("public safety is the highest law") and connect it with Roman dictatorship, with the canonistic principle that *necessity does not acknowledge any law*, with the *comités de salut publique* (Committee of Public Safety) during the French Revolution, and finally with Article 48 of the Weimar Republic, which was the juridical ground for the Nazi regime. Such a genealogy is certainly correct but I do not think that it could really explain the functioning of the security apparatuses and measures which are familiar to us.

While the state of exception was originally conceived as a provisional measure, which was meant to cope with an immediate danger in order to restore the normal situation, the security reasons today constitute a permanent technology of government. When in 2003 I published a book in which I tried to show precisely how the state of exception was becoming a normal system of government in Western democracies (Agamben 2005), I could not imagine that my diagnosis would prove so accurate. The only clear precedent was the Nazi regime. When Hitler took power in February 1933, he immediately proclaimed a decree suspending the articles of the Weimar Constitution concerning personal liberties. The decree was never revoked so the entire Third Reich can be considered as a state of exception which lasted twelve years.

What is happening today is still different. A formal state of exception is not declared and we see instead that vague non-juridical notions—like the security reasons—are used to install a stable state of creeping and fictitious emergency without any clearly identifiable danger. An example of such non-juridical notions, which are used as emergency producing factors, is the concept of crisis. Besides the juridical meaning of judgment in a trial, two semantic traditions converge in the history of this term, which, as is evident, comes from the Greek verb *crino*, a medical and a theological one. In the medical tradition, *crisis* means the moment in which the doctor has to judge, to decide if the patient will die or survive. The day or the days in which this decision is taken are called *crisimoi*, the decisive days. In theology, *crisis* is the Last Judgment pronounced by Christ in the end of times.

What is essential in both traditions is the connection with a certain moment in time. In the present usage of the term, it is precisely this connection which is abolished. The crisis, the judgment, is split from its temporal index and coincides now with the chronological course of time so that—not only in economics and politics but in every aspect of social life—the crisis coincides with normality and becomes, in this way, just a tool of government. Consequently, the capability to decide once and for all disappears and the continuous decision-making process decides nothing. To state it in paradoxical terms we could say that, having to face a continuous state of exception, the government tends to take the form of a perpetual *coup d'état*. By the way, this paradox would be an accurate description of what happens here in Greece as well as in Italy, where to govern means to make a continuous series of small *coups d'état*.

GOVERNING THE EFFECTS

This is why I think that in order to understand the peculiar governmentality under which we live the paradigm of the state of exception is not entirely adequate. I will therefore follow Michel Foucault's suggestion and investigate the origin of the concept of security in the beginning of modern economy, by François Quesnay and the Physiocrats, whose influence on modern governmentality could not be overestimated (Foucault 2012, 14,

16). Starting with the Westphalia Treaty, the great absolutist European states begin to introduce, in their political discourse, the idea that the sovereign has to take care of its subjects' security. But Quesnay is the first to establish security (*sûreté*) as the central notion in the theory of government—and this in a very peculiar way.

One of the main problems governments had to cope with at the time was the problem of famines. Before Quesnay, the usual methodology was trying to prevent famines through the creation of public granaries and forbidding the exportation of cereals. Both these measures had negative effects on production. Quesnay's idea was to reverse the process: instead of trying to prevent famines, he decided to let them happen and to be able to govern them once they occurred, liberalizing both internal and foreign exchanges. "To govern" retains here its etymological cybernetic meaning: a good *kybernes*, a good pilot, can't avoid tempests, but if a tempest occurs he must be able to govern his boat using the force of waves and winds for navigation. This is the meaning of the famous motto *laisser faire, laisser passer*. It is not only the catchword of economic liberalism, it is a paradigm of government which conceives of security (*sûreté* in Quesnay's words) not as the prevention of troubles, but rather as the ability to govern and guide them in the right direction once they take place.

We should not neglect the philosophical implications of this reversal. It means an epochal transformation in the very idea of government, which overturns the traditional hierarchical relation between causes and effects. *Since governing the causes is difficult and expensive, it is safer and more useful to try to govern the effects.* I would suggest that this theorem by Quesnay is the axiom of modern governmentality. The *ancien régime* aimed to rule the causes; modernity pretends to control the effects. This axiom applies to every domain, from economy to ecology, from foreign and military politics to the internal measures of police. We must realize that European governments today have given up any attempt to rule the causes; they only want to govern the effects. And Quesnay's theorem also makes understandable a fact which seems otherwise inexplicable: the paradoxical convergence today of an absolutely liberal paradigm in the economy with an unprecedented and equally absolute paradigm of state and police control. If government aims for the effects and not the causes, it will be obliged to extend and multiply control. Causes demand to be known, while effects can only be checked and controlled.

One important sphere in which the axiom is operative is that of biometric security apparatuses which increasingly pervade every aspect of social life. When biometric technologies first appeared in the eighteenth century in France with Alphonse Bertillon and in England with Francis Galton, the inventor of fingerprinting, they were obviously not meant to prevent crimes but only to recognize recidivist delinquents.[2] Only once a second crime has occurred, can one use the biometric data to identify the offender. Biometric technologies, which had been invented for recidivist criminals, remained for a long time their exclusive privilege. In 1943, the US Congress refused the *Citizen Identification Act*, which was meant to introduce an identity card with fingerprints for every citizen (Lyon 2009, 32). But according to a sort of fatality or unwritten law of modernity, the technologies which have been invented for animals, for criminals, strangers, or Jews, will finally be extended to all human beings. Therefore, in the course of the twentieth century, biometric technologies have been applied to all citizens and Bertillon's identification photographs and Galton's fingerprints are currently in use everywhere for ID cards.

THE DEPOLITICIZATION OF CITIZENSHIP

But this extreme step has been taken only recently and it is still in the process of full realization. With the development of new digital technologies with optical scanners which can easily record not only fingerprints but also the retina or the eye's iris structure, biometric apparatuses tend to move beyond the police stations and immigration offices and spread into everyday life. In many countries, the access to students' restaurants or even to schools is controlled by a biometric apparatus on which the student just puts his or her hand. The European industries in this field, which are quickly growing, recommend that citizens get used to this kind of control from their early youth. The phenomenon is really disturbing, because the European Commissions for the development of security (like the European Security Research Programme) include among their permanent members the representatives of the big industries in the field, which are just the old armaments producers like Thales, Finmeccanica, EADS, and BAE Systems, which have converted to the security business.

It is easy to imagine the dangers represented by a power that could have the unlimited biometric and genetic information of all its citizens at its disposal. With such a power at hand, the extermination of the Jews, which was undertaken on the basis of incomparably less efficient documentation, would have been total and incredibly swift. But I will not dwell on this important aspect of the security problem. The reflections I would like to share concern rather the transformation of political identity and of political relationships that are involved in security technologies. This transformation is so extreme that we can legitimately ask not only if the society in which we live is still a democratic one, but also if this society can still be considered *political*.

Christian Meier has shown how in the fifth century a transformation of the conceptualization of the political took place in Athens which was grounded on what he calls a "politicization" (*politisierung*) of citizenship. While until that moment the fact of belonging to the *polis* was defined by a number of conditions and social statuses of different kinds—for instance belonging to nobility or to a certain cultural community—to be a peasant or merchant, a member of a certain family, etc.—from now on citizenship became the main criterion of social identity:

> The result was a specifically Greek conception of citizenship, in which the fact that men had to behave as citizens found an institutional form. The belonging to economic or religious communities was removed to a secondary rank. The citizens of a democracy considered themselves as members of the *polis* only in so far as they devoted themselves to a political life. *Polis* and *politeia*, city and citizenship, constituted and defined one another. Citizenship became in that way a form of life, by means of which the *polis* constituted itself in a domain clearly distinct from the *oikos*, the house. Politics became therefore a free public space as such opposed to the private space, which was the reign of necessity. (Meier 1979)

According to Meier, this "specifically Greek" process of politicization was transmitted to Western politics, where citizenship remained the decisive element.[3]

The hypothesis I would like to propose is that this fundamental political factor has entered an irrevocable process which we can only define as a process of increasing *de*politicization. What was in the beginning a way of living, an essentially and irreducibly active condition, has now become a purely passive juridical status in which action and inaction, the private and the public, are progressively blurred and become indistinguishable. This process of the depoliticization of citizenship is so evident that I will not dwell on it.

RISE OF THE SECURITY STATE

I will rather try to show how the paradigm of security and the security apparatuses have played a decisive role in this process. The growing extension of technologies which were conceived for criminals, to all citizens inevitably has consequences for the political identity of the citizen. For the first time in the history of humanity identity is no longer a function of the social personality and its recognition by others, but rather a function of biological data which cannot bear any relation to it such as the arabesques of the fingerprints or the disposition of the genes in the double helix of DNA. The most neutral and private thing becomes the decisive factor of social identity, which loses therefore its public character.

If my identity is now determined by biological facts that in no way depend on my will and over which I have no control, then the construction of something like a political and ethical identity becomes problematic. What relationship can I establish with my fingerprints or my genetic code? The new identity is an identity without the person, as it were, in which the space of politics and ethics loses its sense and must be thought again from the ground up. While the classical Greek citizen was defined through the opposition between the private and the public, the *oikos*, which is the place of reproductive life, and the *polis*, the place of political action, the modern citizen seems rather to move in a zone of indifference between the private and the public, or, to quote Hobbes' terms, the physical and the political body (Hobbes 2008, 3).

The materialization in space of this zone of indifference is the video surveillance of the streets and the squares of our cities. Here again an apparatus that had been conceived for the prisons has been extended to public places. But it is evident that a video-recorded place is no longer an *agora* and becomes a hybrid of public and private, a zone of indifference between the prison and the forum. This transformation of the political space is certainly a complex phenomenon that involves a multiplicity of causes, and among them the birth of biopower holds a special place. The primacy of biological identity over political identity is certainly linked to the politicization of bare life in modern states.

But one should never forget that the leveling of social identity on body identity began with the attempt to identify recidivist criminals. We should not be astonished if today the normal relationship between the state and its citizens is defined by suspicion, police filing, and control. The unspoken principle which rules our society can be stated like this: *every citizen is a potential terrorist*. But what is a state ruled by such a principle? Can we still define it as a democratic state? Can we even consider it as something political? In what kind of state do we live today?

Michel Foucault, in his book *Surveiller et Punir* (1975) and in his courses at the Collège de France (Foucault 2004), sketched a typological classification of modern states. He shows how the state of the *ancien régime*, which he calls the territorial or sovereign state and whose motto was *faire mourir et laisser vivre*, evolves progressively into a population state and into a disciplinary state, whose motto reverses now into *faire vivre et laisser mourir*, as it will take care of the citizen's life in order to produce healthy, well-ordered, and manageable bodies.

The state in which we now live is no longer a disciplinary state. Gilles Deleuze (1992, 4) suggested to call it the *État de contrôle*, or "societies of control", because what it wants is not to order and to impose discipline but rather to manage and to control. Deleuze's definition is correct because management and control do not necessarily coincide with order and discipline. No one has said it as clearly as the Italian police officer who, after the Genoa riots in July 2001, declared that the government did not want the police to maintain order but to *manage disorder*.

American political scientists who have tried to analyze the constitutional transformation involved in the Patriot Act and in the other laws that followed September 2001 prefer to speak of a "security state."[4] But what does "security" mean here? It was during the French Revolution that the notion of security—*sûreté*, as they used to say—was linked to the definition of police. The laws of 16 March 1791 and 11 August 1792 introduced thus into French legislation the notion of *police de sûreté* (security police), which was doomed to have a long history in modernity. In the debates that preceded the vote on these laws, the police and security defined one another, but no one among the speakers (Jacques Pierre Brissot, Marie-Jean Héraut de Séchelles, Armand Gensonné) was able to define police or security by themselves.

The debates focused on the situation of the police with respect to justice and judicial power. Gensonné maintained that they are "two separate and distinct powers", yet, while the function of judicial power is clear, it is impossible to define the role of the police. An analysis of the debate shows that the place and function of the police is undecidable and must remain undecidable because if it were really absorbed in the judicial power the police could no longer exist. This is the discretionary power which still defines the actions of police officers who, in a concrete situation of danger, act as a sovereign, so to speak, for public security. But even when he exerts this discretionary power, the police officer does not really take a decision nor prepare, as is usually stated, the judge's decision. Every decision concerns the causes, while the police act on effects, which are by definition undecidable.

The name of this undecidable element is no more today, as it was in the seventeenth century, the *raison d'État*, or state reason.[5] It is rather "security reasons." The security state is a police state but, again, in juridical theory, the police are a kind of black hole. All we can say is that when the so-called "science of the police" first appears in the eighteenth century, the term "police" is brought back to its etymology from the Greek *politeia* and opposed, as such, to "politics." But it is surprising to see that "police" then coincides with the true political function while the term politics is reserved for foreign policy. Thus Von Justi, in his treatise on *Polizeywissenschaft*, calls *Politik* the relationship of a state with other states, while he calls *Polizei* the relationship of a state with itself. It is worthwhile to reflect upon this definition: "police is the relationship of a state with itself."

The hypothesis I would like to suggest here is that, placing itself under the sign of security, the modern state has left the domain of politics to enter a no man's land, whose geography and whose borders are still unknown. The security state, whose name seems to refer to an absence of cares (*securus* from *sine cura*) should, on the contrary, make us worry about the dangers it involves for democracy because, in it, political life has become impossible, while democracy means precisely the possibility of a political life.

DESTITUENT POWER

But I would like to conclude—or better to simply end this essay (in philosophy, as in art, no conclusion is possible, you can only abandon your work)—with something which, as far as I can see now, is perhaps the most urgent political problem. If the state we have in front of us is the security state I have described, we have to think anew the traditional strategies of political conflicts. What shall we do, what strategy shall we follow?

The security paradigm implies that each form of dissent, each more or less violent attempt to overthrow the order, becomes an opportunity to govern these actions into a profitable direction. This is evident in the dialectics that tightly bind together terrorism and state in an endless vicious spiral. Starting with the French Revolution, the political tradition of modernity has conceived of radical changes in the form of a revolutionary process that acts as the *pouvoir constituant*, the "constituent power", of a new institutional order. I think that we have to abandon this paradigm and try to conceive of something as a *puissance destituante*, a purely "destituent power", which cannot be captured in the spiral of security.

It is a destituent power of this sort that Benjamin has in mind in his essay *On the Critique of Violence* (Benjamin 1977, 202), when he tries to define a pure violence which could "break the false dialectics of lawmaking violence and law-preserving violence", an example of which is Sorel's proletarian general strike. "On the breaking of this cycle", he writes at the end of the essay, "maintained by mythic forms of law, on the destitution of law with all the forces on which it depends, finally therefore on the abolition of state power, a new historical epoch is founded."[6] While a constituent power destroys law only to recreate it in a new form, destituent power—insofar as it deposes the law once for all—can open a really new historical epoch.

To think such a purely destituent power is not an easy task. Benjamin wrote once that nothing is so anarchical as the bourgeois order. In the same sense, Pasolini in his last movie has one of the four Salò masters saying to their slaves: "true anarchy is the anarchy of power" (Pasolini and Citti 1975). It is precisely because power constitutes itself through the inclusion and the capture of anarchy and anomy that it is so difficult to have an immediate access to these dimensions; it is so hard to think today of something as a true anarchy or a true anomy. I think that a praxis which would succeed in exposing clearly the anarchy and the anomy captured in the governmental security technologies could act as a purely destituent power. A really new political dimension becomes possible only when we grasp and depose the anarchy and the anomy of power. But this is not only a theoretical task: it means first of all the rediscovery of a form of life, the access to a new figure of that political life whose memory the security state tries to cancel at any price.

NOTES

1   This essay is based on a speech given in Athens on 16 November 2013.
2   For a discussion of the work of Bertillon and Galton, see Rhodes (1956) and Galton (1892).

3 For a revised and expanded version of this article in English, see Christian
   Meier. 1990. Changing Politicosocial Concepts in the Fifth Century B.C. In
   The Greek Discovery of Politics. Trans. David McLintock, 157–85. Cambridge, MA: Harvard University Press.
4 See, for example, Madar (2012).
5 See, for example, Thau (2000).
6 For an English version of this essay, see Walter Benjamin. 1986. Critique of
   Violence. In *Reflections: Essays, Aphorisms, Autobiographical Writings*. Ed.
   P. Demetz. Trans. Edmund Jephcott, 277–300. New York: Schocken Books.

REFERENCES

Agamben, Giorgio. 2005. *State of Exception*. Trans. Kevin Attell. Chicago: Univer-
   sity of Chicago Press.
Benjamin, Walter. 1977. Zur Kritik die Gewalt. In *Gesammelte Schriften* II, 1.
   Frankfurt am Main: Suhrkamp.
Deleuze, Gilles. 1992. Postscript on the Societies of Control. *October 59*: 3–7.
Foucault, Michel. 1975. *Surveiller et punir: Naisance de la prison*. Paris: Gallimard.
Foucault, Michel. 2004. *Sécurité, territoire, population: Cours au Collège de France,
   1977–1978*. Paris: Gallimard.
Foucault, Michel. 2012. *On the Government of the Living: Lectures at the Collège
   de France, 1979–1980*. Ed. Michael Senellart. Trans. Graham Burchell. Hound-
   mills, Basingstoke: Palgrave Macmillan.
Galton, Francis. 1892. *Finger Prints*. London: Macmillan.
Hobbes, Thomas. 2008. *Leviathan: Or the Matter, Forme and Power of a Common-
   wealth Ecclesiastical and Civil*. Ed. Michael Oakeshott. New York: Touchstone.
Lyon, David. 2009. *Identifying Citizens: ID Cards as Surveillance*. Malden, MA:
   Polity Press.
Madar, Chase. 2012. Le president Obama, du prix Nobel aux drones. *Le Monde
   diplomatique*, October.
Meier, Christian. 1979. Der Wandel der politisch-sozialen Begriffswelt im 5. Jahr-
   hundert v. Chr. In *Historische Semantik und Begriffsgeschichte*. Ed. Reinhart
   Koselleck. Stuttgart: Klett-Cotta.
Pasolini, Pier Paolo and Sergio Citti. 1975. *Saló, or the 120 Days of Sodom*. Directed
   by Pier Paolo Pasolini. Beverley Hills, CA: United Artists.
Rhodes, Henry T.F. 1956. *Alphonse Bertillon: Father of Scientific Detection*. New
   York: Abelard-Schuman.
Thau, Étienne. 2000. *Raison d'Etat et pensée politique à l'époque de Richelieu*.
   Paris: Albin Michel.
Von Justi, Johann Heinrich. 1816. *Grundsätze der Polizeywissenschaft*. Vienna:
   Rehm.

2 Posthuman Affirmative Politics

Rosi Braidotti

INTRODUCTION

My argument in this paper is that the concept of the biopolitical, in its classical Foucauldian inception, is currently challenged by the joint impact of contemporary neo-materialism (or rather: "matter-realism") and the posthuman turn. The paper will explain these two notions and go on to argue that they call for a novel approach that challenges the idea of the biopolitical on both conceptual and political grounds. The current conditions of advanced capitalism push the logic of the biopolitical beyond anthropocentrism and pay renewed attention to the necropolitical dimensions of the politics of "Life." I will conclude by emphasizing affirmative ethics as a force capable of redefining politics as living complexity.

ON THE POLITICAL

I concur with Thomas Lemke (2011, 1) that the notion of the biopolitical has become somewhat of a "buzzword." In its original Foucauldian inception, however, this concept aimed to sharpen the edges of the political analyses and to move them beyond dialectical thinking in the turbulent context of 1970s Europe. In their conversation on intellectuals, for instance, Deleuze and Foucault (1977) offer a multifaceted rendition of power as both restrictive or coercive (*potestas*) and empowering or productive (*potential*). This approach is intended as a critique of the traditional Marxist idea of politics, that is to say as a move away from binary dialectical schemes and their resolutely negative vision of power. The core conceptual issue is precisely that of the negative and its relation to politics.

Foucault and Deleuze then go on to postulate an equally crucial distinction between politics (*LA politique*) and the political (*LE politique*). Politics focuses on the management of civil society and its institutions, the political on the transformative experimentations with new arts of existence and ethical relations. Politics is made of progressive emancipatory measures, predicated on chronological continuity, whereas the political is the radical self-styling that requires the circular time of critical praxis. Both Foucault and Deleuze emphasize the difference between the centralized—i.e., majoritarian and agonistic—character of politics and the minor or minoritarian, dynamic, affirmative character of the political. Activism as an affirmative political praxis consists in connecting critical theory not so much to *LA politique*—i.e., organized or majoritarian politics, or "politics as usual"—as to *LE politique*—i.e., the political in its nomadic and transformative forms of becoming.

On the political issue, as on that of rethinking subjectivity after dialectics, however, Deleuze goes much further than Foucault. The qualitative distinction between politics and the political is replicated at the level of the philosophy of time and the form of relational affectivity that a time-continuum may engender. Politics is postulated on *Chronos*—the linear time of institutional deployment of norms and protocols. It is a reactive and majority-bound enterprise that is often made of flat repetitions and predictable reversals that may alter the balance but leave the structure of power basically untouched. The political, on the other hand, is postulated on the axis of *Aion*—the non-linear time of becoming and of affirmative critical practice. It is minoritarian and it aims at the counter-actualization of alternative states of affairs in relation to the present. Nomad thought as a *zoë*-centered (Braidotti 2006) form of material vitalism (Deleuze and Guattari 1987) sets the desire for transformations in the sense of becoming ethico-political at the centre of the theoretical agenda.

In so doing, nomadic neo-materialism (Braidotti 1991; 2011a; 2011b), also known as "matter-realism" (Fraser, Kember and Lury 2006) or "vital materialism" (Bennett 2010, 17), has produced also a critique of classical biopolitics. This critique is multiple. To start with it is worth remembering that Foucault's biopolitical work was essentially an analysis of the political economy of liberal democracies, of the welfare state and the notion of moralized liberal individualism they combined to produce. To what extent this analysis can be extended to cover advanced capitalism is a critical question.

Foucault attacked the liberal individualistic vision of the subject not only in terms of a critique of possessive individualism (MacPherson 1962), but also more conceptually. He proposes a biopolitical analysis that concentrates not only on the production of discourse and the multidirectional circulation of discursive practices, but also on the material grounds of such production and the social and institutional structures that sustain them. The focus is firmly held throughout on the constitution of subjectivity as a discursive and material process that combines a number of heterogeneous elements. A pertinent example of this combinatory process is the correlation firmly established between the care dispensed by the welfare system to its citizens and the mechanisms that control, regulate, and monitor them. Techniques of embodied and embedded discipline and punishment, surveillance and incarceration, constitute a crucial element of an allegedly benevolent political economy. The embodied subject, or rather the subject as embodied matter, emerges at the intersection of these mechanisms of control, which tend to target essential vital functions—such as reproduction, sexuality, health—medical and mental—and hygiene, as the main objects of discursive and material control.

In this regard, Foucault's biopolitical analysis explicitly critiques a liberal vision of the political subject that is assumed to function according to the universalist, humanistic idea of inbuilt rationality, moral goodness, and self-regulating judgment. Foucault articulates his suspicion toward humanism in his masterful *The Order of Things* (1970) in order to declare the humanist project historically, politically, and ethically over. This skepticism goes hand-in-hand with the suspension of belief in the intrinsic value of Enlightenment-based rationality and the self-correcting powers of human reason. This complex anti-humanist argument constitutes for me one of the roots of contemporary posthumanism (Braidotti 2013).

A second and equally important line of criticism, however, runs through Foucault's analysis of the biopolitical: it concerns the role of violence in relation to politics and the political. As a thinker and an activist on the left of the political spectrum, Foucault resisted the naturalization of violence traditionally proposed by right-wing ideologies. He was also critical of the allegedly peace-loving disposition of humanism and openly discussed the compatibility between humanistic reason and the uses of terror and violence (Foucault 1977). Historical examples are, for instance, colonialism, but also the social campaigns against deviants, vagabonds, delinquents, and other undesirables. This critical analysis of humanism is clearly linked to the critique of universal reason.

Foucault, like Deleuze, was also critical, however, of the dogmatism and the authoritarian tendencies of the political left, throughout the 1970s but also beyond. This second line of attack criticizes the aspiration to revolutionary purity of the left, the utopian drive which historically resulted in totalitarian regimes like the USSR and the People's Republic of China and in genocides like Pol-Pot's in Cambodia. Deleuze and Foucault targeted the utopian elements of the Marxist-Leninist and also of the Maoist projects and did so on two main and interrelated grounds. The first is precisely the location of violence within these political projects and the second is the definition of the role of intellectuals as the alleged representatives of the masses.

As to the former, Foucault is an anti-metaphysical thinker who defines power not as an ontological precondition of the political, but rather as a complex strategic situation we all inhabit. As such, power analyses require context-specific, historicized accounts of how such strategic situations were constituted in the first place and by which discursive and material conditions they became structured. The genealogical method, in other words, makes for highly specific and historically grounded analyses of how certain power formations have come into being and how they impact upon our self-representation as subjects. By extension this means that no universalizing generalization about power is possible for Foucault; he stresses instead the need for a change of scale, to unveil power relations where they are most effective and invisible: in the specific locations of one's own discursive and social practice. One has to start from micro-instances of the embodied and embedded self and the complex web of social relations that compose the self. By extension this means that there is no transhistorical political ontology and therefore also no logical necessity for political violence, or for instrumental violence, as the defining feature of the political. It is rather the other way around, namely that ontology itself is, as Oksala (2012, 6) cogently puts it, "the outcome of political practice: it is politics that has forgotten itself."

This position against the ontological necessity for violence also affects Foucault's redefinition of the role of intellectuals. Wisely removed from grandiose visions of the philosopher as revolutionary leader, Foucault actively promoted instead the "specific" intellectual, as distinct also from the Hegelian universal philosopher and the Gramscian organic intellectual. He foregrounds instead the genealogical structure of intellectual work in terms of discourse analysis, which in turn stresses the relational nature and the political responsibility of intellectuals. Power analyses entail both critique and creativity and therefore are not solely oppositional. The critical thinker is neither a transcendental consciousness nor an atomized entity, but rather a non-unitary relational subject, which in my terms is nomadic, accountable, and outward-bound (Braidotti 2013). This vision of subjectivity leads to an increased awareness of the shared vulnerability of embodied subjects, in so far as they are all caught in strategic relations of power, and it results in subtler and more effective analyses of how power works in and through the body. This double emphasis on fragility on the one hand and the critique of despotic power relations on the other is crucial to a nomadic vision of the political thinker and her actions.

ZOĒ/POSTHUMAN LIFE AND MONISTIC MATERIALISM

Deleuze and Guattari push this argument even further and go beyond the biopolitical premises laid out by Foucault, embracing Spinozist monism fully. That radical immanent thought promotes both the necessity of creativity as the counterpart of critique (Braidotti 2006) and also a kind of ontological pacifism, which sustains a democratic move. It moreover calls for post-anthropocentric approaches that move beyond the assumption of transcendental consciousness as the key to human exceptionalism. More on both these concepts later. Contemporary neo-Spinozist monism goes beyond Foucault's idea of the biopolitical in that it implies a notion of subjectivity as vital and self-organizing matter, an embedded form of "matter-realism" that is intrinsically connected to the posthuman definition of Life as *zoē*, or a dynamic and generative non-human force. Radical monistic relationality stresses the ethical aspects of subjectivity and allows us to bypass both the pitfalls of binary thinking and the ontologization of political violence.

Moreover, monistic neo-materialism is a practice of affirmation, not of negativity, and this commitment to the positive constitutes not only its core ethical value, but also its political force. Neo-Spinozist monism places a different emphasis on the affective elements of human subjectivity under advanced capitalism and on the process of political subject-formation. Rejecting the Lacanian conceptual structure and terminology, vital neo-materialist thinkers stress the generative importance of affects and connect them to a positive view of desire as plenitude, not as Lack (Braidotti 2006). The unconscious drives, instead of being played back upon a sort of negative filter linked to the "black box" of desire as Lack with its corollary of negative passions like envy, resentment, and perennial frustration, are approached affirmatively. Affects are the autonomous visceral elements of our allegedly rational belief system (Connolly 1999). What they express is the profoundly relational nature of human subjectivity and its constitutive drive for the freedom of expression of its powers (*potentia*).

By way of contrast, the Hegelian-Marxist school of dialectics of consciousness equates critical political subjectivity with negative, oppositional, or "unhappy" consciousness. Such reactive vision of the subject banks on negativity and even requires it, because it builds on the assumption that the critical position consists in analyzing negative social and discursive conditions, in order to better overthrow them. In other words it is the same conditions that construct the negative moment—for instance the experience of oppression, marginality, injury, or trauma—and also the possibility of overturning them. The same analytic premises provide both the damages and the possibility of positive resistance, counteraction, or transcendence (Foucault 1977). The "wounded attachments" (Brown 1993; 2006) that trigger and at the same time are engendered by this process of vulnerability and resistance constitute the paradoxical core of oppositional consciousness.

As an alternative, Deleuze and Guattari construct a non-Hegelian, monistic, and vital-materialist account of the genesis of political subjectivity that foregrounds the relational, negotiation-driven, and affirmative elements of this process. The political is sustained by a relational affirmative ethics that aims to cultivate collectively and produce the conditions of its own expression: it is an auto-poietic praxis based on a positive definition of the subject as a process-driven "di-vidual". A subject's ethical core is clearly not her moral intentionality, as much as it is the effects of power (as repressive—*potestas*—and positive—*potentia*) her actions are likely to have upon the world. It is a process of engendering empowering modes of becoming (Braidotti 2006; Deleuze 1968).

Here is the punchline of contemporary *zoē*/posthuman neo-Spinozist materialist politics: affirmative ethics defines our politics. Given that the ethical good is equated with radical relationality, aiming at affirmative empowerment, the ethical ideal is to increase one's ability to enter into modes of relation with multiple others. Oppositional consciousness as a reactive mode is replaced by affirmative praxis and political subjectivity is redefined as a process or assemblage that actualizes this ethical propensity. This position aspires to the creation of affirmative alternatives by working through the negative instances so as to collectively transform them into affirmative practices. The drive toward affirmation is a key feature of neo-Spinozist nomadic political subjects.

This view of subjectivity does not condition the emergence of the subject on negation but on creative affirmation, not on loss but on vital generative forces. The rejection of the dialectical scheme also implies a shift of temporal gears. It means that the conditions for political and ethical agency are not dependent on the current state of the terrain; they are not oppositional and thus not tied to the present by negation. Instead they are projected across time as affirmative praxis, geared to creating empowering relations aimed at possible futures. Ethical relations create possible worlds by mobilizing resources that have been left untapped in the present, including our desires and imagination. They are the driving forces that concretize actual, material relations and can thus constitute a network, web, or rhizome of interconnection with others.

Zoë/posthuman monistic vitalism stresses a constitutive sense of intimacy with the world and a sense of entanglement in a web of immanent and ever-shifting relations and perpetual becoming. Georges Bataille's agnostic spirituality is of great inspiration for nomadic thought, in that it leads to a nontheistic form of naturalism that rejects all transcendental mystifications (Bataille 1988) and honors what Bryden calls "a dynamism of the void" (2001, 5). The idea that we are all "part of nature", as Lloyd put it (1994; 1996), generates not only vital monism, but also alternative visions of how matter and mind interact and join forces to co-create affirmative becomings. Intimacy with the world speaks of our ability to re-collect it and re-connect to it and hence of our capacity to find our "homes" within it, in the pursuit of nomadic sustainable relations (Braidotti 2006). Relational nomadic subjects engage in transversal connections with—Haraway speaks of "becoming-with" (2007)—multiple human and non-human others. Such webs of connections and negotiations define belonging not as attachment to static identity lines but as dynamic transversal moves across ecosophically interconnected categories. Relationality consists of a deep sense of negotiations with the multiple ecologies—social, environmental, and psychic (Guattari 2000), that constitute us. A sense of familiarity with the world flows from the simple fact that we are the products of such ecological interconnections and notably of the nature-culture continuum (Haraway 1997) which marks our era.

Theoretically and politically, neo-Spinozist material vitalism stands against the emphasis on political theology that, adapted from Carl Schmitt (1996), shaped the thinking of Leo Strauss and the American neocons through the Bush Jr years (Norton 2004). The difference between the two approaches is that political theology in its classical enunciation as well as in the contemporary reinterpretation by Agamben (1998) reduces modern political theories to the secularized version of theological concepts. This fundamentally authoritarian reduction overemphasizes the ruthlessly dichotomous ("friend or enemy") and polarizing nature ("you are with us or against us") of the political relation. By stressing the antagonistic dimension as the defining core of politics (Mouffe 2005), this approach ends up endorsing negativity and the necessity of violence. It also expresses an indictment of Western modernity and the democratic process as being structurally flawed.

Materialist vital ethics, on the other hand, while being resolutely atheistic, is ontologically pacifist. Deleuze's concept of the univocity of being and the immanence of matter is a vitalist anti-theology. The recognition of our intimacy with the world provides the conceptual grounds to assert a non-unitary ethical subject immersed in the intelligent and self-organizing structure of Life itself. It therefore infuses affect and endurance at the heart of the embodied and embedded materialism of the subject and of matter itself as a nature-culture continuum. The proposed methodology is not social constructivism, but rather neo-Spinozist expressionism (Braidotti 2006; 2013). That is to say that events, phenomena, and subject-formations are approached as actualizations of differential modes of becoming within a monistic universe. The univocity of being means that we have to deal with one matter, which is intelligent, embedded, embodied, and affective. It requires a subtler analysis of differential variations in the process of subjectivation in order to account for the actualization of transversal subject formations, also known as "assemblages" (Deleuze and Guattari 1980).

The Deleuzian position shares the same commitment to overturning the dialectical model of intersubjectivity as the linguistic tradition of semiotics, psychoanalysis, and deconstruction, but takes a different road. It assumes the defamiliarization or relative deterritorialization of established values and habits of thought as a starting point to explore and experiment with alternative forms of subjectivity. This qualitative shift engages our collective imaginings (Gatens and Lloyd 1999) and desire (Braidotti 2011b)—in response to world-historical structural transformations. The nomadic subject is a materially embodied and historically embedded "di-vidual" in that it is a bound instantiation of a common and ever-shifting matter. Each singular self is an actualized and temporarily bound expression of the ongoing process of becoming. Matter is intelligent and self-organizing; specific forms of individuation are carved out of this vital material, according to the monistic vision of matter. In the specific case of the human organism, it implies the embrainment of the body as well as the embodiment of the mind (Marks 1998). Neo-Spinozist vital materialism defies the oppositional character of dialectical thought and posits a pacifist ontology of mutual specification as the motor of processes of individuation (Simondon 2012) and auto-poietic self-styling.

DIFFERENT CARTOGRAPHIES OF ADVANCED CAPITALISM

As I suggested earlier, the drawback of Foucault's anatomy of the biopolitical is that it describes a system of governmentality at the apex of its evolution and thus ends up producing an analytic of the present conditions, which accounts only partially for the actual situation. This feature becomes all the more salient in the aftermath of Foucault's pioneering work.

Several distinct trends can be detected in contemporary thinking about the biopolitical management of life and death. To start with, a school of biopolitical citizenship has emerged, with emphasis on the ethical implications of "biopower" as an instance of governmentality that is as empowering as it is confining (Esposito 2008; Rabinow 2003; Rose 2007). This school of thought locates the political moment in the relational and self-regulating accountability of a bioethical subject that takes full responsibility for her biological and genetic existence, including illness, depression, and other aspects of one's embodiment. This position allows for a residual type of Kantianism to emerge around the last phase of Foucault's work, with emphasis on individual responsibility for the self-management of one's health and lifestyle. The advantage of this position is that it calls for a higher degree of lucidity about posthuman bio-organic existence, which means that the naturalist paradigm is definitely abandoned. The disadvantage of this position, however, is that it redirects the notion of responsibility toward individualism; in a political context of neoliberal dismantling of the British National Health Service, a pillar of the welfare state, and increasing privatization, bioethical citizenship indexes access to and responsibility for the cost of basic social services like health care to an individual's manifest ability to act responsibly by reducing the risks and exertions linked to the wrong lifestyle. In other words, here bioethical agency means taking adequate care of one's own genetic capital. The recent government campaigns against smoking, excessive drinking, and obesity constitute evidence of this neoliberal normative trend that supports hyperindividualism.

This approach raises in my eyes serious theoretical questions about the notion of biopower itself. Considering the fast rate of progress and change undergone by contemporary biotechnologies and the challenges they throw to the status of the human, Foucault's work has been criticized, notably by Haraway (1997), for relying on an outdated vision of contemporary technology. Haraway suggests that Foucault's biopower provides the cartography of a world that no longer exists, in so far as we have now entered the age of the informatics of domination. Other critical theories come closer to the target, notably feminist and queer (Barad 2003; Braidotti 2002; Butler 2004; Grosz 2004), environmentalist (Shiva 1997), and race theorists (Gilroy 2000), who have addressed the shifting status of embodiment and difference in advanced capitalism in a manner that reflects the complexity of global social relations.

A second school of contemporary biopolitical reflection led by Giorgio Agamben (1998) has taken what I call a "forensic turn" in social theory and addresses the contemporary status of the human as *Anthropos*. Agamben brings back the *bíos-zoē* distinction, that is to say the distinction between human and non-human life, but retains a negative definition of the latter. Focussing on the dehumanizing effects of biopower as the result of the lethal intervention of sovereign power onto the embodied subject, Agamben argues that the subject is reduced to "bare life", that is to say an inhuman status of extreme vulnerability, bordering on extinction. Biopower for Agamben means "thanatopolitic"—a distinction which Foucault himself had already introduced in the early phases of his biopolitical analyses. For Agamben, however, the purpose of this distinction is to sustain an indictment of the project of industrialized modernity in view of its dehumanizing effects. There is more than a residual dose of Heideggerian suspicion of technology in this argument. The victims of concentration camps and of colonial plantations are the prototype of this murderous political economy. The enslaved, dehumanized human, exemplified today in illegal migrants and asylum seekers, is almost the epitome of the bare life of *Homo Sacer* (Agamben 1998). This insight results in drawing intrinsic links between modernization and violence, modernity and terror, sovereignty and murder.

The inhuman for Agamben, not unlike Lyotard, is the effect of modernization, but he also learned from Hannah Arendt (1951) to look at phenomena of totalitarianism as the ultimate denial of the humanity of the other. Arendt, however, constructed a powerful alternative to these political extremes by stressing the necessity of human rights for all, even and especially the dehumanized "others". In Seyla Benhabib's brilliant formulation, Arendt is "a reluctant modernist" (1996), but a creative one, whereas I would argue that Agamben is less innovative. He perpetuates the philosophical habit that consists in ontologizing the relation between politics and violence and in taking mortality, or finitude, as the transhistorical horizon for discussions of "life". For him, "bare life" does not express generative vitality, but rather the constitutive vulnerability of the human subject, which sovereign power can kill. "Life" is that which makes the body into disposable matter in the hands of the despotic force of unchecked power. This is linked to Heidegger's theory of being as deriving its force from the annihilation of animal life. Finitude is introduced as a constitutive element within the framework of subjectivity, which also fuels an affective political economy of loss and melancholia at the heart of the subject.

I am perturbed by this fixation on *Thanatos* that Nietzsche criticized over a century ago, and which is still very present in critical debates today. It often produces a gloomy and pessimistic vision not only of power, but also of the technological developments that propel the regimes of biopower. My understanding of "life" as *zoē*-ethics of sustainable transformations (Braidotti 2006) differs considerably from what Agamben calls "bare life" or negative *zoē*. I reject the habit that favors the deployment of the problem of *zoē* on the horizon of death, or on liminal states of nonlife. This over-emphasis on mortality and perishability, which is characteristic of the forensic strand of contemporary social and cultural theory, reinserts the specter of human extinction and is haunted by the limitations of the project of western modernity. I find the emphasis on violence and death as the basic term of reference inadequate to the vital politics of our era. I therefore want to turn to another significant community of scholars who work within a Spinozist framework,[1] and prefer to emphasize the politics of life itself as a relentlessly generative force. This requires an interrogation of the shifting interrelations between human and non-human forces.

Moreover, speaking from the position of an embodied and embedded female subject, capable of reproducing the future and the species, I find the metaphysics of finitude to be a myopic way of asking about the limits of what we call "life". At the heart of my research project lies an ethics that respects vulnerability while actively constructing social horizons of hope. I shall return to this in my conclusion.

CAPITALISM AND SCHIZOPHRENIA

In the previous section I have argued that Foucault's biopolitical analysis describes a system at the moment of its implosion and thus does not fully confront the contradictions of our historicity. A Deleuzian analysis, based on the radical immanence of vital matter-realism, on the other hand, empowers us to analyze the perverse political economy of advanced capitalism in ways that move beyond the anatomies of biopolitical powers. Let me outline now the defining features of this system.

Firstly, advanced capitalism functions by a schizoid logic that defies the principle of excluded middle and sustains the simultaneity of internally contradictory social effects. The growing disparities in access to resources—of ecological, technological, social, and financial kinds—are the most obvious aspect of the structural inequalities engendered by the global economy. On the one hand we see the worldwide spread of economic and cultural processes, which engender increasing conformity in lifestyle, telecommunication, and consumerism. On the other hand, we also witness the fragmentation of these processes, with the concomitant effects of increased structural injustices, the marginalization of large sections of the population, and the resurgence of regional, local, ethnic, and cultural differences not only between the geopolitical blocks, but also within them (Eisenstein 1998).

Advanced capitalism has also instilled a one-way political message at the discursive level (Touraine 2001), by celebrating the so-called "end of ideologies" as one of the strongest ideological formations of our times. It triumphantly asserts the end of the quest for social justice in contemporary neoliberal societies and thus fulfils the conservative fantasy of an immutable "human nature", which allegedly coincides with the ethos of advanced capitalism itself (Fukuyama 2002). This political neoconservatism gets compensated in public discourse by overemphasis on moral issues, which in turn produces an escalating notion of the range of social services for which individuals are expected to take financial responsibility. Muehlebach (2012) calls this ideological discourse "the moral neoliberal." Thus, the potentially innovative, deterritorializing impact of genuinely new developments, in society as in technology, are reterritorialized and tuned down by the reassertion of the gravitational pull of old consumeristic values: interactivity has just become another word for shopping.

Secondly, and as a consequence of the above, advanced capitalism is a differential engine in that it promotes the quantitative proliferation of multiple options in consumer goods. It is a multiplier of deterritorialized differences, a spinning machine that actively produces differences for the sake of commodification. As Eugene Holland (2011) points out, advanced capitalism displays clear entropic and self-destructive tendencies in that it erodes the very foundations that sustain it. It consequently exposes and endangers the very sources of its wealth and power in ways that are unmatched by other economic systems, which kept their resources hidden or protected. Advanced capitalism operates on contemporary decoded or deterritorialized flows of change and reterritorializes or stratifies them for the sake of profit.

Given that the political economy of global capitalism consists in multiplying and distributing differences for the sake of profit, it produces ever-shifting waves of genderization and sexualization, racialization and naturalization of multiple "others". It has thus effectively disrupted the traditional dialectical relationship between the dominant subject and the empirical referents of Otherness—which historically are women and LGBT (sexualized others), indigenous or native populations (racialized others), and animals, plants, and the earth-based organisms (naturalized others) (Braidotti 2002; 2006).

Once this dialectical bond between the dominant subject and his "others" is unhinged, advanced capitalism looks like a system that has the capacity to evacuate meaning from most signifiers. It promotes feminism without women, racism without races, natural laws without nature, reproduction without sex, sexuality without genders, multiculturalism without ending racism, economic growth without development, cash flow without money, ecology without nature. Capitalism as schizophrenia is not as playful as this list may suggest; it is rather a ruthless system of recodification of signifying systems in the service of commodification (Deleuze and Guattari 1977; 1980).

Thirdly, in advanced capitalism, time is structurally out of joint, in keeping with the overemphasis on short-term profits alone. Our temporality is determined by the perverse logic of commodity fetishism, which short-circuits the present. The saturation of the social space with commodities results in immobility and sedentary accumulation. The speedy turnover of available commodities, however, induces a state of jetlag, or temporal disjunction. Capitalism induces a perverse logic of desire based on deferral of fulfilment of pleasure to the "next generation" of technological commodities and gadgets, or "infotainment" and popular culture packages based on sequels and instalments that become obsolete at the speed of light. These are legalized but forceful forms of addiction that titillate without providing release and induce dependency without taking responsibility. This mixture of dependency and dissatisfaction constitutes power as a nucleus of negative passions, such as resentment, frustration, envy, and bitterness. The commodity's function as both attractor and perpetual threat to its fulfilment encapsulates a sort of contraction of space and time: it is therefore caught in the spectral economy of the presence-absence of fulfilment, which is addictive and as such it haunts us. The commodity embodies futurity, as Massumi (1992) argues, following Deleuze, and it has become coextensive with the inner space of subjectivity, as well as the outer space of the market economy.

It follows, therefore, that advanced capitalism is an unsustainable "future eater" (Flannery 1994), driven by the all-consuming entropic energy of addictive and enjoyment-frustrating consumerism. Devoid of the capacity for genuine creativity, which would require higher degrees of self-criticism, global capital promotes an addictive logic that creates hunger where it most feeds, thus erecting the entropy of Lack to the level of a Law that wraps us up in persistent anxiety about the future. Various brands of discourses about extinction are for instance circulating today, in a context of economic and ecological crises. In a schizophrenic double pull of euphoria and paranoia, which confirms Deleuze and Guattari's analyses (1977; 1980), the consumerist political economy highjacks our desires and indexes them on the pursuit of commodities. The vitalistic ecosophy proposed by Deleuze and Guattari critiques capitalist consumerism and the greedy consumption of resources.

Fourthly, advanced capitalism functions through tightly controlled mobility, or a "striated" social space subjected to constant surveillance. It works like the great nomad, the organizer of the mobility of commodified products. A generalized practice of "free circulation" pertains almost exclusively to the domain of goods and commodities, data and capital. People do not circulate nearly as freely (Braidotti 2011a; 2011b). Real-life mobility through migration, for instance, or diasporic movements, is kept in check by relations of class, ethnicity, citizenship, gender, and age, to name but a few crucial variables. The global system of the postindustrial world produces scattered and polycentered, profit-oriented power relations. It is therefore crucial to expose the perverse nomadism of a logic of economic exploitation that equates capitalist flows and flux with profit-minded circulation of commodities, and to provide accurate political cartographies of qualitatively different lines of nomadic flows (Braidotti 2006).

Fifthly comes the technologically mediated structure of schizoid advanced capitalism. It is built on the convergence between different and previously differentiated branches of technology, notably biotechnologies and information technologies. The opportunistic political economy of biogenetic capitalism has also turned Life/zoē—that is to say human and non-human intelligent matter—into a commodity for trade and profit. Advanced capitalism both invests in and profits from the scientific and economic control and the commodification of all that lives. I have argued (Braidotti 2006; 2013) that this context produces a paradoxical and rather opportunistic form of post-anthropocentrism on the part of market forces which happily trade on Life itself (see also Rose 2001).

More specifically, what the neoliberal market forces are after, and what they financially invest in, is the informational power of living matter itself. The capitalization of living matter produces a new political economy, which Melinda Cooper (2008) calls "Life as surplus." It introduces discursive and material political techniques of population control of a very different order from the administration of demographics, which preoccupied Foucault's work on biopolitical governmentality. Today, we are undertaking "risk analyses" not only of entire social and national systems, but also of whole sections of the population in the world risk society (Beck 1999). Databanks of biogenetic and neural information about individuals are the true capital today (Braidotti 2013).

This does not mean, however, that traditional patterns of exploitation and oppression are resolved, far from it. With reference to Cooper and Waldby's *Clinical Labor: Tissue Donors and Research Subjects in the Global Bioeconomy*, I would like to coin the term bio-labor as latching onto the corporeal matter of contemporary bodies, marking them off for menial and exploitative tasks. The mechanisms for capture of these bio-laborers, also known as the digital proletariat, follow the classical lines of anthropomorphic difference: the sexualized and racialized "others", as mentioned above, constitute the core of these new underclasses. Think for example of the global chain of care (Hochschild 2000) and other, more extreme cases of bodily commerce, in sexuality and sex-work, reproduction and surrogacy, medical and health practices, organ transplants and other forms of "clinical labour" (Cooper and Waldby 2014). This combination of high-tech advances and low-life survival is one of the most problematic political aspects of advanced capitalism.

Vital matter-realism alters our understanding of embodied or corporeal matter as well. Patricia Clough, for instance, provides an impressive list of the concrete techniques employed by "cognitive capitalism" (Moulier-Butang 2012) to test and monitor the capacities of affective or "bio-mediated" bodies: DNA testing, brain fingerprinting, neural imaging, body heat detection, and iris or hand recognition. These are the contemporary forms of control that go beyond the sites of confinement that Foucault analyzed in the political economy of the nineteenth and early twentieth century techniques of *Discipline and Punish*. Contemporary surveillance techniques hang on "the cloud" that scan the essence of our informational capital at posthuman speed.

All these are also immediately operationalized as surveillance techniques both in civil society and in the War on Terror. What Deleuze and Guattari teach us is that the "virtual" character of technologically mediated power relations today is not ethereal but materially grounded and hence embodied and embedded. Deleuze's speculations on the "control societies" he saw emerging, and Guattari's writings on the post-media age and on Integrated World Capitalism, which were all written at the end of the 1980s and the beginning of the 1990s, already told us that there is no such thing as a purely virtual cyberspace (Guattari 2000; Deleuze 1992; Braidotti 2002).

Last but not least comes the axiomatic character of advanced capitalism. An axiomatic system, as Toscano (2005) pointed out, refuses to provide definitions of the terms it works with, but prefers to order certain domains into existence with the addition or subtraction of certain norms or commands, their objects being treated as purely functional. Axioms operate by emptying flows of their specific meaning in their coded context and thus by decoding them. As Protevi (2009) puts it, through processes of overcoding, preexistent regimes of signs are decoded and subjected to the aims of a centralizing hierarchical machine that turns activity into labor, land into territories, and surplus value into profit.

Being fundamentally meaningless, the decoded flows of capitalism are purely operational modes of regulation. They can get attached to any type of social organization—slave plantations as well as factories—and to different state structures—socialism as well as liberal democracies. As such, the axioms of capitalism are extremely adaptable, capable of great internal variation, and structured around a perverse sort of opportunism. Such flexibility and multiple realizability constitute a formidable apparatus of domination or capture. In the same vein, nomadic theory argues that no freedom is possible within capitalism because the axiom of money and profit knows no limit. Advanced capitalism never attains absolute deterritorializations and always engenders social subjection.

Nomadic theory opposes to the political economy of axiomatic despotism the diagrammatic process of nomadic becoming, which encourages flows without the insertion of axioms. It rejects the ways in which capitalism axiomatizes and captures subjectivity, in order to subject it to the imperatives of surplus value, and defines political praxis as the construction of alternative models of subjectivity. Deleuze's ecosophy of radical immanence and intensive transformative subjects is an affirmative answer to the unsustainable logic and internal contradictions of advanced capitalism. The Deleuzian body is in fact an ecological unit. This *bíos-zoē-technos*-body is marked by the interdependence with its environment, through a structure of mutual flows and data-transfer that is best configured by the notion of viral contamination, or intensive interconnectedness. This ecology of belonging is complex and multilayered. This environmentally bound intensive subject is a collective entity, an embodied affective and intelligent entity that captures, processes, and transforms energies and forces. Being environmentally bound and territorially based, a rhizomatic embodied entity is immersed in fields of constant flows and transformations. Philosophy therefore needs to create forms of ethical and political agency that reflect this high degree of complexity: we need to learn to think differently about who we are in the process of becoming.

BEYOND BIOPOLITICS

Let us start again from the insight that the politics of Life itself in advanced capitalism mobilizes not only generative forces, but also new and subtler degrees of death and extinction. My argument is that a focus on the vital and self-organizing powers of Life/*zoë* composes the notion of *zoë* as a posthuman yet affirmative lifeforce. Vitalist materialism and its monistic political ontology engender a transversal relational ethics to counteract the inhuman(e) aspects of our predicament. This entails significant changes in the status and structure of what counts as the human, dead and alive. Biopolitical analysis is central to this discussion, but in the current context it has moved beyond the premises articulated by Foucault himself.

The central discrepancy between Foucault's notion of biopower and contemporary posthuman political structures has to do with the displacement of anthropocentrism. I argued that the biogenetic structure of advanced capitalism reduces bodies to carriers of vital information, which get invested with financial value and capitalized. They provide the material for new classifications of entire populations on the basis of the genetic predispositions and vital capacities for self-organization. There is a structural isomorphism between economic and biological growth, which makes the power relations of contemporary neoliberal capitalism rawer and cruder than in the Fordist era (Cooper 2008).

Because genetic information, like psychological traits or neural features, is unevenly distributed, this system is not only inherently discriminatory but also racist at some basic level of the term. Patricia Clough (2008) explores this aspect of the contemporary political economy by analyzing the public debates on the availability of pharmaceutical drugs against HIV, or large-scale vaccines against malaria, to mention just a few contemporary examples of posthuman management of Life. A whole underclass of genetically overexposed and socially underinsured disposable bodies is engendered, both in the western world and within the emerging global economies. This kind of population control goes beyond Foucault's analysis of the biopolitical, as it does not function by techniques of discipline and control, but rather by biogenetic farming of data, and by "biopiracy" (Shiva 1997). As Mark Halsey puts it:

> Where once the sole objective was to control the insane, the young, the feminine, the vagrant and the deviant, the objective in recent times has

been to arrest the nonhuman, the inorganic, the inert—in short, the so-called "natural world." (2006, 15)

This is posthuman *zoē*-politics, not biopolitical governmentality.

Again, monistic posthuman philosophy is of great assistance to think through these challenging new historical conditions. Reading Deleuze through the lenses of Massumi, Clough studies the new mechanisms of capture, not of liberal individuals, but bio-genetic "dividuals"

> statistically configured in populations that surface as profiles of bodily capacities, indicating what a body can do now and [what capacities it might be able to unfold] in the future. The affective capacity of bodies, statistically simulated as risk factors, can be apprehended as such without the subject, even without the individual subject's body, [this results in] bringing forth competing bureaucratic procedures of control and political command in terms of securing the life of populations. (2008, 18)

The new interconnections between forms of political control and the estimation of genetic risk factors constitute a technique that Foucault defined as racism, as it configures—it engenders as "raced"—entire populations in a hierarchical scale, this time not determined by pigmentation, but by other genetic characteristics. Because the aim of this political exercise is to estimate a given population's chance of survival or of extinction, the biopolitical management of the living is not only transversal across species and *zoē*-driven, but also inherently linked to death. This is the death-bound or necropolitical face of post-anthropocentrism and the core of its inhuman(e) character: "it permits the healthy life of some populations to necessitate the death of others, marked as nature's degenerate or unhealthy ones" (Clough 2008, 18).

The political management of embodied subjects nowadays can no longer be understood within the visual economy of biopolitics in Foucault's (1978) sense of the term. The representation of embodied subjects is not visual in the sense of being scopic, as in the post-Platonic sense of the simulacrum. Nor is it specular, as in the psychoanalytic mode of redefining vision, within a dialectical scheme of oppositional recognition of self and/as other. The representation of embodied subjects has been replaced by simulation and has become schizoid, or internally disjointed. Contemporary representation also tends to be spectral: the body doubles as the potential corpse it has always been, and is represented as a self-replicating system that is caught in a visual economy of endless circulation (Braidotti 2002). The contemporary social imaginary has immersed carnal matters—bodies and their derivatives—in a logic of boundless circulation, and thus suspended them somewhere beyond the life and death cycle of the image itself. The biogenetic economy has consequently become forensic in its relationship to the body as virtual corpse and in the quest to control a life that cannot be contained within anthropomorphic parameters. Contemporary embodied subjects have to be accounted for in terms of their surplus value as biogenetic containers on the one hand, and as visual commodities circulating in a global media circuit of cash flow on the other hand. They are therefore doubly mediated by biogenetic and by informational codes. The central insight of Foucault's political anatomy remains valid: biopower also involves the management of dying. In other words, the question of the governance of life contains that of extinction as well. In order to deploy the full ethical and political potential of this brilliant insight, however, we do need to move beyond Foucault.

NECROPOLITICS

Vital politics shifts the boundaries between life and death and consequently deals not only with the government of the living, but also with practices of dying. Most of these are linked to inhuman(e) social and political phenomena linked to advanced capitalism, such as poverty, famine, and homelessness, which Zillah Eisenstein aptly labels "global obscenities" (1998). Vandana Shiva (1997) stresses the extent to which biopower has already turned into a form of "biopiracy", which calls for very grounded and concrete political analyses. Thus, the bodies of the empirical subjects who signify difference (woman and LGBT; indigenous or native; animal, earth, or natural "others") have become the disposable bodies of the global economy.

Contemporary capitalism is indeed "biopolitical" in that it aims at controlling all that lives, but because Life is not the prerogative of humans only, it opens up a *zoë*-political or post-anthropocentric dimension. This inaugurates both a negative or reactive form of panhuman planetary bond, which recomposes humanity around a commonly shared bond of vulnerability, and also new modes of connection between humans and the other species. If solidarity and mutual dependence are the key terms for the latter, death and mutual destruction are the common denominators for the former assemblage.

Let me give you some examples of contemporary ways of dying, to illustrate this necropolitical economy. The posthuman aspects of globalization encompass many phenomena that, while not being a priori inhumane, still trigger significant destructive aspects. The postsecular condition, with the rise of religious extremism in a variety of forms, including Christian fundamentalism, entails a political regression of the rights of women, homosexuals, and all sexual minorities. Significant signs of this regression are the decline in reproductive rights and the rise of violence against women and LGBT people. The effect of global financial networks and unchecked hedge funds has been an increase in poverty, especially among women and the young, affected by the disparity in access to the new technologies. The status of children is a chapter apart; from forced labor to the child-soldier phenomenon, childhood has been violently inserted into infernal cycles of exploitation. On a different score, bodily politics has shifted, with the simultaneous emergence of cyborgs on the one hand and renewed forms of vulnerability on the other. Thus, next to the proliferation of pandemics like SARS, Ebola, HIV, bird flu, and others, more familiar epidemics have also returned, notably malaria and tuberculosis, so much so that health has become a public policy issue as well as a human rights concern.

The point is that Life/*zoē* can be a threatening force, as well as a generative one. A great deal of health and environmental concerns, as well as geopolitical issues, simply blur the distinction between life and death. In the era of biogenetic capitalism and nature-culture continuum, *zoē* has become an infrahuman force and all the attention is now drawn to the emergency of disappearing nature. For instance, the public discourse about environmental catastrophes or "natural" disasters—the Fukushima nuclear plant and the Japanese tsunami, the Australian bushfires, hurricane Katrina in New Orleans, etc.—accomplishes a significant double-bind: it expresses a new ecological awareness, while reinserting the distinction between nature and culture. As Protevi argues (2009), this results in the paradoxical renaturalization of our biotechnologically mediated environment. The geopolitical forces are simultaneously renaturalized and subjected to the old hierarchical power relations determined by the dominant politics of the anthropomorphic subject. Public discourse has become simultaneously moralistic about the inhuman forces of the environment and quite hypocritical in perpetuating anthropocentric arrogance. This position results in the denial of the manmade structure of the catastrophes that we continue to attribute to forces beyond our collective control, like the earth, the cosmos or "nature". Our public morality is simply not up to the challenge of the scale and the complexity of damages engendered by our technological advances. This gives rise to a double ethical urgency: firstly, how to turn anxiety and the tendency to mourn the loss of the natural order into effective social and political action. Secondly, how to ground such an action in the responsibility for future generations, in the spirit of social sustainability which I have also explored elsewhere (Braidotti 2006).

Contemporary politics has more than its fair share of cruelty to account for. New scholarship has concentrated on the brutality of today's wars and the renewed expressions of violence targeted not only at the government of the living, but also multiple practices of dying. Biopower and necropolitics are two sides of the same coin, as Achille Mbembe (2003) brilliantly argues. The explosion of discursive interest in the politics of life itself, in other words, affects also the geopolitical dimension of death and of killing. Mbembe expands Foucault's insight in the direction of a more grounded analysis of the biopolitical management of survival. Aptly renaming it "necropolitics", he defines this power essentially as the administration of death: *"the generalized instrumentalization of human existence and the material destruction of human bodies and populations"* (Mbembe 2003, 14). And *not* only human, I might add, but also planetary.

The post-Cold War world has seen not only a dramatic increase in warfare, but also a profound transformation of the practice of war as such. New forms of warfare entail simultaneously the breath-taking efficiency of "intelligent", unmanned, technological weaponry on the one hand, and the rawness of dismembered and humiliated human bodies on the other. Posthuman wars breed new forms of inhumanity. The implications of this approach to necro-power are radical: it is not up to the rationality of the Law and the universalism of moral values to structure the exercise of power, but rather the unleashing of the unrestricted sovereign right to kill, maim, rape, and destroy the life of others. This political economy structures the attribution of different degrees of "humanity" according to hierarchies that are disengaged from the old dialectics and unhinged from biopolitical logic. They fulfil instead a more instrumental, narrow logic of opportunistic exploitation of the life in each body, which is generic and not only individual.

Contemporary necropolitics has extended the politics of death on a global scale. The new forms of industrial-scale warfare rest upon the commercial privatization of the army and the global reach of conflicts, which deterritorialize the use of and the rationale for armed service. Reduced to *"infrastructural warfare"* (Mbembe 2003, 29), and to a large-scale logistical operation (Virilio 2002), war aims at the destruction of all the services that allow civil society to function: roads, electricity lines, airports, hospitals, and other necessities. The old-fashioned army has now mutated into "[u]rban militias, private armies, armies of regional lords, private security firms, and state armies all claim[ing] the right to exercise violence or to kill" (Mbembe 2003, 32). As a result, as a political category, the "population" has also become disaggregated into: "rebels, child soldiers, victims or refugees, or civilians incapacitated by mutilation or massacred on the model of ancient sacrifices, while the 'survivors', after a horrific exodus, are confined to camps and zones of exception" (Mbembe 2003, 34). Many contemporary wars, led by western coalitions under the cover of "humanitarian aid," are often neocolonial exercises aimed at protecting mineral extraction and other essential geophysical resources needed by the global economy. In this respect, the "new" wars look more like privatized conflicts and guerrilla or terrorist attacks, than the traditional confrontation of enlisted and nationally indexed armies.

Arjun Appadurai has also provided incisive analyses of the new "ethnocidal violence" of the new forms of warfare which involve friends, kinsmen, and neighbors, and expresses his horror at the indignity of these conflicts

> involving mutilation, cannibalism, rape, sexual abuse, and violence against civilian spaces and populations. Put simply, the focus here is on bodily brutality perpetrated by ordinary persons against other persons with whom they may have—or could have—previously lived in relative amity. (1998, 907)

Chomsky (quoted in Davies 2008, 134) broadens this analysis by looking at the configuration of contemporary global wars, and he comments shrewdly on a political economy that he labels "the new military humanism" of the humanitarian interventions:

> armed with the technology of global devastation and the jargon of pulp fiction, tabloid headlines and Playstation games: the War on terror, the Clash of Civilisations, the Axis of Evil, Operation "Shock and Awe." Those adventures set out to save the civilised world (*"homo humanus"*) from its enemies (*"homo barbarus"*) under the venerable banners of liberty, decency and democracy.

This deployment of technologically mediated violence—through Western new drone technology for instance—cannot be adequately described in terms of disciplining the body, fighting the enemy, or even as the techniques of a society of control. We have rather entered the era of orchestrated and instrumental massacres, a new "semiosis of killing", leading to the creation of multiple and parallel *"death-worlds"* (Mbembe 2003, 37, 40). These necropolitical modes of governance also circulate as infotainment in global media circuits, according to the logic of double mediation, which combines the body-politics of control and physical elimination with unprecedented degrees of media exposure.

War and surveillance technologies operate without direct human intervention and, in this respect, can be seen as post-anthropocentric; they are also reshaping the practice of surveillance in the social field. Border controls of immigration and the smuggling of people are major aspects of the contemporary inhuman condition and central players in the necropolitical game. Diken (2004) argues that refugees and asylum seekers become another emblem of the contemporary necropower, because they are the perfect instantiation of the disposable humanity that preoccupies Agamben and thus constitute the ultimate necropolitical subject. The proliferation of detention and high-security camps and prisons within the once civic-minded space of European cities is an example of the inhuman face of Fortress Europe.

Duffield (2008, 149) pushes the necropolitical sociopolitical analysis further and makes a distinction between developed or insured humans and underdeveloped or uninsured humans: "Developed life is sustained primarily through regimes of social insurance and bureaucratic protection historically associated with industrial capitalism and the growth of welfare states." The distinction and the tensions between these two categories constitute the terrain for the "global civil war", which is Duffield's definition of globalized advanced capitalism. The link to colonialism is clear: decolonization created nation-states whose people, once subjected to colonial rule, in some cases enslaved and generally exploited, are now free to circulate globally.

These people constitute a large proportion of the unwanted immigrants, refugees, and asylum seekers who are contained and locked up across the developed world. In a twist not deprived of ironical force, world migration is perceived as a particular threat in Europe precisely because it endangers Europe's main social infrastructure: the welfare state. The growing range of warfare weapons and killing techniques raises critical questions about the status of death as an object of contemporary political analysis.

The broad range of ways of dying and the changing techniques of killing, which now combine sophisticated technological mediation with necropolitical brutality, indicate that death as a concept remains caught in a contradiction. On the one hand, death is central to political theory and practice, as exemplified by the new forms of surveillance, confinement, and killing that are at work within a fast-expanding technological context. Death is also, on the other hand, understated and underexamined as a term in critical theory; as a concept, it tends to be stuck in a metaphysical block, while the repertoire of new ideas and political insights around Life and biopower proliferates and diversifies our understanding.

Fortunately, new posthuman theory is filling this vacuum and making important contributions to rethinking the instance of death. Patrick Hanafin, for example, suggests that renewed interest in necropolitics, coupled with a transversal vision of posthuman subjectivity, may help us provide a political and ethical counternarrative to "the imposed bounded subject of liberal legalism" (2010, 132). For Hanafin, this involves a move from the traditional location of mortality as the defining, quasi-metaphysical horizon of being. The majoritarian masculine legal social contract is built on the desire to survive. This is not a politics of empowerment, but one of entrapment in an imagined natural order that in our system translates into a biopolitical regime of discipline and control of bodies. What this means is that we are recognized as full citizens only through the position of victims' loss and injury and the forms of reparation that come with it. Posthuman necropolitical political and legal theory raise the question of what political theory might look like if it were not based on the negative instances of wound and loss.

Hanafin proposes to take the necropolitical dimension seriously by shifting away from thinking of legal subjectivity as death bound to thinking about singularities without identity who relate intimately to one another and the environment in which they are located. This insight points towards a posthuman critical politics of rights. We see here how another fundamental binary of western philosophical thinking gets uncoupled: that of a political life qualified by death, as opposed to a political and legal philosophy which valorizes our mortal condition and creates a politics of survival. This is a post-identitarian position which encourages us, following Virginia Woolf, to adopt a mode of thinking "as if already gone", that is to say to think with and not against death. The emphasis on the death-life continuum may, according to Hanafin, constitute the ultimate threat to a legal system built on the confining horizon of the metaphysics of mortality.

William Connolly's "politics of becoming" (1999, ch. 2, "Suffering, Justice, and the Politics of Becoming") argues a similar case: against necropolitical destruction, we need to develop an "ethos of engagement" (ch. 6, "An Ethos of Engagement") with existing social and political givens—including the horrors of our times—in order to bring about counter-effects, that is to say, unexpected consequences and transformations. Critical theory needs to engage with the present, becoming "worthy of the times", while resisting the violence, horror, and injustices of the times (Braidotti 2013).

AFFIRMATIVE POLITICS

Affirmative politics is my answer to these challenges and contradictions. It indicates the process of transmuting negative passions into productive and sustainable praxis, which does not deny the reality of horrors, violence, and destruction of our times but proposes a different way of dealing with them. What is positive in the ethics of affirmation is the belief that negative affects can be transformed. This implies a dynamic view of all affects, even the traumas that freeze us in pain, horror, or mourning. The slightly depersonalizing effect of the negative or traumatic event involves a loss of ego boundaries, which is the source of both pain and potentially energetic reactions. Multilocality and multidirectional memory (Rothberg 2009) are the affirmative translation of this negative sense of loss. Let me illustrate this controversial point with an example drawn from diasporic subjects.

Following Glissant (1990), "becoming-nomadic" marks the process of positive transformation of the pain of loss into the active production of multiple forms of belonging and complex allegiances. Every event contains within it the potential for being overcome and overtaken—its negative charge can be transposed. The moment of the actualization is also the moment of its neutralization. The ethical subject is the one with the ability to grasp the freedom to depersonalize the event and transform its negative charge. Affirmative ethics puts the motion back into *e-motion* and the active back into activism, introducing movement, process, becoming. This shift makes all the difference to the patterns of repetition of negative emotions. It also reopens the debate on secularity, in that it actually promotes an act of faith in our collective capacity to endure and to transform.

What is negative about negative affects is not a normative value judgment but rather the effect of arrest, blockage, rigidification, that comes as a result of a blow, a shock, an act of violence, betrayal, a trauma, or just intense boredom. Negative passions do not merely destroy the self, but also harm the self's capacity to relate to others—both human and non-human others—and thus to grow in and through others. Negative affects diminish our capacity to express the high levels of interdependence, the vital reliance on others, that are the key to both a non-unitary vision of the subject and to affirmative ethics. Again, the vitalist notion of Life as *zoē* is important here because it stresses that the Life I inhabit is not mine, it does not bear my name—it is a generative force of becoming, of individuation and differentiation: a-personal, indifferent, and generative. What is negated by negative passions is the power of life itself—its *potential*—as the dynamic force, which unfolds through vital flows of connections and becoming. And this is why they should neither be encouraged nor should we be rewarded for lingering around them too long. Negative passions are black holes.

This is an antithesis of the Kantian moral imperative to avoid pain, or to view pain as the obstacle to moral behavior. It displaces the grounds on which Kantian negotiations of limits can take place. The imperative not to do unto others what you would not want done to you is not rejected as much as enlarged. In affirmative ethics, the harm you do to others is immediately reflected on the harm you do to yourself, in terms of loss of *potentia*, positivity, capacity to relate, and hence freedom. Affirmative ethics is not about the avoidance of pain, but rather about transcending the resignation and passivity that ensue from being hurt, lost, and dispossessed. One has to become ethical, as opposed to applying moral rules and protocols as a form of self-protection: one has to endure. Endurance is the Spinozist code word for this process. Endurance has a spatial side to do with the space of the body as an enfleshed field of actualization of passions or forces. It produces affectivity and joy, as in the capacity for being affected by these forces, to the point of pain or extreme pleasure. Endurance points to the struggle to sustain the pain without being annihilated by it and hence opens up to a temporal dimension, or duration in time.

Affirmative ethics is based on the praxis of enduring by constructing positivity, thus propelling new social conditions and relations into being out of injury and pain. It actively constructs energy by transforming the negative charge of these experiences, even in intimate relationships where the dialectics of domination is at work (Benjamin 1988). For Deleuze and Guattari, the timeline for this political activity is that of *Aion*, the continuous tense of becoming, which is different from working within or against the *Chronos* sequence of the hegemonic political order. We need to actively and collectively work toward a refusal of horror and violence—the inhuman aspects of our present—and to turn this into the construction of affirmative alternatives. Such an approach aims to bring affirmation to bear on undoing existing arrangements, so as to actualize productive alternatives.

As critical thinkers we are always trying to be worthy of the times, to interact with them, in order to resist them, that is to say to differ from them. It is a form of *amor fati*, a way of living up to the intensities of life, so as to be worthy of all that happens to us—to live out our shared capacity to affect and to be affected. Beyond negative dialectics, we need to disengage the process of subject formation from negativity to attach it to affirmative otherness. This involves a change of conceptual references: reciprocity is no longer defined dialectically as the struggle for recognition, but rather auto-poietically as mutual definition or specification. Violence is bypassed by the ontological pacifism of a system based on monistic vital materialism and on the processes of differing that rest upon it.

Amor fati is not passive fatalism, but a pragmatic and liable engagement with the present in order to collectively construct conditions that transform and empower our capacity to act ethically and produce social horizons of hope, or sustainable futures. The ethical cultivation of positivity, moreover, does not exclude, either logically or practically, situations of antagonism or conflict. If we follow the Spinozist rule and de-psychologize the discussion about affirmation and negativity, to cast it instead in terms of an ethics or an ethology of forces, it follows that some of the relations we are likely to establish with others may well be of the antagonistic kind. What matters—and this is the shift of perspective introduced by affirmative ethics—is to resist the habit of inscribing antagonistic relations in a logic of dialectical negativity. The transcendence of dialectics, in other words, has to be enacted in the inner structure of relations—of the interpersonal as well as the non-human kind. Antagonism need not be inscribed in the lethal logic of the dialectical struggle of consciousness. This habit of thought needs to be resisted and recoded away from the necessity to establish negativity as the precondition for the process of subject-formation.

In other words, the "worthiness" of an event—that which ethically compels us to engage with it, is not its intrinsic or explicit value according to given standards of moral or political evaluation, but rather the extent to which it contributes to conditions of becoming. It is a vital force to move beyond the negative. Protevi argues (2009) that in this nomadic view, the political is the nonreactive and the non-habitual response of reactive engagement with the events of one's life that can reshape one's becoming. A sort of creative disorganization of the negative aims at keeping life immanent, non-unitary and non-reified according to dominant codes and hegemonic traditions of both life and thought.

My ethical stance is that there is no logical necessity to link political subjectivity to oppositional consciousness and reduce them both to violence and negativity. Political activism can be all the more effective if it disengages the process of consciousness–raising from negativity and connects it instead to creative affirmation and the actualization of virtual potentials. Because these are by definition not contained in the present conditions, and cannot emerge from them, they have to be brought about or generated creatively by a qualitative leap of the collective praxis and of our ethical imagination.

NOTE

1 This includes Deleuze and Guattari (1977; 1987), Guattari (1995), Glissant (1997), Balibar (2002), and Hardt and Negri (2000).

REFERENCES

Agamben, Giorgio. 1998. *Homo Sacer: Sovereign Power and Bare Life*. Stanford, CA: Stanford University Press.

Appadurai, Arjun. 1998. Dead Certainty: Ethnic Violence in the Era of Globalization. *Development and Change* 29(4): 905–25.

Arendt, Hannah. 1951. *The Origins of Totalitarianism*. New York: Harcourt.

Balibar, Etienne. 2002. *Politics and the Other Scene*. London: Verso.

Barad, Karen. 2003. Posthumanist performativity: Toward an understanding of how matter comes to matter. *Signs* 28 (3): 801–31.

Bataille, Georges. 1988. *The Accursed Share: An Essay on General Economy, Vol. 1*. Trans. Robert Hurley. New York: Zone Books.

Beck, Ulrich. 1999. *World Risk Society*. Cambridge: Polity Press.

Benhabib, Seyla. 1996. *The Reluctant Modernism of Hannah Arendt*. Thousand Oaks, CA: Sage.

Benjamin, Jessica. 1988. *The Bonds of Love: Psychoanalysis, Feminism, and the Problem of Domination*. New York: Pantheon Books.

Bennett, Jane. 2010. *Vibrant Matter: A Political Ecology of Things*. Durham, NC: Duke University Press.

Braidotti, Rosi. 1991. *Patterns of Dissonance*. Cambridge: Polity Press.

Braidotti, Rosi. 2002. *Metamorphoses: Towards a Materialist Theory of Becoming*. Cambridge: Polity Press.

Braidotti, Rosi. 2006. *Transpositions: On Nomadic Ethics*. Cambridge: Polity Press.

Braidotti, Rosi. 2011a. *Nomadic Subjects: Embodiment and Sexual Difference in Contemporary Feminist Theory*, 2nd ed. New York: Columbia University Press.

Braidotti, Rosi. 2011b. *Nomadic Theory. The Portable Rosi Braidotti*. New York: Columbia University Press.

Braidotti, Rosi. 2013. *The Posthuman*. Cambridge: Polity Press.

Brown, Wendy. 1993. Wounded Attachments. *Political Theory* 21 (3): 390–410.

Brown, Wendy. 2006. *Regulating Aversion: Tolerance in the Age of Identity and Empire*. Princeton, NJ: Princeton University Press.

Bryden, Mary, ed. 2001. *Deleuze and Religion*. London: Routledge.

Butler, Judith. 2004. *Undoing Gender*. London: Routledge.

Clough, Patricia. 2008. The Affective Turn: Political Economy, Biomedia and Bodies. *Theory, Culture & Society* 25 (1): 1–23.

Connolly, William E. 1999. *Why Am I Not a Secularist?* Minneapolis, MN: University of Minnesota Press.

Cooper, Melinda. 2008. *Life as Surplus: Biotechnology & Capitalism in the Neoliberal Era*. Seattle, WA: University of Washington Press.

Cooper, Melinda and Catherine Waldby. 2014. *Clinical Labor: Tissue Donors and Research Subjects in the Global Bioeconomy*. Durham, NC: Duke University Press.

Davies, Tony. 2008. *Humanism*. Abingdon, Oxfordshire: Routledge.

Deleuze, Gilles. 1968. *Spinoza et le problème de l'expression*. Paris: Minuit.

Deleuze, Gilles. 1992. Postscript on the Societies of Control. *October* 59 (Winter): 3–7.

Deleuze, Gilles and Michel Foucault. 1977. Intellectuals and Power. In *Language, Counter-Memory and Practice: Selected Essays and Interviews by Michel Foucault*. Ed. Donald F. Bouchard, 205–17. Ithaca, NY: Cornell University Press.

Deleuze, Gilles and Félix Guattari. 1977. *Anti-Oedipus: Capitalism and Schizophrenia*. Trans. Robert Hurley, Mark Seem, and Helen R. Lane. New York: Viking Press.

Deleuze, Gilles and Félix Guattari. 1980. *A Thousand Plateaus: Capitalism and Schizophrenia*. Trans. Brian Massumi. 1987. Minneapolis, MN: University of Minnesota Press.

Diken, Bülent. 2004. From Refugee Camps to Gated Communities: Biopolitics and the End of the City. *Citizenship Studies* 8 (1): 83–106.

Duffield, Mark. 2008. Global Civil War: The Non-Insured, International Containment and Post-Interventionary Society. *Journal of Refugee Studies* 21(2): 145–65.

Eisenstein, Zillah. 1998. *Global Obscenities: Patriarchy, Capitalism, and the Lure of Cyberfantasy*. New York: New York University Press.

Esposito, Roberto. 2008. *Bíos: Biopolitics and Philosophy*. Trans. Timothy Campbell. Minneapolis, MN: University of Minnesota Press.

Flannery, Tim. 1994. *The Future Eaters: An Ecological History of the Australasian Lands and People*. New York: Grove Press.

Foucault, Michel. 1970. *The Order of Things: An Archaeology of the Human Sciences*. New York: Pantheon Books.

Foucault, Michel. 1977. *Discipline and Punish: The Birth of the Prison*. Trans. Alan Sheridan. New York: Pantheon Books.

Foucault, Michel. 1978. *The History of Sexuality, Vol. 1*. Trans. Robert Hurley. New York: Pantheon Books.

Fraser, Mariam, Saraha Kember and Celia Lury, eds. 2006. *Inventive Life: Approaches to the New Vitalism*. London: Sage.

Fukuyama, Francis. 2002. *Our Posthuman Future: Consequences of the BioTechnology Revolution*. New York: Picador.

Gatens, Moira and Genevieve Lloyd. 1999. *Collective Imaginings: Spinoza, Past and Present*. New York: Routledge.

Gilroy, Paul. 2000. *Against Race. Imaging Political Culture beyond the Colour Line*. Cambridge, MA: Harvard University Press.

Glissant, Édouard. 1990. *Poetics of Relation*. Trans. Betsy Wing. 1997. Ann Arbor, MI: University of Michigan Press.

Grosz, Elizabeth. 2004. *The Nick of Time*. Durham, NC: Duke University Press.

Guattari, Félix. 1995. *Chaosmosis: An Ethico-Aesthetic Paradigm*. Trans. Paul Bains and Julian Pefanis. Sydney: Power.

Guattari, Félix. 2000. *The Three Ecologies*. Trans. Ian Pindar and Paul Sutton. London: Athlone Press.

Halsey, Mark. 2006. *Deleuze and Environmental Damage*. London: Ashgate.

Hanafin, Patrick. 2010. On reading *Transpositions*: A response to Rosi Braidotti's *Transpositions: On Nomadic Ethics*. *Subjectivities* 3 (2): 131–5.

Haraway, Donna J. 1997. *Modest_Witness@Second_Millenium.FemaleMan©_Meets_OncoMouseTM: Feminism and Technoscience*. New York: Routledge.

Haraway, Donna J. 2007. *When Species Meet*. Minneapolis, MN: University of Minnesota Press.

Hardt, Michael and Antonio Negri. 2000. *Empire*. Cambridge, MA: Harvard University Press.

Hochschild, Arlie, R. 2000. Global Care Chains and Emotional Surplus Value. In *On The Edge: Living with Global Capitalism*. Ed. Will Hutton and Anthony Giddens, 130–46. London: Jonathan Cape.

Holland, Eugene W. 2011. *Nomad Citizenship: Free-Market Communism and the Slow-Motion General Strike*. Minneapolis, MN: University of Minnesota Press.

Lemke, Thomas. 2011. Biopolitics: *An Advanced Introduction*. Trans. Eric Frederick Trump. New York: New York University Press.

Lloyd, Genevieve. 1994. *Part of Nature: Self-Knowledge in Spinoza's Ethic*. Ithaca, NY: Cornell University Press.

Lloyd, Genevieve. 1996. *Spinoza and the Ethics*. London: Routledge.

MacPherson, Crawford Brough. 1962. *The Political Theory of Possessive Individualism: Hobbes to Locke*. Oxford: Oxford University Press.

Marks, John. 1998. *Gilles Deleuze: Vitalism and Multiplicity*. London: Pluto Press.

Massumi, Brian. 1992. *A User's Guide to Capitalism and Schizophrenia*. Boston: MIT Press.

Mbembe, Achille. 2003. Necropolitics. Trans. Libby Meintjes. *Public Culture* 15 (1): 11–40.

Mouffe, Chantal. 2005. *On the Political*. New York: Routledge.

Moulier-Boutang, Yann. 2012. *Cognitive Capitalism*. Cambridge: Polity Press.

Muehlebach, Andrea. 2012. *The Moral Neoliberal: Welfare and Citizenship in Italy*. Chicago: University of Chicago Press.

Norton, Anne. 2004. *Leo Strauss and the Politics of American Empire*. New Haven, CT: Yale University Press.

Oksala, Johanna. 2012. *Foucault, Politics, and Violence*. Evanston, IL: Northwestern University Press.

Protevi, John. 2009. *Political Affect: Connecting the Social and the Somatic*. Minneapolis, MN: University of Minnesota Press.

Rabinow, Paul. 2003. *Anthropos Today: Reflections of Modern Equipment*. Princeton, NJ: Princeton University Press.

Rose, Nikolas. 2001. The Politics of Life Itself. *Theory, Culture & Society* 18 (6): 1–30.

Rose, Nikolas. 2007. *The Politics of Life Itself: Biomedicine, Power, and Subjectivity in the Twenty-First Century*. Princeton, NJ: Princeton University Press.

Rothberg, Michael. 2009. *Multidirectional Memory: Remembering the Holocaust in the Age of Decolonization*. Stanford, CA: Stanford University Press.

Schmitt, Carl. 1996. *The Concept of the Political*. Trans. George Schwab. Chicago: University of Chicago Press.

Shiva, Vandana. 1997. *Biopiracy: The Plunder of Nature and Knowledge*. Boston, MA: South End Press.

Simondon, Gilbert. 2012. *Du mode d'existence des objets techniques*, Paris: Aubier.

Toscano, Alberto. 2005. Axiomatic. *The Deleuze Dictionary*. Ed. Adrian Parr, 17–8. Edinburgh: Edinburgh University Press.

Touraine, Alain. 2001. *Beyond Neoliberalism*. Cambridge: Polity.

Virilio, Paul. 2002. *Desert Screen: War at the Speed of Light*. London: Continuum.

3 Rethinking Biopolitics
The New Materialism and the Political Economy of Life

Thomas Lemke

Forty years ago, the French philosopher and historian Michel Foucault first pronounced in a lecture the semantic merger of life and politics that would shape his subsequent work and the ensuing theoretical debates (Foucault 2000a, 137).[1] His notion of "biopolitics" points to a historical shift at the threshold of modernity. According to Foucault, biopolitics marks a discontinuity in political practice since it places life at the center of political rationalities and technologies. He distinguishes historically and analytically between two dimensions of biopolitics: the disciplining of the individual body and the social regulation of the population. Furthermore, Foucault's concept signals a theoretical critique of the sovereign paradigm of power. According to this model, power is exercised as interdiction and repression in a framework of law and legality. In contrast, Foucault stresses the productive capacity of power, which cannot be reduced to the ancient sovereign "right of death". While sovereignty seized hold of life in order to suppress it, the new life-administering power is dedicated to inciting, reinforcing, monitoring, and optimizing the forces under its control (Foucault 1980; 2003).

There has been a remarkable interest in the Foucauldian notion of biopolitics since his death in 1984. It is possible to discern several distinct lines of reception. First there are theoretical proposals that seek to "update" Foucault's work for an analytics of contemporary societies. Gilles Deleuze and Donna Haraway have explicitly challenged the logics of discipline and the idea of the body as an integral natural entity by pointing to new mechanisms of control and changing medical and scientific concepts of the body (Deleuze 1995; Haraway 1991). Secondly, we can identify endeavors that suggest an alternative genealogy of biopolitics. Giorgio Agamben and Achille Mbembe have emphasized that biopolitical interventions are not limited to Western modernity; they can be traced back to Greek antiquity (Agamben 1998) and extend beyond the Western hemisphere into the (post-)colonial past (Mbembe 2003). The third area of inquiry concentrates on the mode of the political: what rationality or logic characterizes biopolitical practices and what counterforces do these practices mobilize? How does biopolitics differentiate itself analytically and historically from other eras and from other political formations? Here the writings of Michael Hardt and Antonio Negri, and the work of Roberto Esposito are of crucial interest. While the former identify a new form of biopolitical production, characterized by the tight interlocking of economic structures with juridico-political arrangements, Esposito draws our attention to a "paradigm of immunization" which, he argues, allows the two opposing dimensions of biopolitics—advancement and development of life, on the one hand, and its destruction and elimination, on the other—to be conceived of as two constitutive aspects of modern political thought (Hardt and Negri 2000; 2004; Esposito 2008). The fourth line of reception originates in science and technology studies, medical sociology, and anthropology, as well as in feminist theory and gender studies. The main focus here is on the matter of life. If, as a consequence of bio-scientific innovations, the living body is regarded today less as an organic substratum than as molecular software that can be read and rewritten, then the question as to the foundations, means, and ends of biopolitics needs to be posed in a different manner. In this area of research the writings of Paul Rabinow and Nikolas Rose are of vital importance (Rabinow 1992; Rabinow and Rose 2006; Rose 2007).[2]

My contribution explores promising areas of research that have so far received little attention in the work on biopolitics. I will focus especially on two of them: first on the so-called new materialism, which proposes to reconsider and reevaluate materiality by conceiving of matter as active, forceful, and plural rather than passive, inactive, and unitary (Alaimo and Hekman 2008; Bennett 2004; Colebrook 2008; Coole and Frost 2010b); and secondly on work on the "bioeconomy", which investigates the systematic relations between (neoliberal) capitalism on the one hand and changing concepts of life and the emergence of a biotech industry on the other (for an overview see Helmreich 2008; Birch and Tyfield 2013). As I will argue, both lines of research open up new directions in the analysis of biopolitics by inquiring into the mode of politics and the matter of life. Both go beyond Foucault's original formulation of biopolitics, which was centered around the poles of population and individual, and seek to address a different topography of power. While the new materialism questions the traditional concept of life that takes for granted the distinction between organic and inorganic, or between matter and life, in order to investigate "vibrant matter" (Bennett 2010), theorists of the bioeconomy shift the analytical interest from the state, population policies, and national governance to economic processes and capitalist strategies that take the preservation or enhancement of vitality and well-being as their object (see e.g., Cooper 2008; Rajan 2006).

After briefly presenting the two areas of research, I will explore the perspectives of Foucault's work vis-à-vis the two challenges. I will argue for a closer realignment of the concepts of biopolitics and governmentality as a way of addressing the two challenges and overcoming the (anthropocentric) limitations of Foucault's work. The focus is on an alternative concept of government that Foucault only briefly discusses in his lectures on governmentality: the idea that "to govern means to govern things" (2007, 97).

I suggest that the idea of a "government of things" makes it possible to arrive at a different concept of matter and of the political economy of life.

1. CHALLENGE I: FROM BIOPOLITICS TO BIOECONOMY

Biotechnological innovations and biomedical developments have often generated high expectations and hopes. They have been associated with the idea that new markets, services, and products will emerge to profoundly change and revolutionize societies and economies. This vision has been taken up in ambitious political action plans by the Organisation for Economic Cooperation and Development (OECD) and the European Union, and in national initiatives, and postulates that the boundaries and the substance of the economic have to be redefined. The economy, according to this projection, will soon transform itself into a "bioeconomy".

In 2006, the OECD published *The Bioeconomy to 2030: Designing a Policy Agenda.* "Bioeconomy" is defined in this programmatic text as a society's sum total of economic operations which use the potential value of biological products and processes in order to create new growth and prosperity for citizens and nations (OECD 2006, 3). At approximately the same time as the OECD document appeared, the European Commission (EC) adopted a plan with a similar goal. The EC stressed the potential of a "knowledge-based bioeconomy" (KBBE) that would simultaneously strengthen European competitiveness in international markets, help to protect the environment, and develop more sustainable forms of energy, food, and biological renewable materials (European Commission 2005).

Both the EC's and the OECD's programs are meant to promote new products and services derived from bio-scientific innovations, as are national initiatives like the creation of the Bioeconomy Council in Germany in 2009 to recommend strategies and action plans for a sustainable bioeconomy.[3] Central to this vision, therefore, is the constitution and regulation of markets rather than a fundamental realignment of the economy, as the term "bioeconomy" tends to suggest.

This enlarged meaning of the word appears in academic works, which, in contrast to the political programs, observe a decisive and structural transformation of economic relations. In the past decade, scholars have proposed a variety of new terms and concepts to critically evaluate the articulation of biotechnological innovations and transformations in economic structures and contemporary capitalist regimes. These propositions include: "recombinant capital" (Thacker 2005), "bio-informational capitalism" (Peters 2012), "biomedical mode of reproduction" (Thompson 2005), "biovalue" (Waldby 2000), "biocapital" (Sunder Rajan 2006), "lively capital" (Sunder Rajan 2012), and "life as surplus" (Cooper 2008).

The concepts often differ significantly, and they all merit an extensive discussion—something that is beyond the scope of this chapter. Here,

I will focus on one of the most influential accounts conceptualizing the link between capitalism and the life sciences: the concept of biocapital in Kaushik Sunder Rajan's book *Biocapital: The Constitution of Postgenomic Life* (2006). While Sunder Rajan did not coin the term "biocapital",[4] his book has done much to popularize the concept by providing an insightful account of the relationship between bioscientific innovations and transformations in contemporary capitalism. From a theoretical standpoint, Sunder Rajan links Foucault's concept of biopolitics to Marx's critique of political economy, situating both within his anthropological analysis (ibid., 3–15, 78–79). His empirical thesis is that the emergence of the biosciences marks a new form and a new phase of capitalism (3). He argues that the constitution of biocapital can be mapped through a dual perspective:

> On the one hand, what forms of alienation, expropriation, and divestiture are necessary for a "culture of biotechnology innovation" to take root? On the other hand, how are individual and collective subjectivities and citizenships both shaped and conscripted by these technologies that concern "life itself"? (Sunder Rajan 2006, 78)

The book is based on a multiplicity of field studies, observations, and interviews with scientists, physicians, entrepreneurs, and government representatives in the United States and India. It combines detailed ethnographic research with comprehensive theoretical reflection. Although the book's subject matter is broad, the empirical focus of its analysis is centered on the development of pharmaceuticals and, especially, how genomic research has transformed their production. An important aspect of contemporary pharmaceutical research is that it aims to create "personalized medicine", that is to say, medicine the production of which is based on the genetic traits of the patient.

Sunder Rajan argues that the scientific production of knowledge can no longer be separated from the capitalist production of value. Two risk discourses permeate each other in this area of pharmaceutical research: the medical risk that current and future patients have of suffering from a major illness and the financial risk of pharmaceutical companies whose great investment in research and development will, it is hoped, ultimately result in commodities. Sunder Rajan describes this branch of industry as a special form of capitalism: a speculative capitalism that is based less on the manufacture of concrete products than on hopes and expectations, bringing together, into an "organic" synthesis, the hope of patients that new medical treatments will be developed, and the zeal of risk capitalism for future profits.

An important aspect of the book and its analysis of biocapital is the emphasis it puts on the speculative dimension of biocapital. Sunder Rajan stresses the importance of visions, hype, and promises. From this starting point, he is able to make the argument that, since the circulation of capital cannot be separated from expectations and hope, the production of economic value is also tied to moral values and ethical questions (ibid., 41, 56).

Sunder Rajan's work on biocapital is part of a more general theoretical reconsideration. In a very influential and insightful article, Stefan Helmreich has identified two distinctive clusters of theories or "species of biocapital": first, a Marxist-feminist cluster including the work of social scientists like Sarah Franklin, Margaret Lock, and Charis Thompson, which is concerned with (re-)production and focuses on the analysis of biological matter (Franklin 2003; 2006; Franklin and Lock 2003a; Lock 2001; Thompson 2005); second, a Weberian-Marxist cluster interested in questions of meaning and focusing on how "relations of production are described alongside accountings of ethical subjectivity" (Helmreich 2008, 471). In addition to Sunder Rajan's work, the second cluster contains contributions by Eugene Thacker (2005), Mike Fortun (2008), and others. Helmreich rightly notes that all "species of biocapital" present a specific fusion of Marx's political economy and Foucault's concept of biopolitics, and that there are often overlaps between the two clusters of theories (2008, 471).

However, the exact theoretical relation between Marx and Foucault, between "bioeconomy" and "biopolitics," often remains unclear. While some scholars seem to subscribe to the idea of a "new phase of capitalism", engendered by the emergence of a biotech industry and new biotechnological innovations (e.g., Cooper 2008), others prefer to distance themselves from Marxist theory and to diagnose a straightforward shift from biopolitics to bioeconomics (e.g., Rose 2001, 15). In a critical review of the literature on "bioeconomy", "biovalue", and "biocapital", Kean Birch and David Tyfield have identified several inconsistencies and ambiguities in this area of research (2013, 307–13). First, there is an issue with how to link "vitality" and value, especially in the concept of biovalue originally formulated by Catherine Waldby (2000, 33). The concept seems to be based on the idea of an already present and ever-abundant vitality that pertains to biological material—an idea that conflicts or even contradicts the insight that health and well-being are, at least in part, socially mediated or constructed. Birch and Tyfield stress that biological matter cannot be the source of value in itself; rather, it is the knowledge that makes it possible to transform cells, tissue, and genes into commodities that is valuable. Secondly, they argue that the analytic distinction between economic value and ethical values tends to collapse in many works on the bioeconomy. According to Birch and Tyfield, many writings tend to downplay the importance of political-economic processes by overemphasizing ethical values and subjectivities. Finally, they argue that Marxist concepts like surplus, capital, and value are only selectively adopted without adequately addressing their original formulations in Marxism, especially in the labor theory of value. Birch and Tyfield hold that this unclear appropriation of Marxist terminology "renders these STS theories of the bioeconomy opaque" (2013, 307).

Following Birch and Tyfield, one might also critically inquire into the general danger of fetishizing the notion of "bio" (2013, 309). It is necessary to investigate the (analytical and critical) value of notions like biovalue, biocapital, and bioeconomy; otherwise, critical analysis runs the risk of either essentializing biological processes as something original and vital that is captured and exploited by capitalism, or taking the rhetoric of the bioeconomy as a new area of production and industry at face value.

Despite all the critical points one might want to raise concerning theoretical ambiguities or inconsistencies, the fact remains that we can give credit to work on the political economy of life for extending the traditional focus on political institutions and actors to address economic structures and dynamics. This work also goes beyond the Foucauldian poles of individuals and populations, to take into account the non-human or life-forms beyond the human: genes, cells, embryos, organs, and tissue (Helmreich 2008, 464). At this point, the theoretical interest in the bioeconomy links up with the debate about the new materialism. While scholars working on the "bioeconomy" address living matter in their analyses, new materialists tend to define matter as living.

2. CHALLENGE II: FROM BIOPOWER TO THINGPOWER

Recently, social and political theory has demonstrated a renewed theoretical interest in matter and materiality. The "new materialism", as it is sometimes called (e.g., Ahmed 2008; Coole and Frost 2010a; Dolphijn and van der Tuin 2012; Hird 2004), does not represent a homogeneous style of thought or a single theoretical position but encompasses a plurality of different approaches and disciplinary perspectives, ranging from science and technology studies, via feminist theory and political philosophy, to geography (Alaimo and Hekman 2008; Bennett 2010; Braun and Whatmore 2010a; Latour and Weibel 2005). The new materialist scholarship shares the conviction that the "linguistic turn" or primarily textual accounts are insufficient for an adequate understanding of the complex and dynamic interplay of meaning and matter. New materialists often stress that the focus on discourse, language, and culture not only leads to impoverished theoretical accounts and conceptual flaws but also results in serious political problems and ethical quandaries, as it fails to address central challenges facing contemporary societies, especially economic change and the environmental crisis.[5]

The "material turn" (Bennett and Joyce 2010) criticizes the idea of the natural world and technical artifacts as a mere resource or raw material for technological progress, economic production, or social construction. It aims at a new understanding of ontology, epistemology, ethics, and politics, to be achieved by overcoming anthropocentrism and humanism, the split between nature and culture, linguistic or discursive idealism, social constructivism, positivism, and naturalism. Central to this movement is the extension of the concept of agency and power to the non-human, thereby also calling into question conventional understandings of life.

One of the most important representatives of the new materialism is Jane Bennett. Her starting point in *Vibrant Matter: A Political Ecology of Things* (2010) is the assumption that matter must be addressed as an active part of a political process that has so far been dominated by human subjectivity. The aim of the book is to rethink the traditional distinctions between matter and life, inorganic and organic, passive object and active subject (2004, 353–4). Instead, Bennett invites us to conceive of a "vitality of matter" (2010, vii), a concept that disturbs and upsets this conventional mindset:

> By "vitality" I mean the capacity of things—edibles, commodities, storms, metals—not only to impede or block the will and designs of humans but also to act as quasi agents or forces with trajectories, propensities, or tendencies of their own. (Bennett 2010, viii)

Seen in this light, agency is no longer an exclusive property of human beings. Rather, the force of non-human actors in events needs to be acknowledged. Bennett employs and synthesizes a heterogeneous bunch of theoretical concepts and ideas, from Lucretius, Spinoza, Adorno, Latour, Thoreau, Bergson, Dewey, and Deleuze and Guattari, to arrive at a different concept of agency. First, she argues that agency needs to be "distributed across a wider range of ontological types" (ibid., 9) that cuts across the human-non-human divide so that things like food and minerals can be reconceptualized as actants. Secondly, she moves beyond the focus on individual bodies and their borders to propose a concept of action that is conceived of as the effect of certain configurations of human and non-human forces. Bennett coins the term "thing power" to account for the ability of inanimate things to produce effects by operating in conjunction with other material bodies.

In relation to the problem of biopolitics, one might diagnose a certain ambiguity in Bennett's work. She claims that "everything is, in a sense, alive" (2010, 117). However, this position is only partly convincing. While it is certainly right to conceive of life not as a property that pertains to specific bodies but as a process or rather the outcome of certain materializations, it might be more accurate to distinguish between differently composed materialities and various complexities of conjunctions between bodies—in which the distinction between animate and inanimate bodies may play a crucial role. Bruce Braun and Sarah Whatmore rightly ask whether it might be more pertinent to attend to "the *specificity* of the matter at hand, as opposed to a generic analogy to 'life' that could be described as a metaphysics" (Braun and Whatmore 2010b, xxix, emphasis in original; see also Braun 2008, 675–7).[6]

However, the new materialism does not represent a homogeneous strand of thought or a single theoretical perspective, but encompasses a plurality of different approaches that converge on at least one point: the claim that matter is not a stable and given property but rather the fluid and contingent effect of practices. Its theoretical strength lies in the fact that it makes it possible to question the seemingly stable borderline between the organic and the inorganic and the idea of life as a fundamental biological fact.

To sum up, the research on the bioeconomy and the new materialism poses important challenges and seriously affects the concept of biopolitics. First, it seems that Foucault's concept of power does not provide a dynamic concept of materiality that takes account of the materialization of human as well as non-human bodies (Barad 2007, 200). His anthropocentrism fails to see that biopolitical interventions encompass the "'making-up' of *both* people and *things*" (Rutherford 1999, 44; emphasis in original). In Foucault's writings, and in most work in his legacy, the notion of biopolitics remains intimately linked to the constitution and transformation of human bodies, defining "the entry of phenomena peculiar to the life of the human species into the order of knowledge and power" (Foucault 1980, 141–2). Secondly, Foucault's concept of biopolitics is apparently unprepared to adequately address the political economy of life. It focuses on political rationalities and strategies at the expense of analyzing the complex of economic processes, bio-scientific innovations, and technological developments that allow access to "life itself" (Franklin 2000). It seems that Foucault's concept of biopolitics is incapable of accounting for how the politicization of life is intertwined with its economization, and cannot provide the conceptual tools for a critical analysis of the bioeconomy.

In the following, I propose to resituate the concept of biopolitics within an analytics of government, as a way of overcoming some of the limitations of Foucault's original work and addressing the critical points raised by work on the bioeconomy and in the new materialism.

3. THE GOVERNMENT OF THINGS

In his writings on discipline in the mid-1970s, Foucault repeatedly pointed out that the power of the economy was vested in a prior "economy of power" (2008, 65), because the accumulation of capital presumes technologies of production and forms of labor that make it possible to put to use a multitude of human beings in an organized and profitable manner. By this theoretical reorientation, Foucault hoped to complement and extend Marx's critique of political economy to a critique of "political anatomy" (1977, 221). In his lectures of 1978 and 1979 at the Collège de France, Foucault takes this form of analysis one step further, combining the "microphysics of power" (2007, 28) with the macropolitical question of the state. He shows that the economy defines a distinctive dimension or plane of existence under quite specific historical and institutional conditions that are linked to the emergence of liberal government. As a result, the economy cannot be conceived of as a given object independent of and external to political regulation. Quite on the contrary, the "art of government" (Foucault 2007, 92) is not limited to the field of politics as separated from the economy; instead the constitution of a conceptually and practically distinguished space, characterized by autonomous laws and a proper rationality, is itself an internal differentiation of "economic" government: "a form of schematization peculiar to a particular technology of government" (Foucault 2008, 319).

There is a second theoretical displacement in the lecture series on governmentality in relation to Foucault's previous work, especially the work on biopolitics. When in the lectures of 1978 and 1979 Foucault defines "liberalism as the general framework of biopolitics" (2008, 22), this results from the self-critical insight that his analysis, until then, had been one-sided and unsatisfactory, because it focused mainly on processes involving population regulation and the corporeal disciplining of human bodies. In *Discipline and Punish* (Foucault 1977) and *The History of Sexuality, Volume 1* (Foucault 1980), the investigation of subjectification processes essentially limits itself to subjugation and corporeal dressage, hence to the dimension of *zoē*, with techniques of self-formation receiving little attention. With the problem of government, the perspective broadens and the question of moral and political existence also emerges: the problem, then, of *bíos*.[7] Beyond technologies of bodily disciplining and the regulation of the population, attention is now also drawn to the self-constitution of individual and collective subjects—what Foucault in his later work came to call "political technologies of individuals" and "technologies of the self" (Foucault 1997a, 223; 2000b, 404).[8]

Taken together, these two theoretical shifts result in a closer realignment of the concepts of governmentality and biopolitics. The idea of biopolitics as an "art of government" makes it possible to address the challenges of the new materialism and of work on the "bioeconomy". Foucault's revised perspective on biopolitics enables us to study the interplay of moral values and economic value. It investigates the entanglement of concepts of life, vitality and health on the one hand, and the constitution of markets and products on the other. The new analytical focus results in the concept of a "vital politics". This idea stems from Alexander Rüstow, one of the most important representatives of postwar German liberalism, whom Foucault briefly touches on in the 1979 lecture (Foucault 2008, 148, 157). By "vital politics", Rüstow means a form of politics "that considers all factors upon which happiness, well-being, and satisfaction in reality depend" (1955, 70). This politics is, he indicates, by no means limited to action by the state, but "is politics in the broadest possible sense . . . all social measures and experimental arrangements" (1957, 235); it relies on social ties and spiritual cohesion and reactivates moral values and cultural traditions, its goal being to insert an "ever more dense net and weave of living ties [lebendiger Bindungen] into the entire social realm" (ibid., 238). This is a task of innovation that needs to take in all societal elements and levels, while simultaneously acknowledging their self-directing competencies.

Foucault's analytics of government takes account of these vital political ambitions of (neo-)liberal governmental practice, tying the analysis of physical-biological being to an examination of subjectification processes and moral-political modes of existence. This theoretical interest becomes clear when he discusses Moheau's *Recherches et considerations sur la population de la France* in his lectures, describing the author as "the first great theorist of what we could call biopolitics" (Foucault 2007, 22).[9] Foucault no longer refers to the biological or physical dimension of biopolitics alone, but cites approvingly Moheau's idea that government means to "govern the physical *and* moral existence of their subjects" (Moheau, cited in ibid., 23, emphasis by Lemke).

The concept of a "government of things" that Foucault introduces in the lectures of 1978 and 1979 at the Collège de France makes it possible to respond to the critical points new materialist scholars raise vis-à-vis Foucault's concept of biopolitics. It not only investigates the relations between the physical and the moral, but combines this analysis with a focus on the interactions between the natural and the artificial.[10] The idea of a "government of things" was originally proposed by Guillaume de la Perrière in an early modern tract on the art of government.[11] Here, government is conceived of as "the right dispositions of things arranged so as to lead to a suitable end" (de la Perrière, cited in Foucault 2007, 96). According to Foucault, de la Perrière's notion of a "government of things" does not constitute an additional domain of government apart and separate from the government of men. Rather than restaging "an opposition between things and men", it relies on "a sort of complex of men and things" (de la Perrière, cited in Foucault 2007, 96). It is worth quoting the whole passage:

> The things government must be concerned about, La Perrière says, are men in their relationships, bonds, and complex involvements with things like wealth, resources, means of subsistence, and, of course, the territory with its borders, qualities, climate, dryness, fertility, and so on. "Things" are men in their relationships with things like customs, habits, ways of acting and thinking. Finally, they are men in their relationships with things like accidents, misfortunes, famine, epidemics, and death. (Ibid.)

Following Foucault's interpretation, the "art of government" does not conceive of interactions between two stable and fixed entities—"humans" and "things". Rather, Foucault employs a relational approach. This is why "things" appears in inverted commas. In fact, the qualification "human" or "thing" and the political and moral distinction between them is itself an instrument and effect of the art of government, and does not constitute its origin or point of departure. Thus, the "government of things" does not rely on a foundational sorting of subjects and objects. Quite on the contrary, Foucault questions the idea that contrasts active subjects with passive objects. He employs the term "subject-object" (2007, 44, 77) to address the phenomenon of the population as on the one hand a material body, "on which and towards which mechanisms are directed", and on the other "a subject, since it is called upon to conduct itself in such and such a fashion" (ibid., 42–3). In this perspective, the art of government determines what is defined as subject and object, as human and non-human. It establishes and enacts the boundaries between socially relevant and politically accepted existence and "pure matter", something that does not possess legal-moral protection and is "reduced" to "things".[12]

The idea of a "government of things" takes a more concrete form when Foucault analyzes the concept of the milieu (2007, 20–23, 77–78). He argues that the milieu defines an "intersection between a multiplicity of living individuals working and coexisting with each other in a set of material elements that act on them and on which they act in turn" (ibid., 22). Here, Foucault quite clearly accepts the idea that agency is not exclusively a property of humans; rather, agential power originates in relations between humans and non-human entities. Also, the milieu articulates the link between the natural and the artificial without systematically distinguishing between them. According to Foucault, the milieu provides the "point of articulation" (ibid.) between the "natural" and the "artificial", the physical and the moral. Here we note that Foucault does not take non-human nature for granted, but is interested in how it is articulated within practices—practices that are here conceived as "more-than-human" practices.[13]

The idea of a government of things helps to enact a different understanding of biopolitics that no longer exclusively addresses "the basic biological features of the human species" (Foucault 2007, 2). This important theoretical shift entails two important dimensions. First, we see a move beyond a concept of biopolitics as limited to the physical and biological existence to a "government of things" that takes into account the "intrication of men and things" (ibid., 97), the natural and the artificial, the physical and the moral. Secondly, the biological can only play out in a certain "milieu". In the perspective of a government of things, neither nature nor life is a self-evident and stable entity or property. Foucault refers to "a multiplicity of individuals who are and fundamentally and essentially only exist biologically bound to the materiality within which they live" (ibid., 21). In this perspective, life is not a given but depends on conditions of existence within and beyond life processes.[14]

4. THE HISTORICAL ONTOLOGY OF LIFE

New materialists sometimes claim that Foucault has to be subsumed under the category of social constructivism and anthropocentrism (see e.g., Braun 2008, 668). The charge is that Foucault's work remains within the "traditional humanist orbit" (Barad 2007, 235), restricting agency to human subjects without taking into consideration the agential properties of non-human forces. In this light, some argue that Foucault's notion of bio-power "might have been adequate to eighteenth- and nineteenth-century practices, but not contemporary ones" (Barad 2007, 200; see also Haraway 1997, 12).

This chapter suggests a more nuanced assessment based on elements in Foucault's work on governmentality. The concept of a "government of things" does not take life as an essence or a pre-given that at some point in history enters "into the order of knowledge and power" (Foucault 1980, 141–2); quite on the contrary, it inquires into the conditions of the emergence of "life" as a distinctive domain of practice and thought. The "historical nominalism" Foucault proposes in the governmentality lectures (2008, 318), in analyzing "politics" and "the economy", is also useful for investigating the matter of life. According to Foucault, these entities are "things that do not exist and yet which are inscribed in reality and fall under a regime of truth dividing the true and the false" (ibid., 20). Like "politics" and "the economy", "life" is not an object that is always already there, nor can it be reduced to an (illusionary or ideological) effect brought about by scientific practices. Rather, it has to be conceptualized as a "transactional reality" (see ibid., 297), that is to say, a dynamic ensemble of matter and meaning that finally makes it possible to account for the "historical ontology" (Foucault 1997b, 315) of life.

The "relational materialism" (Mol 2013, 381) of a "government of things" might prove more fruitful in exploring the material and technical conditions that produce "life", dependent on, and operating in, historical specific conjunctions with other bodies, than the idea of an all-encompassing "vitality of matter" and an original "force of things" (Bennett 2004). Furthermore, it might be helpful in evading the "fetishisation of the 'bio'" (Birch and Tyfield 2013, 309) in much of the literature on biocapital, bioeconomy, and so on, as it does not endorse the idea of a singular and stable substance and an originary force of life. This perspective also invites us to reconsider the notion of resistance, or rather a certain concept of resistance, that tends to reinscribe passivity or rearticulate the opposition of activity versus passivity, instead of conceiving of the agential properties of human and non-human bodies and their eventfulness and indeterminacy (see Braun and Whatmore 2010b, xx–xxii).

Finally, in Foucault's account, the "bioeconomy" is not a "new face, and a new phase of capitalism" (Sunder Rajan 2006, 3); rather, the economy has been a political economy of life from its very beginnings. Already in his early work, Foucault reminds us that the eighteenth century emergence of political economy cannot be separated from the beginnings of modern biology. Liberal concepts of autonomy and freedom are closely connected to biological concepts of self-preservation and self-regulation that came to prevail over the previously dominant physical-mechanistic model for investigating the body. Originating by 1800, biology was based on an organizational principle understanding the visible phenomena of life as emerging essentially at random, without a set plan. Internal organization thus replaced an external order corresponding to the plans of a higher authority beyond life, with "life" functioning as an abstract and dynamic principle equally inherent in all organisms. Categories such as self-preservation, reproduction, and development now came to characterize living bodies, placed at a greater distance from artificial creations than had been the case before (Foucault 1970).[15]

It has to be stressed that Foucault's writing did not so much systematically pursue as offer promising suggestions for the analytic perspective outlined here. He never concretized his remarks on the relation between biopolitics and (neo-)liberalism, though this was meant to stand at the center of the 1979 lecture (see Foucault 2008, 21–2, 78). Nor did he ever actively develop the idea of government liberated from its anthropocentric connotations. However, while Foucault chose not to directly address the problem of human and non-human relations, or to extend his ideas on the political economy of life, the concept of a "government of things" addresses many critical points subsequently raised by new materialists and scholars of the bioeconomy in their readings of his work—and this analytical perspective might be helpful in overcoming some of their own shortcomings and limitations.

NOTES

1 I would like to thank Katharina Hoppe, who helped me with the work on the manuscript, and Gerard Holden, who copyedited the text.
2 For a more extensive discussion of the different lines of reception, see Folkers and Lemke (2014).
3 Bioeconomy Council. 2014. What is the Bioeconomy Council? Accessed March 3, 2014. http://biooekonomierat.de/home-en/bioeconomy-council.html.
4 See e.g., Franklin and Lock (2003b, 8).
5 In their edited volume *New Materialisms: Ontology, Agency, and Politics*, Diana Coole and Samantha Frost delineate three distinctive themes or topics in new materialist scholarship: (1) an "ontological reorientation" that takes up, or is even based on, developments in the natural sciences; (2) "consideration of a raft of biopolitical and bioethical issues concerning the status of life and of the human"; (3) "a critical and nondogmatic reengagement with political economy" (Coole and Frost 2010b, 6–7; see also Dolphijn and van der Tuin 2012).
6 As Bryan E. Bannon commented in his review of *Vibrant Matter*:

> It is unproblematic to assert that all existing bodies are affective and susceptible to affectation, and one need not equate this two-sided capacity with life, even the asubjective life of metal that Bennett describes. If life is a *field of intensities* in the way Bennett describes, then, far from being a property, it is a particular way of *relating* to the affections that surround an assemblage. Thus, on Bennett's own account, it is possible to assert that matter itself is not alive per se, but that life denotes a particular intricacy of responsiveness with complex alliances between smaller constituent assemblages. (Bannon 2011, 3; emphasis in original)

7   I here take up Giorgio Agamben's distinction between *zoē* and *bíos* as two forms of life (Agamben 1998).

8   For a more extensive argument on this theoretical shift, see Lemke 2011. It is not only Foucault's concept of biopolitics that changes after *The History of Sexuality, Volume 1*; his view of liberalism itself undergoes a shift of emphasis. Whereas in a text from 1977 he still understands political economy rather traditionally as an external limitation on power by law, in the lectures of 1978 and 1979 it stands for an inner self-limitation on power (Senellart 2004).

9   The book was first published in 1778 in Paris (for bibliographical information and the debate on the contested identity of the author, see Foucault 2007, 27 n. 39).

10 For a more detailed account of the idea of a "government of things" in Foucault's work and how it might be linked to the debate on the new materialism, see Lemke 2014.

11 Foucault is referring to the book *Le Miroire politique, œuvre non moins utile que necessaire à tout monarches, roys, princes, seigneurs, magistrats, et autres surintendants et gouverneurs de Republicques* (Lyon 1555).

12 See Agamben's work (1998), and especially his notion of "bare life" in this respect.

13 The term "more-than-human" was coined by Braun and Whatmore (2010b, xii). On Foucault's interpretation of the "naturalism" (2008, 61) of liberalism and its focus on the "market milieu" (ibid., 259) as a self-regulating matter of government, see Folkers (2013); see also Terranova (2009).

14 The philosopher of biology John Dupré has recently suggested that "functional biological wholes, the entities that we primarily think of as organisms, are in fact cooperating assemblies of a wide variety of lineage-forming entities" (Dupré 2012, 126). Dupré rejects the assumption that all cells in an organism belong to the same species. Quite on the contrary, "living things" according to this account are "extremely diverse and opportunistic compilations of elements from many distinct sources" (Dupré 2012, 126). Dupré argues for a redefinition of "organisms" as "cooperating assemblies". In this perspective, human life only exists as the effect of symbiotic systems linking "human" and "non-human" life:

> A functioning human organism is a symbiotic system containing a multitude of microbial cells—bacteria, archaea, and fungi—without which the whole would be seriously dysfunctional and ultimately non-viable. Most of these reside in the gut, but they are also found on the skin, and in all body cavities. In fact about 90 per cent of the cells that make up the human body belong to such microbial symbionts and, owing to their great diversity, they contribute something like 99 per cent of the genes in the human body. (Ibid., 125)

For a "thing materialism" which points to the mineral conditions for the emergence of biological entities (bone), see Bennett (2004, 360)

15 Citing Foucault (2008, 16), Brian Massumi rightly stresses the importance of nature for the workings of (neo-)liberal governmentality:

> Nature is something that runs under, through, and in the exercise of governmentality. It is, if you like, its indispensable hypodermis. It is the other face of something whose visible face, visible for the governors, is their own action. . . . The laws of nature are the formal mirroring, for the system, of the system's own pattern of actions. They are the form in which the system internalizes the activity of its outside as its own. (Massumi 2009, 165–6)

REFERENCES

Agamben, Giorgio. 1998. *Homo Sacer: Sovereign Power and Bare Life*. Trans. Daniel Heller-Roazen. Stanford: Stanford University Press.

Ahmed, Sara. 2008. Open Forum Imaginary Prohibitions: Some Preliminary Remarks on the Founding Gestures of the "New Materialism". *European Journal of Women's Studies* 15: 23–39.

Alaimo, Stacy, and Susan Hekman, eds. 2008. *Material Feminisms*. Indianapolis: Indiana University Press.

Bannon, Bryan E. 2011. Review of "Vibrant Matter: A Political Ecology of Things", by Jane Bennett. *Environmental Philosophy* 5: 1–4.

Barad, Karen. 2007. *Meeting the Universe Halfway: Quantum Physics and the Entanglement of Matter and Meaning*. Durham, NC: Duke University Press.

Bennett, Jane. 2004. The Force of Things: Steps Toward an Ecology of Matter. *Political Theory* 32 (3): 347–72.

Bennett, Jane. 2010. *Vibrant Matter: A Political Ecology of Things*. Durham NC: Duke University Press.

Bennett, Tony and Patrick Joyce, eds. 2010. *Material Powers: Cultural Studies, History and the Material Turn*. London: Routledge.

Birch, Kean and David Tyfield. 2013. Theorizing the Bioeconomy: Biovalue, Biocapital, Bioeconomics or . . . What? *Science, Technology & Human Values* 38 (3): 299–327.

Braun, Bruce. 2008. Environmental Issues: Inventive Life. *Progress in Human Geography* 32 (5): 667–79.

Braun, Bruce and Sarah Whatmore, eds. 2010a. *Political Matter: Technoscience, Democracy and Public Life*. Minneapolis: University of Minnesota Press.

Braun, Bruce and Sarah Whatmore. 2010b. The Stuff of Politics: An Introduction. In *Political Matter: Technoscience, Democracy and Public Life*. Ed. Bruce Braun and Sarah Whatmore, ix–xl. Minneapolis: University of Minnesota Press.

Colebrook, Claire. 2008. On Non Becoming Man: The Materialist Politics of Unactualized Potential. In *Material Feminisms*. Ed. Stacy Alaimo and Susan Hekman, 52–84. Bloomington, IN: Indiana University Press.

Coole, Diana and Samantha Frost, eds. 2010a. *New Materialisms: Ontology, Agency, and Politics*. Durham NC: Duke University Press.

Coole, Diana and Samantha Frost. 2010b. Introducing the New Materialisms. In *New Materialisms: Ontology, Agency, and Politics*. Ed. Diana Coole and Samantha Frost, 1–43. Durham, NC: Duke University Press.

Cooper, Melinda. 2008. *Life as Surplus: Biotechnology and Capitalism in the Neoliberal Era*. Seattle: University of Washington Press.

Deleuze, Gilles. 1995. Postscript on Control Societies. In *Negotiations, 1972–1990*. Trans. Martin Joughin, 177–82. New York: Columbia University Press.

Dolphijn, Rick, and Iris van der Tuin. 2012. *New Materialism: Interviews & Cartographies*. Ann Arbor, MI: Open Humanities Press.

Dupré, John. 2012. *Processes of Life: Essays in the Philosophy of Biology*. Oxford: Oxford University Press.

Esposito, Roberto. 2008. *Bíos: Biopolitics and Philosophy*. Minneapolis: University of Minnesota Press.

European Commission. 2005. *New Perspectives on the Knowledge-Based Bio-economy: Conference Report*. Brussels: European Commission.

Folkers, Andreas. 2013. After Virtù and Fortuna: Foucault on the Government of Economic Events. *Distinktion: Scandinavian Journal of Social Theory* 15 (1): 88–104.

Folkers, Andreas and Thomas Lemke. 2014. Einleitung. In *Biopolitik. Ein Reader.* Ed. Andreas Folkers and Thomas Lemke, 7–61. Berlin: Suhrkamp.

Fortun, Michael. 2008. *Promising Genomics: Iceland and deCODE Genetics in a World of Speculation.* Berkeley: University of California Press.

Foucault, Michel. 1970. *The Order of Things.* Trans. Alan Sheridan. New York: Random House.

Foucault, Michel. 1977. *Discipline and Punish: The Birth of the Prison.* Trans. Alan Sheridan. New York: Vintage Books.

Foucault, Michel. 1980. *The History of Sexuality, Vol. 1: An Introduction.* Trans. Robert Hurley New York: Vintage Books.

Foucault, Michel. 1997a. Sex, Power, and the Politics of Identity. In *Ethics: Subjectivity and Truth: Essential Works of Michel Foucault, 1954–1984, Vol. 1.* Ed. Paul Rabinow. Trans. Robert Hurley, 163–173. New York: New Press.

Foucault, Michel. 1997b. What Is Enlightenment? In *Ethics: Subjectivity and Truth: Essential Works of Michel Foucault, 1954–1984, Vol. 1.* Ed. Paul Rabinow. Trans. Robert Hurley, 303–19. New York: New Press.

Foucault, Michel. 2000a. The Birth of Social Medicine. In *Power: Essential Works of Foucault, 1954–1984, Vol. 3.* Ed. James D. Faublon. Trans. Robert Hurley, 134–56. New York: New Press.

Foucault, Michel. 2000b. The Subject and Power. In *Power: Essential Works of Foucault, 1954–1984, Vol. 3.* Ed. James D. Faublon. Trans. Robert Hurley, 326–48. New York: New Press.

Foucault, Michel. 2003. *Society Must Be Defended: Lectures at the Collège de France, 1975–1976.* Ed. Mauro Bertani and Alessandro Fontana. Trans. David Macey. New York: Picador.

Foucault, Michel. 2007. *Security, Territory, Population: Lectures at the Collège de France, 1977–78.* Ed. Michel Senellart. Trans. Graham Burchell. London: Palgrave Macmillan.

Foucault, Michel. 2008. *The Birth of Biopolitics: Lectures at the Collège de France, 1978–79.* Ed. Michel Senellart. Trans. Graham Burchell. London: Palgrave Macmillan.

Franklin, Sarah. 2000. Life Itself: Global Nature and the Genetic Imaginary. In *Global Nature, Global Culture.* Ed. Sarah Franklin, Celia Lury, and Jackie Stacey, 188–227. London: Sage.

Franklin, Sarah. 2003. Ethical Biocapital: New Strategies of Cell Culture. In *Remaking Life and Death: Toward an Anthropology of the Biosciences.* Ed. Sarah Franklin and Margaret Lock, 97–128. Santa Fe: School of American Research Press.

Franklin, Sarah. 2006. Bio-Economies: Biowealth from the Inside Out. *Development* 49: 97–101.

Franklin, Sarah and Margaret Lock, eds. 2003a. *Remaking Life and Death: Toward an Anthropology of the Biosciences.* Santa Fe: School of American Research Press.

Franklin, Sarah and Margaret Lock. 2003b. Animation and Cessation: The Remaking of Life and Death. In *Remaking Life and Death. Toward an Anthropology of the Biosciences.* Ed. Sarah Franklin and Margaret Lock, 3–22. Santa Fe: School of American Research Press.

Haraway, Donna J. 1991. *Simians, Cyborgs and Women: The Reinvention of Nature.* New York: Routledge.

Haraway, Donna J. 1997. *Modest_Witness@Second_Millenium. FemaleMan©_Meets_OncoMouse™: Feminism and Technoscience.* New York: Routledge.

Hardt, Michael and Antonio Negri. 2000. *Empire: The New World Order.* Cambridge, MA: Harvard University Press.

Hardt, Michael and Antonio Negri. 2004. *Multitude: War and Democracy in the Age of Empire*. New York: Penguin.

Helmreich, Stefan. 2008. Species of Biocapital. *Science as Culture* 17 (4): 463–78.

Hird, Myra J. 2004. Feminist Matters: New Materialist Considerations of Sexual Difference. *Feminist Theory* 5 (2): 223–32.

Latour, Bruno and Peter Weibel, eds. 2005. *Making Things Public: Atmospheres of Democracy*. Cambridge, MA: MIT Press.

Lemke, Thomas. 2011. Beyond Foucault: From Biopolitics to the Government of Life. In *Governmentality: Current Issues and Future Challenges*. Ed. Ulrich Bröckling, Susanne Krasmann, and Thomas Lemke, 165–84. New York: Routledge.

Lemke, Thomas. 2014. New Materialisms: Foucault and the "Government of Things". *Theory, Culture & Society* 32: 3–25.

Lock, Margaret. 2001. The Alienation of Body Tissue and the Biopolitics of Immortalized Cell Lines. *Body & Society* 7 (2–3): 63–91.

Massumi, Brian. 2009. National Enterprise Emergency: Steps Toward an Ecology of Powers. *Theory, Culture & Society* 26 (6): 153–85.

Mbembe, Achille. 2003. Necropolitics. Trans. Libby Meintjes. *Public Culture* 15 (1): 11–40.

Mol, Annemarie. 2013. Mind Your Plate! The Ontonorms of Dutch Dieting. *Social Studies of Science* 43 (3): 379–96.

OECD. 2006. *The Bioeconomy to 2030: Designing a Policy Agenda*. Paris: OECD.

Peters, Michael A. (2012). Bio-informational Capitalism. *Thesis Eleven* 110 (1): 98–111.

Rabinow, Paul. 1992. Artificiality and Enlightenment: From Sociobiology to Biosociality. In *Incorporations*. Ed. Jonathan Crary and Sanford Kwinter, 234–52. New York: Zone Books.

Rabinow, Paul and Nikolas Rose. 2006. Biopower Today. *BioSocieties* 1: 195–217.

Rose, Nikolas. 2001. The Politics of Life Itself. *Theory, Culture & Society* 18 (6): 1–30.

Rose, Nikolas. 2006. *The Politics of Life Itself: Biomedicine, Power, and Subjectivity in the Twenty-First Century*. Princeton: Princeton University Press.

Rüstow, Alexander. 1955. Wirtschaftsethische Probleme der sozialen Marktwirtschaft. In *Der Christ und die soziale Marktwirtschaft*. Ed. Patrick M. Boarman, 53–74. Stuttgart/Köln: Kohlhammer.

Rüstow, Alexander. 1957. Vitalpolitik gegen Vermassung. In *Masse und Demokratie*. Ed. Albert Hunold, 215–38. Erlenbach-Zürich/Stuttgart: Eugen Rentsch Verlag.

Rutherford, Paul. 1999. The Entry of Life into History. In *Discourses of the Environment*. Ed. Éric Darier, 37–62. Oxford: Blackwell.

Senellart, Michel. 2004. La question du libéralisme. *Le Magazine Littéraire* 435: 55–6.

Sunder Rajan, Kaushik. 2006. *Biocapital: The Constitution of Postgenomic Life*. Durham, NC: Duke University Press.

Sunder Rajan, Kaushik, ed. 2012. *Lively Capital: Biotechnologies, Ethics, and Governance in Global Markets*. Durham, NC: Duke University Press.

Terranova, Tiziana. 2009. Another Life: The Nature of Political Economy in Foucault's Genealogy of Biopolitics. *Theory, Culture & Society* 26 (6): 234–62.

Thacker, Eugene. 2005. *The Global Genome: Biotechnology, Politics, and Culture*. Cambridge, MA: MIT Press.

Thompson, Charis. 2005. *Making Parents: The Ontological Choreography of Reproductive Technologies*. Cambridge: MIT Press.

Waldby, Catherine. 2000. *The Visible Human Project: Informatic Bodies and Posthuman Medicine*. London: Routledge.

4 From Biopolitics to Biophilosophy, or the Vanishing Subject of Biopolitics

Audronė Žukauskaitė

The proliferation of theories of biopolitics raises the question: Who is the subject of biopolitics?[1] If every theory of biopolitics refers to a different subject, then how are these different theories compatible? In this essay I will examine three ways in which the biopolitical opposition between "appropriate bodies" and "improper bodies" has been defined: the division between citizens and non-citizens; the division between persons and non-persons; and the division between humans and non-humans or animals. The first division, between citizens and non-citizens, is discussed as the paradox of human rights, which is revealed in the works of Hannah Arendt, Jacques Rancière, and Giorgio Agamben. The problem here is that so-called human rights are attributed only to those individuals who are defined as citizens of a certain nation-state; by contrast, all other subjects are treated as the objects of humanitarian help. The second division is related to the notion of person that is the founding concept of the Universal Declaration of Human Rights of 1948. By contrast to the 1789 Declaration, which emphasized the concept of citizenship, the Declaration of 1948 praised the dignity of the human person. Unfortunately, the concept of person was immediately accompanied by the creation of a shadow realm of non-persons. This leads us to the third division between humans and non-humans, or between human and animal, discussed in Agamben's and Jacques Derrida's works. Agamben and Derrida insist that the animalization of man and the humanization of animal are two sides of the same biopolitical division. The analysis of these three divisions makes us question the notions of citizen, person, and human, and to rethink the body in Gilles Deleuze and Félix Guattari's terms. Deleuze and Guattari oppose the modernist notions of subjectivity with their theory of becoming-imperceptible, that is to say becoming asubjective, asignifying, and inorganic. Thus the philosophy of the imperceptible erases the division between *zoē* and *bios*, or between naked life and political existence, and creates the conditions for fluid ontology and non-hierarchized being.

ON CITIZENS AND NON-CITIZENS

The Agambenian concept of biopolitics is usually related to the concept of sovereign power and its right to declare a state of exception. In other words, biopolitics is related to exceptional conditions, in which habitual jurisdiction is suspended, such as in a concentration camp in Nazi Germany or a Gulag in Soviet Russia. In this essay I would like not to focus on Agamben's analysis of sovereignty but to discuss the concept of human rights. Agamben reveals the ambivalent nature of human rights, which were formulated to protect biological life from the executions of biopower, but are still dependent on biopolitical decisions. In his analysis of human rights, Agamben is influenced by Arendt's text "We Refugees" (1943). Here Arendt introduces the new political figure of her time—a refugee—who is stripped of any rights and at the same time represents the *avant-garde* of their people. Later in *The Origins of Totalitarianism* Arendt rephrases this question in a chapter entitled "The Perplexities of the Rights of Man" (reproduced in Arendt 2013). Here she reveals the ambivalent status of the rights of man: although the rights of man were created to protect the rights of those who have lost their political status, she argues, "Actually the opposite is the case. It seems that a man who is nothing but a man has lost the very qualities which make it possible for other people to treat him as a fellow-man" (Arendt 2013, 93). In other words, if human rights have any meaning or intention at all, it is to protect precisely this "nakedness of being human" (92). But "nakedness" as such does not belong to the jurisdiction of human rights because human rights are inseparably connected to the rights of the citizen. As the title of the Declaration of 1789—*Déclaration des droits de l'homme et du citoyen*—implies, the right to have rights is attributed only to those human beings who are already citizens of a certain nation-state. Therefore, human rights mean the rights of citizenship and cannot be extended to an abstract human being. As Arendt points out:

> The conception of human rights, based upon the assumed existence of a human being as such, broke down at the very moment when those who professed to believe in it were for the first time confronted with people who had indeed lost all other qualities and specific relationships—except that they were still human. The world found nothing sacred in the abstract nakedness of being human. (2013, 92)

Thus the abstract "nakedness of being human" reduces the stateless people to the status in which they are being regarded as "savages" and even as "beasts" (Arendt 2013, 92). And although Arendt's distinction between political community and "savages" is very problematic in itself, her point is that a person, after losing all signs of belonging to a political community, loses the possibility to participate in a common world and falls back to "the dark background of mere givenness", becoming "different in general, representing nothing but his own absolutely unique individuality" (Arendt 2013, 94–5).

As many commentators have pointed out, Arendt's definition of the political poses many theoretical problems. The main problem is that her account of the political is based on the opposition between the public sphere, where all equal members can act and "build a common world", and the private sphere, based on the "dark background of mere givenness" (Arendt 2013, 94). As Rancière points out, the distinction between the public and the private is itself a political question, depending on where you draw the line separating one from the other. In this sense politics is not about separation, but about putting these worlds together, or creating "the scenes of dissensus" (Rancière 2004, 304). The problem here is that, by claiming human rights for stateless people, Arendt immediately makes this claim at the expense of those who cannot act in the common world—"savages" and "beasts", Another problem relates to her understanding of the public political sphere as based on the law of equality, and of the private sphere as based on the law of difference. In this sense the political community can be created only with equals and between equals. In the "dark background of mere givenness", our unique nature and individuality pose essential limitations to our political activity. The political community succeeds only by reducing the "dark background of differences" to a minimum. Here Arendt herself gets confused when, after describing the sacredness of being human and the need of the human to preserve its unique individuality, she simultaneously renounces this individuality as an obstacle to achieve political equality. As Cary Wolfe argues, here Arendt

> falls back on a classically humanist argument that derives from Aristotle: for the "right to have rights" consists in the ability to enter into relations of reciprocal obligation (or what she calls, a little more lyrically, "a framework where one is judged by one's actions and opinions"). (2013, 7)

Arendt never raises the question about what happens to those "savages" and "beasts" who are so "unique" that they are not invited to create our common world. By prescribing individuality and difference to the sphere of idiosyncratic privacy, she depoliticizes naked life and fails to reframe the sphere of the political in such a way that it will be inclusive rather than exclusive.

Nevertheless, these theoretical problems do not diminish Arendt's insights into the question of biopolitcs, which were elaborated in Agamben's own text "We Refugees" (1995). In this text, which later appears as a chapter entitled "Beyond Human Rights" in his book *Means Without End* (2000), Agamben points out that Arendt's analysis has lost none of its relevance and that we still have to rethink the subject of the political beyond the concept of human rights and the citizen. The figure of the refugee (who is not necessarily stateless but may prefer to be such instead of returning to his/her country) creates a rupture between naked life and the juridical order of the nation-state. As Agamben points out:

> Nation-state means a state that makes nativity or birth [*nascita*] (that is, naked human life) the foundation of its own sovereignty. . . . The fiction that is implicit here is that *birth* [*nascita*] comes into being immediately as *nation*, so that there may not be any difference between the two moments. (2000, 20)

Thus human rights represent the inscription of naked life into the political order of the nation-state. In a paradoxical way the nation-state functions as a biopolitical apparatus, which includes naked life in the political order but with the condition that this naked life can be excluded again at any time.

The figure of the refugee, in Agamben's words, deconstructs the fictional identity between the fact and place of birth and the nation-state, or between nativity and nationality: "Inasmuch as the refugee, an apparently marginal figure, unhinges the old trinity of state-nation-territory, it deserves instead to be regarded as the central figure of our political history" (2000, 21). In other words, the concept of the refugee creates a kind of fluid ontology, which cannot be stabilized either by naturalization or by repatriation. In this condition the nation-state functions as an apparatus of capture, producing two dangerous tendencies. The first tendency is the growing epidemic of ethnocentrism, racism, and xenophobia, when the state apparatus is constantly sorting out the individuals within the nation-state and deciding which of them are "appropriate" and which of them are "improper". The second tendency is so-called humanitarianism, when the concern with "naked life" is depoliticized and replaced outside the nation-state or any juridical order. In Agamben's analysis, humanitarianism represents nothing other than a separation of the rights of man from the rights of the citizen, not in order to give a political status to the "naked life" but in order to expel it beyond the political. As Agamben points out:

> In the final analysis, however, humanitarian organizations—which today are more and more supported by international commissions—can only grasp human life in the figure of bare or sacred life, and therefore, despite themselves, maintain a secret solidarity with the very powers they ought to fight. (1998, 133)

The point here is not to separate the figure of the refugee either within or outside the nation-state, but to invent new categories of rethinking his/her political status.

> The refugee must be considered for what he is: nothing less than a limit concept that radically calls into question the fundamental categories of the nation-state, from the birth-nation to the man-citizen link, and that

thereby makes it possible to clear the way for a long-overdue renewal of categories in the service of a politics in which bare life is no longer separated and excepted, either in the state order or in the figure of human rights. (Agamben 1998, 134)

In other words, the notion of "naked life" or bare life should be rethought in such a way that it could not be eliminated from the political order.

ON PERSONS AND NON-PERSONS

The tension between "naked life" and the political is reconsidered in Roberto Esposito's works. Just as Agamben interprets biopolitics in terms of inclusion and exclusion of bare life in the system of law, Esposito rethinks the laws of life as a defensive mechanism of immunity, which works by including or excluding the pathological element in the immunitary system of an organism. In *Bíos: Biopolitics and Philosophy* (2008) and *Immunitas: The Protection and Negation of Life* (2011), Esposito defines what he calls the "immunitary paradigm", which defines the political realm from the ancient *polis* until late modernity. Esposito redefines the term of immunity taken from the biomedical sphere in such a way that it could be expanded to the political-juridical order:

> Where the term "immunity" for the biomedical sphere refers to a condition of natural or induced refractoriness on the part of a living organism when faced with a given disease, immunity in political-juridical language alludes to a temporary or definitive exemption on the part of [the] subject with regard to concrete obligations or responsibilities that under normal circumstances would bind one to others. (2008, 45)

On the one hand, immunization can be seen as a negative practice of protecting both the organism and the political body (sovereignty, property, liberty) from outside intruders. On the other hand, immunization can be seen as an affirmative practice of incorporation, including the dangerous element within itself. As Esposito points out, "For life to remain as such, it must submit itself to an alien force that, if not entirely hostile, at least inhibits its development" (2011, 8). Esposito himself concentrates on the negative practices of biopolitics, such as Nazi eugenics, or the search for security in relation to the threat of terrorism. In this sense immunity as a reactionary practice limits the emergence of community, or the being-in-common. What remains unthought in Esposito's theory is the way in which the immunitary paradigm could take an affirmative turn. For example, we can only imagine the practices which could break down the defensive mechanisms of the political body and force it to accept the foreign and intrusive element without resistance. But Esposito is very clear about the reasons why the immunitary paradigm is so strong and long-lasting: it is because all political theories are based on the model of the individual and individual body. Even those political theories which presuppose the multitude of individual wills generate a single individual body: "it is because the political body is already inscribed in a single body that its parts can or must be consolidated in an identical figure whose object precisely is the self-preservation of the political organism as a whole" (Esposito 2008, 158). This logic of the individual body also prevailed in the constitution of nation-states, and, as a consequence of this, this logic qualifies as a pathogenic element everything that is not compatible with it.

Thus, in order to break through the defensive mechanism of immunitary protection, the political philosophy has to give up the notion of the individual and the individual body. In his most recent texts Esposito seeks to create a philosophy of the impersonal, which is seen as a counter-strategy to the dominant concept of person (Esposito 2012, 101–2; Esposito 2013, 112–22). Esposito recalls Arendt's argument, made in *The Origins of Totalitarianism*, that the notion of the nation-state opens a gap between a man and a citizen, and argues that it was precisely the concept of the person which, as was expected, would extend the concept of human rights to every individual. After the experience of Nazism it seemed that the concept of person, which became the basis of the Universal Declaration of Human Rights of 1948,[2] was the most appropriate to reconcile the division between the citizen and the man, law and life. Jacques Maritain, who was one of the authors of the 1948 Declaration, pointed out that the human person possesses rights because of the very fact that it is a person capable of self-determination (Esposito 2012, 72–1). Nevertheless, Esposito critically asks if the discourse of personhood is still sufficient to ensure the extension of human rights to all individuals. He argues that, on the contrary, the opposite is true: "despite the rising rhetoric of humanitarian commitment, human life remains largely outside the protection of the law" (Esposito 2012, 73). Esposito radically argues that this outcome becomes possible not in spite of, but precisely because of the notion of personhood:

> As I argue, this outcome occurs *because* of the conceptual lexicon of the person, not in spite of it. My thesis is that the dispositif of the person, intended by the creators of the Declaration on Human Rights to fill in the chasm between man and citizen left gaping since 1789, produced an equally profound gap between rights and life. The very paradigm that appears to be a vehicle for their epochal reunion acts instead as a separation filter, or as a differential diaphragm between two elements that fail to meet up, except in the form of their separation. (2012, 73–4)

In other words, the concept of person acquires its meaning only by distantiating a person from those who are treated as non-persons. For example, in Roman society a slave was treated as a non-person, or as a living thing.

Similarly, the condition of children can be compared to that of the slave. In a recent context another problem arises when we try to define the status of the body, which oscillates between the notion of a person and that of a thing. Recent biomedical practices evoke a wide array of in between states between person and thing: for example, what is the status of embryos, of fetuses, and of clinical waste? In other words, it appears that the notion of person works as a biopolitical apparatus, separating those who are considered to be self-determined persons (with reason, will, and moral sense), and those who come closer to things. As Esposito points out:

> What really qualifies as "person" only occupies the central section: that of adult, healthy individuals. Before and after this lies the no man's land of the non-person (the fetus), the quasi-person (the infant), the semi-person (the elderly, no longer mentally or physically able), the no-longer-person (the patient in a vegetative state), and, finally, the anti-person (the fool, whom Singer puts in the same relation to the intelligent human being as obtains between the animal and the normal human being—albeit with a clear preference for the animal). (2012, 97)

It is clear without saying it that those who are supposed to be persons can exercise their power toward not-yet-persons or already-non-persons, such as children with disabilities or the elderly. In this sense the notion of the person operates as an apparatus of exclusion, deciding which is worthy of life and which is not. This apparatus of exclusion not only comes very close to the practices of eugenics but also reveals that the sublime concept of person always exists at the expense of non-person. As Esposito points out, "Personalization and depersonalization are nothing but different flows of the same process, one that is ancient in origin but whose effects are far from being exhausted" (2012, 99).

Having in mind the biopolitical effect that the notion of person has on individuals, Esposito argues for the need to create the philosophy of the impersonal. Following Simone Weil's ideas on the impersonality of justice, Maurice Blanchot's notion of the impersonality of writing, Emmanuel Levinas' notion of anonymity, and Gilles Deleuze's philosophy of life, Esposito tries to define the philosophy of the impersonal that could break the immunitary barrier between one individual and another. Esposito is right in arguing that Deleuzian philosophical strategies seem to be the most radical in destroying the notion of the person and of the subject. It is the Deleuzian notion of life which is able to destroy the separations and hierarchies between "appropriate" and "improper" forms of life. Esposito also defines the three specific Deleuzian strategies which are crucial for evading the humanist paradigm (2012, 147–51). The first strategy involves replacing the category of possibility with that of virtuality; if the possible can be easily dismissed as not being real, by contrast, the notion of the virtual has the same status as reality. In other words, all life forms, even in their virtual stages (the embryo, for example), are supposed to be equally real. The second strategy replaces the concept of the individual with that of individuation. Inspired by the philosophy of Gilbert Simondon, Deleuze argues that individuals are only temporary stages in the permanent process of individuation. Every individual preserves a remainder which can be developed in a successive phase of individuation. In this sense the transition from one stage to another, or from one dimension (biological, psychic, mental) to another, is more fluid than was imagined (Esposito 2008, 180). From this follows the third strategy, that of becoming. Deleuze refers to different forms of becoming, which may be biological (becoming-animal, becoming-child, becoming-molecule), as well as political (becoming-woman, becoming-minoritarian, becoming-black, etc.). Deleuze also writes about becoming-imperceptible, which can be seen as a more complete version of the philosophy of the impersonal, as I will argue later. For the moment we can conclude that for Esposito the philosophy of the impersonal is a counter-strategy which enables the deconstruction of the divisions between different life forms and ensures that any expression of individuation has an equal chance for existence.

ON HUMANS AND NON-HUMANS

Esposito's analysis of the distinction between persons and non-persons raises a question about the biological dimension of our lives. Between a person and a thing there is a vast area of biological life, including not only animals in the proper sense but also animality produced by a biopolitical apparatus of inclusion and exclusion. In *The Open: Man and Animal* (2004), Agamben defines this division as an "anthropological machine" and argues that this machine functions by animalizing the human and by humanizing the animal (2004, 37). The limit between the human and the non-human is always displaced, either by producing the animal in human form (the slave, the "savage", or the "beast"), or by inventing a human animal. Here we can reconsider two examples. The first example, discussed by Wolfe (2013, 11), relates to The Great Ape Project, inspired by philosophers Peter Singer and Paola Cavalieri in 1994. On 25 June 2008, the Environmental Committee of the Spanish Parliament approved a resolution to grant basic rights to the great apes.[3] As described in the World Declaration on Great Primates, these three basic rights seem to be quite human: these are the right to life, the protection of individual liberty, and the prohibition of torture.[4] The second example is quite recent and it relates to the migrant center on the Mediterranean island of Lampedusa. On 18 December 2013, an amateur video was revealed, showing migrants, stripped naked in the winter cold, being sprayed for disinfection. One of the migrants, identified as Khalid, complained that the detained migrants in the center were treated like "animals."[5] As these examples demonstrate, the "anthropological machine" already is at work here by questioning the limits between the human and

the non-human, or between the human and the animal, either by humanizing animals (The Great Ape Project), or by animalizing humans (the migrant
center in Lampedusa). The "anthropological machine" produces a zone of
indeterminacy or a state of exception which is:

> The place of a ceaselessly updated decision in which the caesurae and
> their rearticulation are always dislocated and displaced anew. What
> would thus be obtained, however, is neither an animal life nor a human
> life, but only a life that is separated and excluded from itself—only a
> *bare life*. (Agamben 2004, 38)

In other words, the "anthropological machine" not only separates man
and animal, but invents a biological life within the human itself, which can
be captured and manipulated by a political decision. As Matthew Calarco
argues, Agamben's work helps us to see that the division between man and
animal is always political and ethical:

> For not only does the distinction create the opening for the exploitation
> of nonhuman animals and others considered not fully human . . ., but
> it also creates the conditions for contemporary biopolitics, in which
> more and more of the "biological" and "animal" aspects of human
> life are brought under the purview of the State and the juridical order.
> (2008, 94)

What these two examples also reveal is that the model of human rights should
be questioned and abandoned instead of trying to apply it to non-humans
or animals. As Derrida points out, "It is preferable not to introduce this
problematic concerning the relations between humans and animals into the
existing juridical framework" by extending some form of human rights to
animals (Derrida and Roudinesco 2004, 74). If the concept of rights is transferred to animals, the animals are subjected to the paradigm of humanization and in this sense experience a conceptual violence. The problem here
is that in order to "liberate" animals, certain activists are using the same
anthropocentric legal and moral framework which actually effectuated the
subjected status of animals. As Calarco points out:

> There is a peculiar irony at work when animal rights theorists and ani-
> mal liberationists employ classical humanist and anthropocentric crite-
> ria to argue for granting animals certain rights or protecting them from
> suffering, for *it is these very criteria that have served historically to
> justify violence toward animals*. (2008, 128)

Another problem here is that in equating humans and animals by using the
same framework of human rights we arrive at morally doubtful conclusions,

for example, when Singer (2001, 162), who questions the fact that a new-born baby has the right to life, so generously attributes the same right to the great apes. It seems that the concept of rights is always working at the expense of others, so that it "is concerned not with justice and compassion but with 'a system of entitlement' and with who gets what within such a system" (Wolfe 2013, 17). Of course, this doesn't mean that the question of the animal should not be raised; but the concepts in which this question could be thought are still missing. As Derrida points out:

> For the moment, we ought to limit ourselves to working out the rules of law [*droit*] such as they exist. But it will eventually be necessary to reconsider the history of this law and to understand that although ani-mals cannot be placed under concepts like citizen, consciousness linked with speech, subject, etc., they are not for all that without a "right". It's the very concept of right that will have to be "rethought". (Derrida and Roudinesco 2004, 74)

Derrida returns to the same question in his book *The Animal That There-fore I Am*, where he refers to the Universal Declaration of Animal Rights of 1989 and asks whether the concept of "rights" is here the appropriate concept (2008, 87–8). Derrida, following Kant, argues that the concept of rights is inseparably connected with a specific notion of the subject, who is not only the subject of rights but also the subject of duties:

> Kant envisages the rupture of such a correlation within the subjectivity of the subject in two cases: that of the serf who is subject to duties but enjoys no rights, and that of God, who has every right but is bound by no duty. (2008, 99–100)

In other words, we cannot attribute the concepts of rights to animals with-out simultaneously subjecting them to duties, therefore making of them rational, liberal, and autonomous subjects. So it's obvious that the whole framework of rights is affiliated with power: to have rights means to have the rights to power. This is why the question of animal rights has to be refor-mulated and, according to Derrida, it was Jeremy Bentham who proposed changing the very form of the question regarding the animal:

> Bentham said something like this: the question is not to know whether the animal can think, reason, or speak, etc., something we still pretend to be asking ourselves (from Aristotle to Descartes, from Descartes, especially, to Heidegger, Levinas, and Lacan, and this question deter-mines so many others concerning *power* or *capability* [pouvoirs] and *attributes* [avoirs] . . .) . . . The *first* and *decisive* question would rather be to know whether animals *can suffer*. (2008, 27)

Thus the framework of power is reformulated in terms of affect or affectivity, or, as Derrida names it, non-power or inability:

> Being able to suffer is no longer a power; it is a possibility without power, a possibility of the impossible. Mortality resides there, as the most radical means of thinking the finitude that we share with animals, the mortality that belongs to the very finitude of life. (2008, 28)

In other words, mortality and suffering is not a privilege of man, but something which humans share with non-humans.

Seen from this perspective we cannot deny the unprecedented violence inflicted on animals, which is compared by Derrida with "the worst cases of genocide". As Wolfe argues, "Derrida (an Algerian Jew) is well aware of the complexities of the analogy here" (2013, 44). Nevertheless, Derrida points out that the annihilation of animals can be described in terms of "animal genocide", which is much more complicated because it relates not only to annihilation but also to "virtually interminable survival" (2008, 26). This process ironically brings to mind the Foucauldian idea that the old sovereign's "right of death" is transformed in the biopolitical paradigm into the right to ensure, maintain, and develop life (Foucault 1998, 136). In other words, certain species of animals are not only annihilated but also subjected to artificially prolonged life and overpopulation. In this sense the "animal genocide" is even more horrible than the historical genocide:

> As if, for example, instead of throwing a people into ovens and gas chambers (let's say Nazi) doctors and geneticists had decided to organize the overproduction and overgeneration of Jews, gypsies, and homosexuals by means of artificial insemination, so that, being continually more numerous and better fed, they could be destined in always increasing numbers for the same hell, that of the imposition of genetic experimentation, or extermination by gas or by fire. (2008, 26)

This reversed fantasy demonstrates what kind of industrial, hormonal, and genetic violence is inflicted on animals. In this sense it is today's animal which becomes the real "subject" of biopolitics.

In this context, as Wolfe argues, we can see the irony of the Spanish Parliament's decision to grant basic rights to the great apes: it seems that those animals, which are "racially" similar to us, can be protected, while all other species can be submitted to an "animal holocaust" (2013, 104). The problem here is neither deciding which species are more human, nor drawing a distinction between human and non-human; the problem is that such a decision is always a biopolitical decision, creating a new framework (and a new declaration) of inclusion and exclusion. As Wolfe points out:

> To live under biopolitics is to live in a situation in which we are all always already (potential) "animals" before the law—not just nonhuman

animals according to zoological classification, but any group of living beings that is so framed. (2013, 10)

Thus reframed—or, more precisely, unframed—the question of the animal should be thought of as part of a continuum of multiple life forms. As Derrida points out:

> Beyond the edge of *so-called* human, beyond it but by no means on a single opposing side, rather than "The Animal" or "Animal Life" there is already a heterogeneous multiplicity of the living, or more precisely . . . , a multiplicity of organizations of relations between living and dead, relations of organization or lack of organization among realms that are more and more difficult to dissociate by means of the figures of the organic and inorganic, of life and/or death. (2008, 31)

These relations are so intertwined and fluid that they cannot be objectified. This means that the so-called human is also a part of this multiplicity of organizations and cannot achieve an autonomous being outside these organizations. Having this in mind, Calarco argues that it is impossible to make "an insuperable division between human and animal" and that "'human beings' belong to this multiplicity of beings and relations" (2008, 142). Formulated in this way, these questions force us to abandon the humanist perspective and to overcome the divisions between humans and non-humans.

ON BECOMING-IMPERCEPTIBLE

Derrida, by insisting on the "heterogeneous multiplicity of the living" (2008, 31), comes very close to Deleuze and Guattari's notion of becoming that is formulated in *A Thousand Plateaus*. In this work Deleuze and Guattari oppose the metaphysical tradition of self-identical being with their fluid ontology of becoming. Deleuze and Guattari refer to different kinds of becomings: becoming-woman, becoming-child, becoming-animal, becoming-molecule, and, finally, becoming-imperceptible. As mentioned earlier, Esposito is relying on Deleuze and Guattari's notion of becoming in trying to define his philosophy of the impersonal. Although this theoretical move is certainly justified, I would argue that Deleuze and Guattari's idea of becoming is much more radical than just an attack on the concept of person or personality. The concept of becoming not only deconstructs the notion of person or subject as such but also defies the centrality of man and human to such an extent that it moves beyond the scope of human perception toward becoming-imperceptible. As I will demonstrate later, the concept of becoming-imperceptible opens up the perspective of non-anthropomorphic, asubjective, and pre-individual life, which goes beyond the division between human and non-human.

The problem with Deleuze and Guattari's notion of life is that it strongly depends on literary examples and seems to be very far away from a political vocabulary. Although Deleuze and Guattari refuse to speak in terms of political theory, they are very well aware of the political importance of their philosophical project. They distinguish between macropolitics, which operates on the level of class or the state, and micropolitics, which functions on the level of minorities, of masses and flows. Macropolitics creates and recreates a majority which, being homogeneous, self-identical, and countable, can be easily hierarchized and controlled. By contrast, micropolitics invents the becoming of minorities, which destabilizes and evades the state's power. In this sense every becoming is becoming-minoritarian: "becoming-minoritarian is a political affair and necessitates a labor of power (*puissance*), an active micropolitics" (Deleuze and Guattari 2004, 322). For example, man is majoritarian *par excellence*, and so there is no becoming-man. Man as a majority is never becoming because majority implies a state of domination, an exercise of rights and power; "in this sense women, children, but also animals, plants, and molecules, are minoritarian" (Deleuze and Guattari 2004, 321). The most important point to understand here is that for Deleuze and Guattari becoming-minoritarian is not about celebrating minorities as certain identities but about opening up the medium of becoming. That means that even minorities have to detach themselves from a certain given identity and enter the process of becoming: even black people must become-black, even women must become-woman, even Jews must become-Jewish (Deleuze and Guattari 2004, 321). Thus, becoming is an active medium in which the oppressed groups can express themselves.

In this sense the process of becoming transgresses the divisions between citizens and non-citizens (becoming-minoritarian), between persons and non-persons (becoming-child), and between humans and non-humans (becoming-animal, but also becoming-molecule and becoming-imperceptible). Instead of the logic of inclusion and exclusion, which is at work in every declaration of rights, the notion of becoming introduces the logic of multiplicity. As Deleuze and Guattari point out, becoming and multiplicity are the same thing (2004, 275). Each multiplicity is heterogeneous and continually transforming itself into other multiplicities, gaining or losing dimensions which are simultaneously changing in kind. As Deleuze and Guattari point out, "A fiber stretches from a human to an animal, from a human or an animal to molecules, from molecules to particles, and so on to the imperceptible" (2004, 275). In this respect multiplicities create a fluid ontology of becoming, where the power over life is overwhelmed by the power of life. If power works by totalizing, the resistance to power comes through multiplication and the creation of multiplicities.

Becoming as multiplicity introduces a different kind of relationship with otherness. If political theories of modernity, as was demonstrated before, are founded on the notion of immunity, Deleuze and Guattari propose the concept of contagion to describe the new forms of connections:

> We oppose epidemic to filiation, contagion to heredity, peopling by contagion to sexual reproduction. Bands, human or animal, proliferate by contagion, epidemics, battlefields, and catastrophes. . . . The difference is that contagion, epidemic, involves terms that are entirely heterogeneous: for example, a human being, an animal, and a bacterium, a virus, a molecule, a microorganism. . . . These combinations are neither genetic nor structural; they are interkingdoms, unnatural participations. (2004, 266–7)

So if immunity means the body's capacity to keep its own identity and to resist otherness, becoming expresses the body's ability to change, to experience differences in kind, to affect and to be affected. The immune system can tolerate the heterogeneous element only in small quantities and only on the condition that this absorbed heterogeneous element will help to protect the body from a much more serious threat in the future; this is the logic of vaccination. By contrast, the logic of contagion or epidemics destroys the body's identity in order to open ways to new connections and "unnatural participations."

As such, becoming destroys any notion of subjectivity or individuality, replacing it with haecceities, subjectless individuations, singularities. In his last text, "Immanence: A Life" (2005), Deleuze introduces the notion of life which refers not to the life of the individual but to impersonal yet singular life which may or may not coincide with the life of an individual (Deleuze 2005, 28, 30). A life is a virtual quality which may be actualized in the life of an individual. As was mentioned before, Deleuze replaces the notion of an individual with the notion of individuation, which means only a temporary point in the process of becoming. The division between individual and individuation here resonates with Arendt's concern that a refugee represents nothing but his own unique individuality. Seen from the Deleuzian perspective, the situation is even more radical, because even individuality must be given away in order to enter the flow of becomings and individuations. In this sense becoming, which opens the lines of escape or flight, is an affirmative reformulation of the status of a refugee, but this time without any leftover identity.

Thus, becoming is moving through different stages toward becoming-imperceptible. But what does becoming-imperceptible mean and why should all the molecular becomings come to becoming-imperceptible? As Deleuze and Guattari formulate, "Becoming-imperceptible means many things. What is the relation between the (anorganic) imperceptible, the (asignifying) indiscernible, and the (asubjective) impersonal?" (2004, 308). Here we can see that becoming-imperceptible works on three different levels. First, it renounces the notions of subject and subjectivity, and replaces them with individuations, haecceities, and becomings. Second, it renounces the structuralist model of signification and interpretation and replaces it with the capacity to affect and to be affected. The third level is much more complicated, because the term "anorganic" here can mean either the renunciation of the organic in favor of the inorganic, or the renunciation of the organism as the principle of organization. In this sense to become imperceptible would mean to go beyond the limits of the organizing principle of human perception. The notion of becoming-imperceptible should be related to the notion of the Body without Organs and the three great strata that it opposes: subjectification, signification, and the organism (Deleuze and Guattari 2004, 176). As far as the organism is the result of organization, renouncing the organism means to oppose the principle of organization, hierarchized domination, and exclusion. Thus becoming-imperceptible implies not the subject's dissolution in a soup of haecceities, but an escape from the structures of domination and power.

FROM BIOPOLITICS TO BIOPHILOSOPHY

Thus the Deleuzian notion of becoming-imperceptible reveals a different strategy to oppose power than those proposed by political theory. Deleuze reveals the same strategy in his reading of Foucault's *The History of Sexuality*: here Deleuze discloses the twofold relationship between life and power. As Deleuze points out, "When the diagram of power abandons the model of sovereignty in favor of a disciplinary model, when it becomes the 'bio-power' or 'bio-politics' of populations, controlling and administering life, it is indeed life that emerges as the new object of power" (2006, 76). But it is precisely at this moment when power takes life as its object that resistance turns life against power. As Foucault concludes, "Life as a political object was in a sense taken at face value and turned back against the system that was bent on controlling it. It was life more than the law that became the issue of political struggles, even if the latter were formulated through affirmations concerning rights" (1998, 145). In other words, resistance comes not with the multiplication of the models of rights but with the assertion of life as a multiplicity. As Deleuze points out, "When power becomes bio-power, resistance becomes the power of life, a vital power that cannot be confined within species, environment or the paths of a particular diagram. . . . Is not life this capacity to resist force?" (2006, 77). Life is understood as a certain vitalism, as a potentiality which comes to resist power over life. Both Foucault and Deleuze argue that this new power of life becomes possible only when it is liberated from the concept of the (hu)man and unleashes forces which are non-human, impersonal, or asubjective. Life can resist biopolitical power only by taking this path of becoming-imperceptible which leads to asubjective, asignifying, and inorganic life forms.

In this regard we can claim that Deleuze is creating a philosophy of life or biophilosophy. The notion of biophilosophy, which appears in Eugene Thacker's essay "Biophilosophy for the 21st Century" (in this volume), marks the distinction between the understanding of life in the classical tradition and in the contemporary thinking of life, inspired by new bio-technologies and the posthumanist materialist paradigm of thinking. In classical philosophy the question of life was discussed in terms of transcendent principle—either it is *eidos*, *telos*, essence, or soul. By contrast, the posthumanist materialist thinking interprets life as an immanent principle which is at work in every living system. As Deleuze argues in "Immanence: A Life", immanence "eludes all transcendence of the subject and of the object" (2005, 26). It is not immanence to something else but immanence to itself—in this sense Deleuze says that life is pure immanence. Life can be accounted or explained not by referring to transcendent principles but revealing its immanent modes of existence. From this, it follows that biophilosophy examines not the essence of life but the multiplicity of life's expressions.

As Manuel De Landa argues in *Intensive Science and Virtual Philosophy*, the notion of multiplicity has replaced the metaphysical concept of essence (2002, 9). The concept of essence defines a thing's identity, which is always conceived as being unified and timeless, or eternal. By contrast, the concept of multiplicity defines something as having a variable number of dimensions, which cannot be subsumed by a higher dimension, that is to say, which cannot be unified or universalized. Moreover, the essence or the identity of a thing is given all at once, whereas the multiplicity is never given all at once and appears in a form of progressive differentiation. As De Landa explains, "The singularities which define a multiplicity come in sets, and these sets are not given all at once but are structured in such a way that they *progressively specify the nature of a multiplicity* as they unfold following recurrent sequences" (2002, 16). This means that life understood as a multiplicity is never a closed or unified system and it unfolds in duration and time. Instead of asking "what is life", or asking the question of essence, we should think about it in terms of multiplicity and follow its lines of progressive differentiation.

Thus biophilosophy can be defined by these characteristics: pure immanence, multiplicity, and progressive differentiation. However, it is important to stress that multiple life forms, which follow the lines of progressive differentiation and temporal duration, do not create an unpredictable chaos but coexist in a virtual continuum. This is why virtuality for Deleuze is one of the most important of life's characteristics. As Deleuze points out, "A life contains only virtuals. It is made up of virtualities, events, singularities. What we call virtual is not something that lacks reality but something that is engaged in a process of actualization" (2005, 31). For Deleuze the virtual is real even if it is not actualized in the states of things or of life. In other words, all life forms co-exist on the virtual plane of immanence. Agamben points out that it is precisely the virtual character of life which makes it incommensurable with biopolitical power. If biopolitics operates by making divisions and exclusions, by making decisions about which life is "worth living", by contrast, biophilosophy follows the logic of multiplicities which is all inclusive and non-hierarchic. As Agamben points out, "A life . . . marks the radical impossibility of establishing hierarchies and separations. The plane of immanence thus functions as a principle of virtual indetermination, in which the vegetative and the animal, the inside and the outside and even the organic and inorganic, in passing through one another, cannot be told apart" (1999, 232). In this sense life as pure immanence replaces exclusive hierarchies of power with the inclusive synthesis of multiplicities and in this way invents an alternative to biopolitical power.

NOTES

1 This research was funded by a grant (No. MIP-105/2012) from the Research Council of Lithuania.
2 See the Universal Declaration of Human Rights, http://www.ohchr.org/EN/UDHR/Pages/Introduction.aspx, accessed 22 January 2015.
3 See Cavalieri and Singer (1993) and the Great Ape Project Website, http://www.projetogap.org.br/en/, accessed 22 January 2015.
4 See the World Declaration on Great Primates, http://www.projetogap.org.br/en/world-declaration-on-great-primates/, accessed 22 January 2015.
5 See "'Like concentration camp': Strip-spraying migrants video shocks Italy". *RT*, 18 December 2013, http://rt.com/news/italy-migrants-lampedusa-video-432/, accessed 22 January 2015.

REFERENCES

Agamben, Giorgio. 1995. We Refugees. Trans. Michael Rocke. *Symposium* 49 (2): 114–9.
Agamben, Giorgio. 1998. *Homo Sacer. Sovereign Power and Bare Life*. Trans. Daniel Heller-Roazen. Stanford: Stanford University Press.
Agamben, Giorgio. 1999. *Potentialities: Collected Essays in Philosophy*. Trans. Daniel Heller-Roazen. Stanford: Stanford University Press.
Agamben, Giorgio. 2000. *Means Without End: Notes on Politics*. Trans. Vincenzo Binetti and Cesare Casarino. Minneapolis: University of Minnesota Press.
Agamben, Giorgio. 2004. *The Open: Man and Animal*. Trans. Kevin Attell. Stanford: Stanford University Press.
Arendt, Hannah. 1943. We Refugees. *Menorah Journal* 31: 69–77.
Arendt, Hannah. 2013. The Perplexities of the Rights of Man. In *Biopolitics: A Reader*. Ed. Timothy Campbell and Adam Sitze, 82–97. Durham, NC: Duke University Press.
Calarco, Matthew. 2008. *Zoographies: The Question of the Animal From Heidegger to Derrida*. New York: Columbia University Press.
Cavalieri, Paola and Peter Singer, eds. 1993. *The Great Ape Project: Equality Beyond Humanity*. New York: St. Martin's Press.
De Landa, Manuel. 2002. *Intensive Science and Virtual Philosophy*. New York: Continuum.

Deleuze, Gilles. 2005. Immanence: A Life. In *Pure Immanence: Essays on A Life*. Trans. Anna Boyman, 25–33. New York: Zone Books.

Deleuze, Gilles. 2006. *Foucault*. Trans. Seán Hand. New York: Continuum.

Deleuze, Gilles and Félix Guattari. 2004. *A Thousand Plateaus: Capitalism and Schizophrenia*. Trans. Brian Massumi. New York: Continuum.

Derrida, Jacques. 2008. *The Animal That Therefore I Am*. Ed. Marie-Louise Mallet. Trans. David Wills. New York: Fordham University Press.

Derrida, Jacques and Elisabeth Roudinesco. 2004. *For What Tomorrow: A Dialogue*. Trans. Jeff Fort. Stanford: Stanford University Press.

Esposito, Roberto. 2008. *Bíos: Biopolitics and Philosophy*. Trans. Timothy Campbell. Minneapolis: University of Minnesota Press.

Esposito Roberto. 2011. *Immunitas: The Protection and Negation of Life*. Trans. Zakiya Hanafi. Cambridge: Polity Press.

Esposito, Roberto. 2012. *Third Person: Politics of Life and Philosophy of the Impersonal*. Trans. Zakiya Hanafi. Cambridge: Polity Press.

Esposito, Roberto. 2013. Toward a Philosophy of the Impersonal. In *Terms of the Political: Community, Immunity, Biopolitics*. Trans. Rhiannon Noel Welch, 112–22. New York: Fordham University Press.

Foucault, Michel. 1998. *The History of Sexuality, Vol. 1: The Will to Knowledge*. Trans. Robert Hurley. London: Penguin Books.

Rancière, Jacques. 2004. Who Is the Subject of the Rights of Man? *South Atlantic Quarterly* 103 (2–3): 297–310.

Singer, Peter. 2001. *Writings on an Ethical Life*. New York: HarperCollins.

Wolfe, Cary. 2013. *Before the Law: Humans and Other Animals in a Biopolitical Frame*. Chicago: The University of Chicago Press.

Part II
Life, Bioethics, and Bioart

5 Chimerism and *Immunitas*
The Emergence of a Posthumanist Biophilosophy

Margrit Shildrick

In its claim to be the master discourse, modernist philosophy has long appeared to float above the messy matters of bioscience, which constitutes, nonetheless, one of the major authoritative discourses that sustain the western obsession with the distinction between self and other, between one body and the other. One powerful validation arises from appeals to the putatively unique and temporally stable genetic signature of each human cell, which in turn determines the precise makeup of the Human Leucocyte Antigens (HLA) that underpin our immune systems.[1] It is extremely rare for two individuals to have the same gene-encoded set of HLA molecules[2]—collectively called a tissue type—and as a result, we consider that the biological distinction between self and non-self is absolute and embodied. In lay terms, and indeed in many current textbooks, immunology itself is described as the science of self/non-self-discrimination. In reality, the purity of that distinction is illusory and what constitutes the proper "me" is already shot through with otherness. We have long known that all human bodies swarm with a multitude of putatively alien others such as the countless bacteria that inhabit our gut (Waldby and Mitchell 2006); while current research on the microbiome indicates that the microbial communities that cohabit in and on our bodies immeasurably exceed the strictly human cell components.[3] We are at very least, as Benezra, DeStefano, and Gordon (2012) put it, *supraorganisms* that are deeply hybrid in nature and display none of the expected distinctions that mark out self from other.

In drawing on my own research around the inherent hybridity of organ transplantation that already unsettles identity to the self and signals new ways of becoming other, I want to look more specifically at the event of chimerism as it contests the discourse of the self's immunity to the other. The word is derived from the Greek myth of the Chimera, a fabulous creature that combined elements of a lion, a goat, and a serpent, and thus broke species boundaries to create something new. In biomedicine, the term most frequently encountered is *micro*chimerism, which refers to a small but significant presence of so-called non-self cells coexisting with a dominant population of self cells in the same body. More extensive chimerism is said to occur when the host cells are outnumbered or even replaced, within a solid organ for example. That the occurrence of chimerism within a supposedly single body presents a serious challenge to one of the fundamental *doxa* of western medicine and specifically contests the definitive principle of the immune system might seem self-evident, yet in the face of a sociocultural imaginary that insists on clear boundaries between self and other, the authorized discourse, of the clinic at least, remains largely unchanged, stressing the importance of securing immunity and assuring us all of our continuing essential singularity. Nonetheless, where the hybridism of solid organs stirs up ontological issues for transplant recipients and philosophers alike, but seems to present no nonclinical problems to practitioners, the question of cell microchimerism is the starting point for a series of reconfigurations. For conventionally trained bioscientists, the search is for a functional explanation as to why and how what is expected at most to be a transitory phenomenon—better yet it did not occur at all—may persist for decades; for me as a critical cultural theorist, the thrill is to bring philosophical speculation to bear on the problematic in a way that opens up to the concept of assemblage in its Deleuzian sense as a better model for organic life, including human life.

Before turning to that side later, I want first to refer to some of the developments over the last 60 years since Peter Medawar initially set out his new understanding of the immune system, which had previously focused on infection. Motivated by the graft rejection experienced by postwar military personnel who needed skin grafts for their injuries, Medawar identified the immune system as the effective destroyer of potentially palliative non-self tissue. His goal therefore was to find ways of securing *induced* immunotolerance.[4] Interestingly, he demonstrated and named the phenomenon of long-term dizygotic twin chimerism in certain nonhuman mammals,[5] and even very rarely in humans, and made the connection to *natural* immunotolerance, but went no further than calling its occurrence a "natural accident" and "astonishing" (Medawar 1960). While several of Medawar's observations have been subsequently superseded, what stuck was the apparently natural antagonism of the self/non-self cellular relation. For some decades, the biomedical imaginary characterized the immune system through a series of militaristic metaphors expressing aggression, invasion, outright warfare, and foreignness being met by a swath of self-defense mechanisms such as the forcefully named Natural Killer (NK) cells. The exploration of this characteristic language of biomedical knowledge production has been taken up by many feminist scholars, including Donna Haraway, for whom notions of the self indicate tellingly that "individuality is a strategic defense problem" (1989a, 15), in terms of maintaining the boundaries between the normal self and the pathological other. Emily Martin (1990), Lisa Weasel (2001), and more recently Susan Kelly (2012) have all commented on the emergence of the specific discourse of immunology—and although the analogies have long been undermined by research findings that clearly could not be fitted to the self-defense model, the very same metaphors still hold sway in popular discourse. The problem at any level is that while the body's immunological counter to the putative threat of otherness in the form of a bacterial infection, or even a carcinoma, might understandably evoke images of steadfast defense, its hostile reaction to many therapeutic interventions such as tissue and organ transplants, or bone marrow implants, creates biomedical as well as metaphorical trouble.[6] In any case, the efficacy and closure of the self/non-self distinction are challenged by the existence of several puzzling anomalies, the most potent of which were the phenomenon of autoimmunity, where the body's own cells are misrecognized and responded to as other, and the apparently *natural* tolerance between a pregnant woman and her fetus. As Martin summarizes it:

> The imagery of aggressive immuno-warfare against the foreign focuses on the body that is all of one kind, all purely self . . . hence the normal woman would destroy her foetus to return to a normal state of internal purity. (1990, 148)

With regard to human reproduction, it is conventionally inexplicable, given the different HLA systems in place, that the maternal body should not reject the fetal material, or vice versa, but perhaps what should most startle us is that the paradox surrounding such a ubiquitous and indeed essential natural event as pregnancy has not resulted in any obvious rethinking of the mutually hostile self/other paradigm. The mechanisms of the maternal-fetal relation remain something of a mystery, but the view persists that the two bodies operate as separate entities, immunologically opposed to one another rather than mutually supportive. While the placenta was recognized as a limited site of exchange between mother and fetus—oxygen, nutrients, and hormones passing in one direction, products of excretion in the other—it was, until the late twentieth century, mainly seen as a protective barrier separating the distinct maternal and fetal bodies. And, as many women have found to their detriment, that supposed mutual distinction leaks over into the biopolitical sphere in the context of abortion politics, for example. At the same time, however, a great deal of research—initially associated with Diana Bianchi's lab[7]—has uncovered strong evidence, now widely accepted, that both maternal and fetal cells cross the placental barrier as a matter of course, effecting a kind of microchimerism within each body. The research is conducted, for convenience, by tracing the incidence of—conventionally out of place—Y-coded (male) cells within the female body, and then generalized to other forms of maternal-fetal chimerism. Whilst few bioscientists would now doubt the existence of microchimerism—which refers to very small proportions of mismatched HLA in the host body—many prefer to see it as always transient and insignificant. Nonetheless, the operative discourse shows a clear change from dominant metaphors of fetal intrusion, where the fetal cells are most certainly considered out of place, to a new language of trafficking and migration that catches the inevitability of the process, albeit with some negative implications in place (Martin 2010). There are also significant questions to address concerning what it might mean for either mother or child to carry within themselves allogenetic antigens: do they persist post-birth? Are they really tolerated by the host body or simply unrecognized? Are the effects neutral, benign, or harmful? Does the aging of maternal cells in the bodies of offspring pose any risk?

Over the last couple of decades, the phenomenon of intercorporeal cellular motility has been mired in controversy, particularly in relation to maternal or fetal/infant ill-health. The notion that fetal stem cells can rebuild damaged tissue, bone, or muscle is by now very familiar, so there is an immediate sense of recognition that a positive claim for cellular chimerism might be made at least with regard to the mother's body. Nonetheless, micro-chimerism has been marked as *both* active in tissue repair and regeneration, and as a contributing cause of mysterious autoimmune diseases—often in relation to the very same diseases—such as rheumatoid arthritis, MS, lupus (Nelson 1996; Bianchi 2007; see also Kelly 2012). What is perhaps confusing is that the evidence presented on either side of the debate is more or less the same with only the speculative interpretation making the difference in evaluation. The common ground is that non-self cells are frequently found in greater concentrations at the site of lesions than in peripheral blood, and in greater concentrations than in healthy control groups. The association then suggests either a causal link in the disease process, or alternately that differential HLA—with its distinct immunological signature—gathers to offer additional protection and repair in the face of damage (Kallenbach, Johnson and Bianchi 2011). For the most part, opinions seem fairly entrenched on either side, with few researchers looking for explanations of the paradoxical nature of the research findings. One exception is J. Lee Nelson who seems to have moved from an initial position of skepticism with regard to any beneficial effects (1996, 2002, 2012) which pitched her against the optimism of Diana Bianchi's lab, to a wide consideration of both the positive and negative implications, and, unusually for a research scientist, even an explicit rejection of the exclusionary self/non-self paradigm of human health.[8] In general, the debate about the putatively destructive or preservative nature of transcorporeal cell mobility shows little sign of being definitively resolved and most researchers have fallen back to saying that chimerism is sometimes beneficial, sometimes not, with most hopes of therapeutic implications shelved. The outcome, as Emily Martin sees it, is that as the clear-cut metaphors of self versus non-self become redundant, interest in and research funding for the area drops away.

Turning to the field of solid organ transplants—livers, kidneys, hearts, and so on—the relation between chimerism and tolerance is a central concern for improving graft acceptance rates. To forestall the rejection of donor organs, which usually excite a massive immunological response that fully exemplifies the self-other paradigm, the holy grail has long been to induce tolerance. In conventional practice that tolerance has devolved on

the strategy of engaging recipients in a usually lifelong regimen of immunosuppressant drugs that allow the donor organ to continuing functioning
free of host versus graft disease (or in some cases graft versus host disease)
that would lead to rejection. Grafts are rarely close HLA matches—which
remember are unique for each of us—but the assumption has been that the
alien cellular matter would stay *in situ*, thus localizing the danger, albeit it
one that could rapidly result in death. Very surprisingly a few early kidney
transplant recipients from the 1960s survived without temporally extensive
immunosuppression, but this was not brought to attention until retrospec-
tive studies done almost 30 years later. The first big jolt to the accepted
doxa was Starzl's systematic reviews of the early 1990s, which indicated
that donor HLA could be found not only in association with the transplant
organ but throughout the recipient body (Starzl et al. 1992). Other researchers indicated that cell mobility was bidirectional insofar as the transplant
organ itself could show signs of incorporating the recipient's existing HLA.
In other words, extensive chimerism had taken place. In Starzl's understanding, such chimerism might solve the problem of rejection by keeping the
immunogenetic effects of the two different populations of cells in balance;
he recommended that, pre-treatment, recipients should be given hematopoietic (stem) cells derived from the donor bone marrow, infused directly
into the peripheral blood, which would obviate the need for highly toxic
programs of immunosuppression. A similar approach has been suggested in
the case of xenotransplantation, which is of enduring interest in the transplant field as it could address the acute shortage of suitable human organs.
The problem is that the cellular immunologic rejection of xenografts is even
stronger than with allografts, but it might be overcome if, for example, prior
chimerism were induced in the human recipient of a pig's heart through the
use of porcine bone marrow (Abe, Qi, Sykes and Yang 2002). Like other
optimistic scenarios, the procedure—for both xeno- and allografts—has not
proven to be as effective as hoped, mainly it seems because such chimerism
appears transient and is eventually overtaken by the resurgence of chronic
rejection.

But is chimerism as ephemeral a phenomenon as many clinical papers
suggest? It would seem from most standard research that it is characteristi-
cally short-term, and limited to the procedures of transplantation, or to the
nexus of cellular exchanges that occur during pregnancy,[9] but some very
different results indicate that it is universal and persistent. One of Bianchi's most startling findings was that pregnancy-generated chimerism could
be detected in women many decades later, and even in women who had
never been pregnant; in other words, Y chromosomes could be found in
the peripheral blood and tissues of women who had never conceived, let
alone carried a male fetus (Bianchi et al. 1996). Nor had such women any
history of blood transfusions or of bone marrow or organ transplants.
Now this observation, confirmed many times over, suggests that chimerism
must have additional explanations. One that appears to fit the available

cellular profiling is that chimerism "handed down" as it were from mother to child could entail the translocation of HLA deriving from a *previous* pregnancy in which fetal markers (effectively traced as male ones) had entered the maternal body. In other words, any subsequent female offspring of the same mother could carry Y-coded HLA, not from any pregnancy of her own, but from the circulation of her own older male sibling's cells in the maternal body (Guettier et al. 2005; Yan et al. 2005). But if, as many basic scientists believe, chimerism is potentially life-long in duration (Maloney et al. 1999; Aractingi and Khosrotehrani 2005), then one can begin to surmise an intergenerational scenario in which each one of us—regardless of pregnancy status—could carry non-self cells from a variety of genetic relations. In principle it is difficult to see why chimerism—if it can indeed be detected at all ages—should not persist indefinitely.[10] It is telling that although there is a similar suggestion in Nelson's quasi-journalistic article "Your cells are my cells" for *Scientific American* (2008), there is not a hint of it in her reports of her extensive laboratory work. Biomedical science is clearly not yet ready for such a radical challenge to one of its central tenets. Yet chimerism is already known to relate to non-irradiated blood transfusions (Nelson 2002), bone marrow transplants, all types of tissue and organ transplant, pregnancy, generational genetic transfer, and human dizygotic fusion.[11] But why stop there? Although no one has yet devised any way of testing the idea, it has also been suggested that lactation and fluid sexual exchanges can also generate microchimerism (Yan et al. 2005), which further raises the question of the implausibility of genetic inviolability. If each body in the normal course of health carries plural and durable populations of differentially active HLA, it is clearly not just our understanding of the immune system that needs to be revised. The very existence of chimerism, and the probability that it is ubiquitous, deeply disorders any notion of the bounded self or of individuality while, at the same time, *reinforcing* the trope of uniqueness by multiplying its specific markers.

I want, then, to explore how the empirical observations that I have been outlining could be further deconstructed by applying a theoretical critique to the sense of the bounded self, protected against the incursions of others. It is worth remembering that the notion of immunity as defense of the body only emerged in biomedicine in the late nineteenth century after many hundreds of years in which it was a purely juridical and political concept (Cohen 2009). My starting point is with an intriguing dissection of the word "immunity" undertaken by the philosopher Roberto Esposito, who sees it as intertwined with what at first glance may seem to be an opposing concept, that of "community". Where the latter refers to something public or held in common, immunity signifies that which is private and particular to myself, but as Esposito (2008) points out, the two terms have a common root in the Latin *munus* which means an obligation or even a gift. Munus is all about obligations of responsiveness to the other, about reciprocity; it is what oils the wheels of community, and what is rejected by immunity. The one who is immune is exonerated from reciprocal gift giving and stands as an autonomous individual, free from the abnegation of self that community demands. This is a highly familiar theme in my own research around heart transplantation, where the acceptance of a donor organ, the so-called gift of life—which is currently reliant on the suppression of the recipient's immune system—inaugurates an enduring obligation within the recipient (Shildrick 2013a). Among the many troubling issues faced by heart recipients as they attempt to restore the texture of their prior lives is the new reality of their undeniably hybrid bodies giving rise to the ontological question of "who am I now?", and the equally confusing sense, or imposition, of kinship with the donor family that follows transplant (Shildrick 2013a; 2013b). A very high proportion of donees experience considerable psychic disturbance, regardless of the medical success of the transplant procedures or the prospect of a much increased life expectancy. Commenting on Esposito's work, Timothy Campbell writes: "Accepting the *munus* directly undermines the capacity of the individual to identify himself or herself as such and not part of the community" (2008, x). And that is precisely the issue with recipients, the majority of whom—prior to accepting the "gift of life"—understand themselves within the normative paradigms of western modernism as autonomous selves, sovereign individuals, with a very clear sense of the corporeal distinction between one self and the other. As Esposito notes, it is logically unthinkable for classical culture to tolerate the two-in-one or the one-that-is-made-two. As a result, organ transplantation, in which the differential DNA of the donor material is never assimilated as such but remains fundamentally other, offers a somewhat paradoxical take on the preservation of the individual life.

The move to bring together the seemingly disparate arenas of political philosophy and the body is at the heart of biopolitics, and it is worth remembering that biology both emerges as a symptom of politico-cultural discourse (Cohen 2009) and is generative of that discourse. The embodied self, then, is always a point of biopolitical production. As such, any cellular nomadism across the supposedly impermeable borders of distinct human organisms challenges immunity at all its levels. The underlying model for Esposito's biopolitics (2008; 2013) is of course biomedical, so when he says that the immune "is the 'nonbeing' or the 'not-having' anything in common" (2008, 51), he directly recalls the scientific definition based on self/non-self distinction. And in both forms, immunity is both protective against the destruction of the individual through what might be called excessive relations and at risk of over-identifying otherness to the point of destroying precisely what would be lifesaving. The conventional understanding of the standard immune response of chronic rejection toward a donor organ or tissue, a response that is managed by lifelong immunosuppression, is a case in point. Should immuno*suppression*—which itself always risks the deadly onset of other uncontrollable diseases—fail, then both the incoming material and the host body, which is wholly dependent on the replacement, are expected to die. As Esposito notes: "The idea of immunity, which is necessary for protecting our life, if carried past a certain threshold, winds up negating life" (2013, 61). And like Derrida (2003), he sees autoimmunity as the characteristic mode of contemporary politics, the inevitable outcome of that same process, where "increased to the point at which it turns into its opposite, the negative protection of life will end up destroying, along with the enemy outside, its own body" (Esposito 2013, 64). What is rejected in the overdetermined immunity against putative risk—in both politics and biology—is the possibility of (mutually) productive reconfigurations that go beyond the oppositional mode of self and other. Esposito writes: "The only way for life to defer death isn't to preserve it as such (perhaps in the immunitary form of negative protection) but rather to be reborn continually in different guises" (2008, 181). What he clearly understands is that biomedical technologies—and indeed political practices—entail both the technological and ontological transmutation of the human body. And while he mentions transplantation only in passing, Esposito does refer to pregnancy as a model for an immunity that does not end up destroying the life it seeks to preserve, that is not simply tolerant but hospitable to, and nourishing of, difference.[12] As I mentioned earlier, there is growing acceptance that HLA chimerism activated by pregnancy is bidirectional and at very least partially protective to both mother and fetus.

This surely calls to mind the kind of reciprocal and embodied ethics that has been developed by the feminist philosopher Rosalyn Diprose. In response to the work of Marcel Mauss, Jacques Derrida, and others on the meaning of the gift, Diprose (2002) has developed the term "corporeal generosity" to describe the unavoidable yet largely overlooked debt of embodied generosity. It evokes an unacknowledged form of community in which the self is not so much jeopardized by the risky contact with the common, as Esposito would put it (2008), but engaged in a mode of giving and receiving that is not dependent on the equity—the reduction to sameness—implied by exchange. Corporeal generosity, for Diprose, can only emerge precisely in the event of difference, which must be both protected and responded to. As she writes:

> Intercorporeal generosity *maintains* alterity and ambiguity in the possibilities it opens . . . generosity is only possible if neither sameness [n]or unity is assumed as either the basis or the goal of an encounter with another. (Diprose 2002, 90–1, my emphasis).

What Diprose means is that all life is dependent on an indeterminate range of openings to other lives that in turn bring into play a fluid array of connections which undermines the notion of an atomistic self. In effect, both the isolation of *immunitas*—the refusal of gifting—and the embrace of *communitas*—the coming together of generosity—may be as risky to the integrity of bodies as they are of opening up new possibilities. But if we understand integrity to intend the propriety—the completion and closure—of individual bodies, then that can only refer to the modernist illusion of embodiment that has long since theoretically crashed. The contemporary bodies of sociocultural politics and of bioscience alike have lost their singularity, and what Diprose's model of corporeal generosity offers is yet another challenge to the viability of the stubbornly entrenched biomedical and clinical narrative of a static and autonomous self.

There is no mention of chimerism or even hybridity in Diprose's account, but in related work Myra Hird (2007) extends the notion of the "affective material offering of our body to the other" (Diprose 2002, 191) to the specific relations of maternity. She suggests that alongside the transfer of DNA, there are a variety of other materials such as viruses, antibodies which boost the fetal immune system, nutrients, bacteria, biochemical substances such as nicotine, and indeed cellular material in transit between the maternal and fetal body, all of which could be seen as instances of corporeal generosity. While she does not explore the notion that the microchimerism of pregnancy is bidirectional, Hird, following Diprose, is insistent that all such embodied gifting is unpredictable and potentially disturbing—an expansive network of incalculable significance—and she references Haraway's view of the subsequent child as a "randomly associated genetic package" (Haraway 1989b, 352).[13] Just as the basic science uncovers both the beneficial and harmful consequences of maternal-fetal microchimerism, so corporeal generosity may both threaten the integrity of bodies and open up new possibilities. What matters here is the perception that neither corporeal generosity nor chimerism are instances of assimilation or merging where a new singular form materializes; bodies and micro-parts of bodies are conjoined in their irreducible difference. Just as the geep is a true chimera in that it is capable of reproducing either a lamb or a kid depending on which cells predominate in the reproductive gametes, the critical point of this strange form of *communitas* is that in overriding the immunological discourses of self and other, it does not transcend what Deleuze would call difference in itself. Esposito himself is always clear that the terms *communitas* and *immunitas* both presuppose what they appear to negate, and that immunization is "the fold that in some way separates community from itself, sheltering it from an unbearable excess" (2008, 52). The ethical task is to ensure some livable and coterminous existence.

I shall not pursue further any of the implications of Esposito's reflections on biopolitics, or his warnings about contemporary thanatopolitics and medical technologies, except to note that although he has much in common with Derrida,[14] Esposito decisively opens up the field with his gestures toward Deleuze. In marking the technological and ontological transmutation of the body, Esposito leaves behind the biomedical trope of tolerance—which in immunological terms refers to a lack of reaction to the other, a kind of passive coexistence—and posits a logic of dynamic multiplicity where variation is mutually affective. As he puts it, "We have to find a way, a form, or a conceptual language that converts the immunitary declension . . . into a singular and plural logic in which differences become precisely what holds the world together" (2013, 65). Undoubtedly, Derrida's notion of hospitality (2000) has already done some of that work in establishing the fundamental interiority of otherness, but I am uncertain that it can offer up the "*affirmative* biopolitics" that Esposito is seeking. What he wants is a way of thinking afresh and constructing more adequate concepts about the events that involve and transform us, which, he points out, is precisely what Deleuze explicitly sees as the primary purpose of philosophy. In appealing to the *im*personal as the only vital and singular mode that goes beyond the conventional semantics that continues to function in relation to the individuality of the person, Deleuze gives recognition to the one in the other, and to the unbounded potentiality of life's becoming. As Esposito notes, "That anything that lives needs to be thought in the unity of life . . . means that no part of it can be destroyed in favor of another: every life is a form of life and every form refers to life" (2008, 194).

I want to end then with some very preliminary speculations on where a Deleuzian approach that seeks to explode the mythology of self-other distinctions might take us in engaging with wholesale chimerism. The radical break pursued by Deleuze not only contests the boundaries of embodiment *per se*, but makes sense of—and to a degree settles—many of the troubling aspects of the question "Who am I?" which are so contested in both pregnancy and transplantation. The fundamental shift is from the conventional paradigm of "self versus other" in the formulation that still dominates immunology, and thus biomedical science more generally, to a view of the normal self as constitutively chimeric.

At the heart of Deleuze's philosophy is a decisive break with the notion of an atomistic subject—the sovereign subject of modernity—that celebrates not static "being", but a state of becoming in which any *individual* subject is always in a process of unraveling (Deleuze and Guattari, 1987). Each of us is caught up in multiple and unpredictable webs of interconnections—what Deleuze calls assemblages—in which life itself is characterized as a non-personal vitalist force that exceeds the unique experiences and composition of each individual (see Braidotti 2006; Shildrick 2013b). Now this idea of assemblages is highly effective for understanding what is at stake in chimerism, which is never about an assimilation that wipes out the differences, but about a coming together of disparate elements that deform and reform each other, yet go on functioning in some kind of new configuration.

In previous research, I have recently been rethinking organ transplantation in terms of parasitism, which equally relies on chimerism, but which inevitably preserves some features of self/other antagonism. A Deleuzian mode, in contrast, stresses that life is marked by the generative power of connection and the unending processes of transformation. Such an approach is highly apposite for our technologized society, and may be particularly useful in rethinking organ transplantation as an *ongoing* project, not only for the recipient but for the donor too. In Deleuzian terms, life is not a discrete essence, actualized in the individual body, but simply an element in the broader cycle of becoming that encompasses all manner of beings,

organisms, and machines. Each human life course *is* clearly marked by discrete events such as pregnancy or transplantation where things change or
transform, but in another sense, events are also incorporeal and atemporal
forces and intensities that are excessive to any given form of embodiment.
In short, the strategic defense of the self that immunity—both biomedical
and political—is supposed to mount gives way to what Esposito might even
call community.

So how could this open up new ways of figuring the chimerism which
so troubles the conventionally exclusionary function of the immune system?
In the liberal humanist context in which only individual identity counts, it
is reasonable in the case of transplantation that recipients should wish to
be restored to the person who preceded the surgery, but in Deleuzian terms
(Braidotti 2006), individual "ownership" of life gives way to the intensity of
continued becoming in a process with neither beginning nor end—and that
should call to mind my earlier speculation about intergenerational chimerism.
In that sense, elements of the donor coexist with the recipient in a new assemblage that contributes to the ongoing flux and flow of life. Indeed postmortem
donation is already strongly Deleuzian insofar as each deceased donor body
will provide on average organs and tissues for around seven recipients, illus-
trating clearly the fluid and crosscutting power of connectivity and assemblage that supersedes the individual death. For Deleuze, the point would not
be about restored functional efficacy for the recipient, or even the hope of
prolonged life, but about what Braidotti has called sustainability, "the very
possibility of the future, of duration, of continuity" (2006, 137). The relation
between recipient and donor in such a model is not one of self and other as
the science of immunology supposes, but an impersonal coming together in
a new and unpredictable assemblage that reflects the actual chimerism that
has taken place. Alongside the anticipated changes that occur as a result
of transplantation, many unexpected transmutations like the emergence of
chimerism—that can be perceived as both positive and negative—disorder
existing material boundaries and temporal limits and move always toward
new possibilities of becoming other than the conventional self.

As an authoritative discourse, biomedicine cannot be separated from the
realm of the sociopolitical, where the concept of immunity speaks to the
modernist desire to protect the illusory purity of the defended self (Cohen
2009) and, in Esposito's terms, undermines the development of positive
community. In accessing the basic science research on gestational chimerism
and its many possible intergenerational offshoots, or on various aspects of
organ and tissue transplantation, including the newly emerging stem cell
transplants for neurological disorders, it is increasingly clear that they are
all forms of biopolitical objects. In short, the *biological* ground of the mutually reinforcing biopolitical trope of immunity as the underpinning of the
distinct identities of self and other is far from certain. With the exposure
of the inherent plasticity of human embodiment, not only in terms of visible body modifications, but at the cellular level, a more adequate mode of

perception might be that of an intermingling of corporeal materials that, in maintaining the notion of irreducible difference, cannot be adequately expressed by the metaphor of hybridity. As our understanding that chimerism and microchimerism are the rule rather than the exception strengthens, it inevitably undoes any unproblematized belief in the illusion of self/other separation and disjunction, and suggests new ways of thinking our existence, not in terms of self-defense, but through dynamic coexistence and the inherently communal form of assemblage. Thought together, chimerism and an immunopolitics might resolve the inherent tensions of the self/other model and intensify the postmodernist insistence on the internal diversity, permeability, and intersection of all bodies.

NOTES

1 In the conventional allopathic model of biomedical discourse, all the cells of the body incorporate the Human Leukocyte Antigen (HLA) that marks them as self. When the immune system encounters cells without these precise combinations—as, for example, in bacterial or viral infections, tumors or transplants—it identifies them as not-self and launches an immune response that treats the unrecognized material as pathogens to be neutralized.

2 There are two major types of MHC (Major Histocompatibility Complex) protein molecules—class I and class II—that span the membrane of almost every cell in an organism. In humans these molecules—HLA—are encoded by several genes all clustered in the same region on chromosome 6.

3 This is far from speculative science with the *Human Microbiome Project* receiving substantial ongoing funding from United States *National Institutes of Health*. Nonetheless, the task of identifying the genetic entanglements of the specific biomass lends itself to profound speculation on the nature of the human.

4 "'Immunological tolerance' may be described as a state of indifference or non-reactivity towards a substance that would normally be expected to excite an immunological response" (Medawar 1960).

5 A very recent report in *Nature*, announcing the newly sequenced genome of the marmoset monkey refers to its "unique" capacity to produce "dizygotic twins that exchange hematopoietic stem cells *in utero*, a process that leads to lifelong chimerism" (Marmoset Genome Sequencing and Analysis Consortium 2014). This is odd as Medawar himself, referring to earlier work by R. D. Owen in 1945, was well aware of the similar process in dizygotic twin cattle (Medawar 1960).

6 In addition, as Susan Kelly (2012) points out, cellular therapies, such as those involving bone marrow, and regenerative medicine in general, already challenge the illusion of the atomistic individual body.

7 See Aryn Martin (2010) for her revealing interviews with Bianchi about the difficulties she faced in getting her research accepted.

8 Nelson is also engaged in the popular understanding of science, and her provocative article for *Scientific American* (2008) displays a plethora of metaphors revealing the changing status of microchimeric cells. She variously calls them "interlopers", "undesirable aliens", "masqueraders", "stowaways", "migrants", "two-way traffic", "émigrés", "adopted cells", and "seeds . . . that take root".

9 The problem is that having once identified maternal-fetal chimerism, few researchers are actively looking elsewhere for it.

10 I'm a little wary here of invoking something equivalent to the eighteenth century doctrine of preformation.
11 Tetragametic chimerism occurs when two zygotes fuse and develop into a single body with two distinct sets of DNA. See Norton and Zehner (2008) for a nonclinical account.
12 In the process of providing placentally-mediated immunity to the fetus, the mother's own immune system also changes. The cell-mediated T-cells, which might otherwise attack the fetus, become weaker, while the B-cells of humoral immunity show an increase.
13 Diprose's account and to a great extent Hird's—like my own in this paper—are still very firmly rooted in a humanist scenario of interhuman transfer, but it is clear that such multilateral movement can involve components of many other living creatures as well as the non-living.
14 I have already explored at length Derrida's understanding of the gift and of an absolute hospitality in the context of rethinking the nature of transplantation (Shildrick 2008; 2013a).

REFERENCES

Abe, Masahiro, Jin Qi, Megan Sykes, Yong-Guang Yang. 2002. Mixed Chimerism Induces Donor-specific T-cell Tolerance Across a Highly Disparate Xenogenic Barrier. *Blood* 99 (10): 3823–9.
Aractingi, Sélim and Kiarash Khosrotehrani. 2005. Microchimerism: Fears and Hopes. *Dermatology* 210: 1–2.
Benezra, Amber, Joseph DeStefano, and Jeffrey I. Gordon. 2012. Anthropology of Microbes. *Proceedings in the National Academy of Sciences* 109 (17): 6378–81.
Bianchi, D. W. 2007. Fetomaternal Cell Trafficking: A Story that Begins with Prenatal Diagnosis and May End with Stem Cell Therapy. *Journal of Pediatric Surgery* 42: 12–8.
Bianchi, D. W., G. K. Zickwolf, G. J. Weil, S. Sylvester, and M. A. DeMaria. 1996. Male Fetal Progenitor Cells Persist in the Maternal Blood for as Long as 27 Years Postpartum. *Proceedings in the National Academy of Sciences* 93 (2): 705–8.
Braidotti, Rosi. 2006. *Transpositions: On Nomadic Ethics*. Cambridge: Polity.
Campbell, Timothy. 2008. *Bíos*, Immunity, Life: The Thought of Roberto Esposito. Translator's introduction to *Bíos: Biopolitics and Philosophy*, by Roberto Esposito, vii–xlii. Minneapolis: University of Minnesota Press.
Cohen, Ed. 2009. *A Body Worth Defending: Immunity, Biopolitics, and the Apotheosis of the Modern Body*. Durham, NC: Duke University Press.
Deleuze, Gilles and Guattari Félix. 1987. *A Thousand Plateaus: Capitalism and Schizophrenia*. Trans. Brian Massumi. Minneapolis: Minnesota University Press.
Derrida, Jacques. 2000. *Of Hospitality: Anne Dufourmantelle Invites Jacques Derrida to Respond*. Trans. Rachel Bowlby. Stanford, CA: Stanford University Press.
Derrida, Jacques. 2003. Autoimmunity: Real and Symbolic Suicides. In *Philosophy in a Time of Terror: Dialogues with Jürgen Habermas and Jacques Derrida*. Ed. Giovanna Borradori, 85–172. Chicago: University of Chicago Press.
Diprose, Rosalyn. 2002. *Corporeal Generosity: On Giving with Nietzsche, Merleau-Ponty, and Levinas*. Albany, NY: State University of New York Press.
Esposito, Roberto. 2008. *Bíos: Biopolitics and Philosophy*. Trans. Timothy Campbell. Minneapolis: University of Minnesota Press.
Esposito, Roberto. 2013. *Terms of the Political: Community, Immunity, Biopolitics*. New York: Fordham University Press.
Guettier, Catherine, Mylène Sebagh, Jérome Buard, Danielle Feneux, Monique Ortin-Serrano, Michele Gigou, Viviane Tricottet, Michel Reynès, Didier Samuel

108   *Margrit Shildrick*

and Cyrille Féray. 2005. Male Cell Microchimerism in Normal and Diseased Female Livers from Fetal Life to Adulthood. *Hepatology* 42 (1): 35–43.
Haraway, Donna. 1989a. The Biopolitics of Postmodern Bodies: Determinations of Self in Immune System Discourse. *Differences: A Journal of Feminist Cultural Studies* 1 (1): 3–43.
Haraway, Donna. 1989b. *Primate Visions: Gender, Race, and Nature in the World of Modern Science.* New York: Routledge.
Hird, Myra. 2007. The Corporeal Generosity of Maternity. *Body & Society* 13 (1): 1–20.
Kallenbach, Lisa R., Kirby L. Johnson, and Diana W. Bianchi. 2011. Fetal Cell Microchimerism and Cancer: A Nexus of Reproduction, Immunology and Tumor Biology. *Cancer Research* 71 (1): 8–12.
Kelly, Susan Elizabeth. 2012. The Maternal-Foetal Interface and Gestational Chimerism: The Emerging Importance of Chimeric Bodies. *Science as Culture* 21 (2): 233–57.
Maloney, Sean, Anajane Smith, Daniel E. Furst, David Myerson, Kate Rupert, Paul C. Evans, and J. Lee Nelson. 1999. Microchimerism of Maternal Origin Persists into Adult Life. *Journal of Clinical Investigation* 104 (1): 41–7.
Marmoset Genome Sequencing and Analysis Consortium. 2014. The Common Marmoset Genome Provides Insight into Primate Biology and Evolution. *Nature Genetics* 46 (8): 850–7.
Martin, Aryn. 2010. Microchimerism in the Mother(land): Blurring the Borders of Body and Nation in Fetomaternal Cell Trafficking. *Body & Society* 16 (3): 23–50.
Martin, Emily. 1990. Toward an Anthropology of Immunology: The Body as Nation State. *Medical Anthropology Quarterly* 4 (4): 410–26.
Medawar, Peter. 1960. Immunological Tolerance. Nobel Lecture. 12 December 1960. Accessed 3 January 2015. http://www.nobelprize.org/nobel_prizes/medicine/lau reates/1960/medawar-lecture.html.
Nelson, J. Lee. 1996. Maternal–fetal Immunology and Autoimmune Disease. Is Some Autoimmune Disease Auto-alloimmune or Allo-autoimmune? In *Arthritis & Rheumatism* 39 (2): 191–4.
Nelson, J. Lee. 2002. Microchimerism: Incidental Byproduct of Pregnancy or Active Participant in Human Health? In *Trends in Molecular Medicine* 8 (3): 109–13.
Nelson, J. Lee. 2008. Your Cells are My Cells. *Scientific American* 298 (2): 73–9.
Nelson, J. Lee. 2012. The Otherness of Self: Microchimerism in Health and Disease. *Trends in Immunology* 33 (8): 421–7.
Norton, Aaron and Ozzie Zehner. 2008. Which Half Is Mommy?: Tetragametic Chimerism and Trans-Subjectivity. *WSQ: Women's Studies Quarterly* 36 (3–4): 106–25.
Shildrick, Margrit. 2008. The Critical Turn in Feminist Bioethics: The Case of Heart Transplantation. *International Journal of Feminist Approaches to Bioethics* 1 (1): 28–47.
Shildrick, Margrit. 2013a. Hospitality and "The Gift of Life": Reconfiguring the Other in Heart Transplantation. In *Embodied Selves*. Ed. Stella Gonzalez-Arnal, Gill Jagger, and Kathleen Lennon, 196–208. London: Palgrave Macmillan.
Shildrick, Margrit. 2013b. Re-imagining Embodiment: Prostheses, Supplements and Boundaries. *Somatechnics* 3 (2): 270–86.
Starzl, T. E., A. J. Demetris, N. Murase, S. Ildstad, C. Ricordi, and M. Trucco. 1992. Cell Migration, Chimerism, and Graft Acceptance. *Lancet* 339 (8809): 1579–82.
Waldby, Catherine, and Robert Mitchell. 2006. *Tissue Economies: Blood, Organs, and Cell Lines in Late Capitalism.* Durham, NC: Duke University Press.
Weasel, Lisa. 2001. Dismantling the Self/Other Dichotomy in Science: Towards a Feminist Model of the Immune System. *Hypatia* 16 (1): 27–44.
Yan, Zhen, Nathalie C. Lambert, Katherine A. Guthrie, Allison J. Porter, Laurence S. Loubiere, Margaret M. Madeleine, Anne M. Stevens, Heidi M. Hermes, and J. Lee Nelson. 2005. Male Microchimerism in Women without Sons. *American Journal of Medicine* 118 (8): 899–906.

6 Resisting Biopolitics, Resisting Freedom
Prenatal Testing and Choice

Catherine Mills

Within the context of biopower, what would resistance be? Is it possible to say clearly what is on the side of resistance, and what is on the side of biopower? That the notion of freedom can itself be mobilized as a modern liberal or neoliberal biopolitical technique makes this question especially perplexing, for then the exercise of freedom entangles one further in the very structures that it may ordinarily be understood to challenge. As Nikolas Rose (2007, 67) argues, freedom is an artifact of government; he writes, "Freedom has been an objective of government, freedom has been an instrument or means of government, freedom has inspired the invention of a variety of technologies of governing." As Rose insists, though, this does not mean that freedom is thereby illusory, or nothing more than a governmental technique that only reveals the depths of oppression. Instead, it points to the possibility of different kinds of freedom, or more specifically, to the possibility of practicing freedom differently. If this is right, then it is worth examining in more detail the contours of freedom as a mechanism of both resistance and reiteration, of liberation and entrapment, in the context of modern biopower. In this essay I want to initiate an inquiry into the ways that notions of individual freedom operate in the context of human reproduction, the management of which I take to be a key axis of biopower.[1]

In contemporary practices and politics of reproduction, the principle of individual liberty or freedom has come to hold sway over public policy debates, as well as the moral decision-making that prospective parents and medical practitioners engage in throughout pregnancy care. Freedom here is most consistently rendered in the terms of individual or parental choice, and the imperative of reducing external interference in that choice. I examine this apparatus of choice as it operates in regards to prenatal testing, since this is a central aspect of almost all pregnancies in developed countries, and increasingly also in developing countries. Further, prenatal testing is inseparable from decisions about selective termination, which are essentially biopolitical decisions about who comes into the world and who does not. As feminist and disability scholars have often pointed out, prenatal testing technologies such as amniocentesis and obstetric ultrasound are central to processes of medical normalization in prenatal care (e.g., Katz Rothman

1993; Rapp 2000; Tremain 2006). Foucault's writings on normalization have thus been significant for analyzing the social operation of prenatal technologies. However, what has remained somewhat obscure is the relation between medical normalization and the apparatus of choice as it operates in relation to prenatal testing in particular and reproductive decision-making more generally. In illuminating this relation, I also try to bring out a different ethos for the formulation and framing of choice.

BIOPOWER AND REPRODUCTION

The concept of biopower first received significant articulation in Foucault's work in his 1976 lectures, later published under the title *Society Must Be Defended* (2003). The most influential statement of what Foucault sought to indicate by this term, though, is his lapidary revision of this lecture included in *Will to Knowledge* (1990) under the subtitle, "Right of Death and Power over Life." In both these discussions, Foucault contrasts biopower with the regime of sovereignty that it came to supplant, and differentiates between a disciplinary power that operates at the level of the individual body, and a biopolitics that concerns itself with the new political subject of the population, or "man-as-a-species" (2003, 243). The notion of bio-*power* therefore combines the earlier work on disciplinary power with a new form of power that Foucault identifies as bio-*politics*. Thus, in *Will to Knowledge*, Foucault (1990, 138) writes that "the ancient right to *take* life or *let* live was replaced by a power to *foster* life or *disallow* it to the point of death." He goes on to argue that this new life-administering power emerged in two basic forms, beginning from the late seventeenth century and extending through to the nineteenth: discipline and biopolitics of population. These two forms of power operate as the poles of biopower, where one focuses on the body in order to individualize and manipulate its forces, and the other is "centered not upon the body but upon life", in which "bodies are replaced by general biological processes" (Foucault 2003, 249). These two poles, Foucault insists, are tied together through a "whole intermediary cluster of relations" at the level not of speculative discourse but of "concrete arrangements that would go to make up the great technology of power in the nineteenth century" (1990, 140). One of the principle mechanisms that tied these two poles together was, in Foucault's view, the deployment of sexuality. Sexuality, Foucault argues, emerged in the nineteenth century as one of the most significant vectors of the new formation of power because of its "privileged position . . . between organism and population, between the body and general phenomena" (2003, 252). He writes, "It was at the pivot of the two axes along which developed the entire political technology of life", tied both to the intensification and subjugation of the forces of the individual body in discipline and applied to populations because of its consequences (1990, 145). "Sex was a means of access both to the life of the body and the life of the species" (1990, 146). Foucault argues that while sovereign power had prioritized the blood relation as one of its fundamental values, the regime of biopower that emerged in the eighteenth and nineteenth century focused instead on sexuality. This shift of focus entailed the emergence of a congeries of concepts such as heredity, progeny, degeneracy, and perversion, within which sexuality was not a symbol of power, but its object and target. Foucault is careful to note, though, that the transition from a society of blood to one of sexuality was not a distinct rupture, but entailed a series of "overlappings, interactions and echoes" (1990, 149). Given this shift, however, Foucault's fundamental point in *Will to Knowledge* is that rather than discourse about sex and sexuality being repressed in the Victorian age, it was continuously produced and incited, requiring infinitesimal detail and giving rise to constant anxiety.

In the initial reception of *Will to Knowledge*, it is interesting that matters of perversion took priority over the other foci of regulation that Foucault identified, probably largely because queer theorists led the way in recognizing the importance of Foucault's text for modern sexual politics. But this has come at the cost of obscuring other aspects of Foucault's analysis, aspects that are now being highlighted in feminist scholarship on the biopolitics of reproduction (Deutscher 2008; 2012). Reproduction was a central political and social problem in the contexts of the social transformations of the eighteenth and nineteenth centuries, such as increasing urbanism, the decline of the aristocracy, the emergence of capital, and the reproduction of the labor force, and imperial colonialism. In his analysis of the deployment of sexuality, Foucault identifies the "socialization of procreative behavior" as one of the "four great strategic unities" that formed the mechanisms of a power focused on sex (1990, 103–4). This socialization took place, Foucault argues,

> via all the incitements and restrictions, the "social" and fiscal measures brought to bear on the fertility of couples: a political socialization achieved through the "responsibilization" of couples with regard to the social body as a whole (which had to be limited or on the contrary reinvigorated), and a medical socialization carried out by attributing a pathogenic value—for the individual and the species—to birth-control practices. (1990, 104–5)

Foucault goes on to argue that the family unit was significant in the deployment of sexuality, for it is "the interchange of sexuality and alliance: it conveys the law and the juridical dimension in the deployment of sexuality; and it conveys the economy of pleasure and the intensity of sensations in the regime of alliance" (1990, 108).

Of course, the management of life through reproduction in biopower has transformed since the eighteenth and nineteenth centuries. While this transformation is impossible to map in any detail here, several interconnected characteristics or trends stand out as significant within reproductive biopolitics today. Since the postwar period of the twentieth century, reproduction has been subject to an ever-increasing technologization and a correlative commercialization, both of which take place against the backdrop of the expansion of neoliberalism and the dismantling of the welfare state. The technologization of pregnancy predates the clinical emergence of in vitro fertilization (IVF) in the late 1970s, but IVF permits ever-greater technological penetration into the processes of human reproduction (see Franklin 2013). The capacity to stimulate conception outside a woman's body and keep an embryo *ex vivo* for very early stages of its development has not only made it possible for technological interventions such as preimplantation genetic diagnosis to become an increasingly routine aspect of clinical reality, but has allowed for—and required—increasing ranges of choices to be made by prospective parents. Furthermore, the externalization of reproductive processes is intimately related to the commercialization of human reproduction. It has permitted the development of vast international markets in reproductive material such as gametes, and ultimately, in the products of reproduction in commercial gestational surrogacy (see Spar 2006; Murphy 2012; Almeling 2011). This trend of commercialization has also seen IVF companies listed on the stock exchange: in 2013, Virtus Health was listed on the Australian Securities Exchange, soon followed by the Monash IVF Group in 2014.

In the context of these transformations of human reproduction and its management in the twentieth and early twenty-first century, individual freedom has become the preeminent moral value and guiding politico-legal principle. While the eugenics typical of nineteenth and early twentieth century biopolitics involved state regulation of reproduction—ranging from enforced sterilization to prevent the birth of those deemed unfit to their outright murder (in the Nazi T4 program for instance)—this is no longer the predominant mode of biopolitical management. Rather, today, individuals themselves are responsible for the enactment of biopolitics in reproduction. Foucault saw this trend emerging in the nineteenth century, where the family was integral to a "political socialization achieved through the 'responsibilization' of couples with regard to the social body as a whole" (Foucault 1990, 104–5). Thus, if late twentieth century biopolitics involves a "flexible eugenics" (Taussig, Rapp and Heath 2005) enacted through individual decision-making, this is not entirely new. But the extent of this individual responsibilization is probably unprecedented.

This kind of individualized responsibility is evident in contemporary bioethics discourse for instance. Bioethical approaches to reproduction often insist on the preeminent value of reproductive freedom, while also elaborating the obligation on the part of parents to have the best child possible, or at least take measures to avoid the birth of the abnormal or anomalous. In general, in this context, individual freedom is understood as a matter of non-interference; the principle of reproductive freedom or liberty requires that neither institutions, such as the state, nor other people limit the exercise of individual freedom, except in circumstances where choices would result in harm to others. Even more prosaically, it is understood as being permitted or enabled to make one's own choices in regards to reproduction. Given this, how might we think about individual choice in the context of contemporary biopower?

NORMALIZING REPRODUCTION

While arguing that sexuality held a privileged position in regards to the individual body and the population, Foucault also goes on to identify another element that circulates between these two poles of biopower: norms. Foucault gives a central role to norms and normalization as the principal form of social and political regulation, suggesting at one point that "[a] normalizing society is the historical outcome of a technology of power centered on life" (1990, 144). While always maintaining this centrality of norms, across his discussions of normalization in various texts, Foucault nevertheless presented various accounts of the specific role and operation of normalization within biopower. For instance, in *Discipline and Punish*, he suggests that disciplinary techniques such as norms have "swarmed" from their originating institutions, such as the prison, to take over the entire social field (Foucault 1977). In *Society Must Be Defended*, though, he rejects this as "a first and inadequate interpretation" and argues instead that "the normalizing society is a society in which the norm of discipline and the norm of regulation intersect along an orthogonal articulation" (Foucault 2003, 253). Later, Foucault refines his view to suggest that normalization works in opposing ways in discipline and a biopolitics of population. In the former, infractions of the norm are produced as a consequence of the application of the norm understood as an idealization; this means that the phenomenal particularity of an individual is identified and calibrated through the application of an ideal. Consequently, as Foucault (1977, 184) states, normalization produces individuals as the necessary mode and counterpart of the operation of norms, that is, as a material artifact of power. In a biopolitics of population, Foucault suggests that norms are mobilized in exactly the opposite way, insofar as "the normal comes first and the norm is deduced from it" (2007, 63). Biopolitics of populations and the apparatuses of security that Foucault identifies as crucial to it, involve:

> a plotting of the normal and the abnormal, of different curves of normality, and the operation of normalization consists in establishing an interplay between these different distributions of normality and [in] acting to bring the most unfavorable into line with the most favorable. (Foucault 2007, 63)

This is a more strictly statistical concept of the norm, in which the norm is derived from empirical phenomena though it may not be strictly identifiable with any particular empirical being.

In light of this, how might we understand prenatal testing, which assays the fetus and, where diverges from norms are found, requires reproducing agents to decide on the continued existence or otherwise of that fetus? To what extent is Foucault's account of normalization in biopower helpful for understanding the operation and social effect of prenatal testing? And can his work guide us in understanding possible modes of resistance to fetal normalization? To address these questions, I focus on the technology of obstetric ultrasound, since it is a kind of "frontline" of normalization, acting as a threshold technology that either provides reassurance of normality on the one hand, or identifies possible and actual malformation and shunts prospective parents into channels for further normalization and potential elimination of the fetus. It is, I think, a particularly interesting technology since it operates in a context of inherent epistemic uncertainty and calculation of risk, while being simultaneously bound to moral and affective economies that shape the very experience of pregnancy, and thus, the lives of women. In its mobilization as a prenatal screening technology, obstetric ultrasound is, to use the helpful phrase proposed by Timothy O'Leary (2010, 172), an "apparatus of experience".

While I think there are various ways in which ultrasound may be construed as a biopolitical technology, one of these is the way in which it contributes to and is integrated within the normalization of the fetus. However, ultrasound is not obviously a disciplinary measure as conceived by Foucault, in that it does not entail the application of an idealized conception of the normal to the fetal body—there is no ideal fetus from which actual fetuses depart in varying degrees. Rather, ultrasound relies upon detailed measurements derived from actual fetuses, which allow for the perception of a more or less flexible normal range for a variety of indicators. For example, the Australasian Society for Ultrasound Medicine (ASUM), which is one of the main professional bodies for the regulation of obstetric ultrasound practice in Australia and which provides guidelines for clinical use of ultrasound in obstetric care, has generated a policy document called the "Statement on Normal Ultrasonic Fetal Measurements" (ASUM 2001). This document sets out charts for fetal measurements based on the Australian population, for use in the reporting of obstetric measurements in Australia and New Zealand. The charts are the outcome of 11,600 measurements of fetal parameters across 3,800 pregnancies. They include normal range measurements for head circumference, abdominal circumference, femur length, and biparietal diameter, to name a few. Clearly, this is more in line with the operation of normalization in a biopolitics of populations, wherein the norm is derived statistically from preexistent bodies. To be sure, there is a sense in which the application of the norm thus derived reveals a phenomenal particularity in the manner suggested by discipline—but the norm in this formation is supposed to be ideal only in the sense of being a statistically identified "empirical ideal".

Further, ultrasound necessarily operates within a context of the medical management of risk and uncertainty on the one hand, and the political management of population wellbeing on the other. For instance, in Australia, one of the main—though certainly not the only—tasks of the first trimester scan is to contribute to the generation of a risk calculation for the likelihood of a fetus having Down Syndrome. This is done through the measurement of the nuchal fold at the back of the fetus' neck, which is subsequently combined with maternal age and blood results to generate a risk assessment. Note that this test is not diagnostic—it cannot say definitively whether the fetus has Down Syndrome, but can only track certain markers for it and give an indication of likelihood. If the risk calculation yielded for any particular fetus is sufficiently high—typically, a greater than 1 in 300 chance—then further diagnostic testing may be recommended. This particular cut-off point trades off various factors such as detection rates, false positive rates, and economic costs (Schwennesen and Koch 2012, 290) to arrive at a way of distributing access to health resources that strives to track, if not limit, the appearance of the abnormal within a population. Both the individual woman and the population are asked to live with a certain risk of abnormality in the context of the distribution of limited health resources, but when that risk is deemed too high, further intervention may be required. The reification of this particular ratio means, though, that it becomes naturalized and the medico-political calculation behind it is almost entirely obscured.

Importantly, the norms that underpin ultrasound practice do not in themselves require the eradication of the abnormal. Rather, the eradication of the abnormal points not to an intrinsic aspect of norms so much as to their location and operation within moral and affective economies of the normal. What makes normalization especially dangerous is what Eva Kittay (2006) has identified as the "desire for the normal." For Kittay, norms that are simultaneously descriptive and prescriptive are analytically desirable, whereas purely descriptive statistical norms become the basis of social norms and institutions that enforce distinctions between the functionally valuable and devalued. Further, following Canguilhem, Kittay speculates that even mere statistical frequency ultimately appears as an expression of value, insofar as it means that a trait is normative within a particular kind of life. This point links the norm with conceptions of health and wellbeing, and more generally, with ideas about the good life. Oftentimes, the eradication of the abnormal, or even anomalous, is justified not by the need to maintain or protect the normal, but by concerns that deviation from the normal will undermine prospects of wellbeing and the capacity to live a good life, that is, to flourish. In this way, concerns about biological normality are made inextricable from concerns about ways of living well.

This brings to the fore difficult issues about the confluence of individual desires for the normal with the biopolitical structuration of reproductive choice. It seems reasonable to suppose that prospective parents using prenatal testing technologies are not typically making decisions based on biopolitical notions of what is good for the social body; rather, their interest lies in ensuring the wellbeing of their own future child (though this is not to say that this is their only interest). Nevertheless, the structuration of choice in the context of normalization means that the aggregation of individual choices coincides with biopolitical aims, such as those of ensuring population wellbeing. Further, the confluence of conceptions of wellbeing and the normal make such decisions unavoidably biopolitical. Several quick points can be made here. First, this suggests that the salient issue is not simply the *choices made* by any given individual or couple, but the "meanings and cultural tensions of the *choices made available* by the increasing number and routinization of prenatal testing" (Press 2000, 217; emphasis added). Second, there is also the matter of the various subtle ways that particular "choice architectures" foster certain attitudes, decisions, and actions, while denying or rendering others improbable if not impossible (Thaler and Sunstein 2008).[2] Finally, as Nancy Press (2000, 225–7) outlines, people may hold quite contradictory beliefs about the desirability of the normal and notions of wellbeing. For instance, they may simultaneously hold positive (often romanticized) notions of disability conditions such as Down Syndrome when considered in general, but be distressed by the prospect of their own child being diagnosed with the condition in prenatal testing. These points reveal a complex nexus of affect, ethics, and normalization surrounding the apparatus of choice in individualized biopolitics. A detailed analysis of this nexus is beyond the scope of this essay, but it clearly provides productive material for further investigation of the ways in which biopower operates today. Rather than extending this analysis here, though, let me return to the original question of resistance.

RESISTING REPRODUCTIVE CHOICE?

Given this context, how might one think about resisting biopolitics? What is it about biopolitics that warrants "resistance", and what would resistance look like? These questions are difficult to respond to in a context that involves desires for the future wellbeing of (possible) others, where wellbeing has become tied in various ways to conceptions of the normal, as well as longstanding political struggles over the scope of women's freedom, overwhelmingly figured in reproduction in terms of autonomous choice. Foucault was clear that the possibility of resistance was intrinsic to relations of power, since otherwise, such relations would constitute domination rather than power *per se*. Power relations, he insisted, required a degree of freedom for those involved in them in order to be considered power relations. Further, in a brief reflection upon efforts to resist the management of life in biopolitics, he argues that such efforts relied upon the very thing in which that power was invested, that is, "on life and man as a living being" (Foucault 1990, 144). This, he suggests, is the case even if those struggles against biopolitics were cast in terms of rights, such as the right to health and to self-expression, for example. While it would not be desirable, or perhaps even possible, to specify models of resistance to reproductive biopolitics, the relation between life and power may point toward a kind of desideratum for a new approach to prenatal testing in the context of reproductive biopolitics.

I argued in the previous section that the notion of reproductive freedom is itself harnessed to the biopolitical management of life through the responsibilization of individuals in reproductive choice. Individual choice then becomes the matrix through which the desire for the normal is channeled, and through which the excise of the abnormal itself becomes increasingly normal. However, it would be too hasty to suppose that as individual choice is part of the operation of biopolitics in reproduction, it should therefore be constrained or eliminated. For one, the notion of choice has been central to feminist efforts in regards to the politics of reproduction, and its mere rejection would entirely miss the important ways that the mobilization of this notion has contributed to achieving gender justice. Further, a simple rejection of choice would align too easily with anti-abortion campaigns that seek to restrict reproductive freedom. What appears to be required, then, is a continued commitment to freedom, but one that insists that freedom be practiced differently than in choices that align with hegemonic norms and conceptions of health and wellbeing, that is, with the normal. Interestingly, there is evidence to suggest that women who choose not to undertake prenatal testing or to not terminate their pregnancy when fetal malformations are detected often find themselves at odds with the institutions of normalization (see, for example, Tankard Reist 2006). This is not to say that these decisions should necessarily be taken as models of resistance to biopolitical normalization—though they might constitute possible forms of such resistance. More particularly, though, there is something in the *spirit* of these choices that may point toward a different practice of reproductive freedom.

Two paths for interpreting matters of resistance to biopolitics that may have some bearing on prenatal testing emerge in Foucault's work—the first evident in Foucault's interpretation of the work of Georges Canguilhem, and the second hinted at in a comment by Canguilhem on the development of Foucault's work. In his essay on Canguilhem, Foucault argues that at the center of the problems which preoccupy Canguilhem resides "a chance occurrence . . . like a disturbance in the informative system, something like a 'mistake,'" in short, "error;" Foucault states, "Life—and this is its radical feature—is that which is capable of error" (1998, 476). He goes on to suggest that, for Canguilhem, the errancy internal to life proves to be the radical contingency around which the history of life and the development of human beings are twined. Further, Foucault locates error at the centre of the relation of life and knowledge as understood by Canguilhem; he writes,

> If one grants that the concept is the reply that life itself has given to that chance process, one must agree that error is the root of what produces

human thought and its history. The opposition of the true and the false, the values that are attributed to the one and the other, the power effects that different societies and different institutions link to that division—all this may be nothing but the most belated response to that possibility of error inherent in life. (1998, 476)

Thus, it is through the notion of error that life is placed in relations of contiguity and contingency with truth and the structures within which it is told. "Error", or the inherent capacity of life to "err", both establishes the relation of life to truth and undermines that relation by disentangling man from the structures of truth and power that respond to the potential for error. Hence, "with man, life has led to a living being that is never completely in the right place, that is destined to 'err' and to be 'wrong'" (Foucault 1998, 476).

In terms of considering resistance in the context of prenatal testing, there is a sense in which this particular view brings out the intrinsically recalcitrant aspect of life insofar as it continually produces errors in the face of normalization. Normalization does not ultimately override the errancy of life, and the continual production of error ensures that life constantly escapes or exceeds the techniques that govern and administer it (Foucault 1990, 143). However, while this may be so, the continual production of error also ensures the continued enforcement of normalization. It is because life is subject to errancy that systems of normalization attain their apparent justification and appeal. Indeed, as I have suggested elsewhere, biopower thus appears as the modern response to the vital possibility of error (Mills 2013, 89; forthcoming). Furthermore, while some theorists have attempted to interpret Canguilhem's valorization of error within the socio-political sphere, it is difficult to see just how it would provide a model for, or even yield insight into, political resistance. For here, it seems that what is at stake is not simply the inherent errancy of life, but the response to that errancy. Can the response to error be anything other than a modality of biopower? Must such a response take the form of normalization? Or, how might one respond to errancy other than through desire for the normal?

This leads us to the second approach to resistance, which plays on the ambiguity of the concept of life to locate resistance in *bios*, or particular forms of life, rather than in biological life itself. In considering the tendency to posit a conceptual break between the two "stages" of the *History of Sexuality* series, Canguilhem was prompted to comment, "It was normal, in the properly axiological sense, that Foucault would undertake the elaboration of an ethics. In the face of normalization and against it, *Le Souci de soi*" (1997, 32). According to Canguilhem, the project of the later volumes of *History of Sexuality* did not present so much a rupture from earlier work, but its "completion, precipitated, perhaps, by a premonitory anxiety." This positing of a more or less strict continuity in the conceptual concerns that

Foucault was working with is worth considering for a moment, for it is not immediately obvious what Canguilhem means in his comments. First, the axiological sense that Canguilhem wants to emphasize brings to the fore the way that a study of ethics is logically an extension of the study of normalization—it is part and parcel of the study of norms and of value. But Canguilhem also says more than this. About what is Foucault supposedly anxious, for instance? In what sense is an ethics "in the face of normalization and against it"? Here, Canguilhem appears to be indicating that the particular formulation of ethics that Foucault offers may be seen as a working out of a form of resistance to normalization. Various versions of this interpretation have been common in commentary on Foucault's work, but I think at the least we can say that it is not the whole story.

For one, with the view that a consideration of ethics is an axiological extension of the study of normalization in mind, there are at least two moments or modalities at issue in the ethics of the self that Foucault proposes. First, an ethics of the self involves the integration of norms that one finds in one's culture into one's own ethical subjectivity. Thus, it may explain the ways that norms and values—such as the value of individual freedom, for instance—are taken up and made the object of ethical experience, that is, how they are integrated into one's own becoming as an ethical subject. In short, even as a biopolitical technique, freedom has normative force; indeed, "the force of individual freedom as a moral norm is inseparable from it being a norm of subjection" (Mills 2011, 50). Second, though, Foucault does seem at times to be reaching for an understanding of how those same norms might be dislodged, lived differently, suspended, or even transformed. In this way, the ethics of the self understood as a practice of freedom can also be seen as moving toward a freedom that is not already delimited by the norms of political and moral liberalism and the version of individual freedom offered therein. His work gestures, at times, to a different way of practicing freedom. In the context of homosexual relations, Foucault pointed toward the possibility of friendship as a model for practicing freedom differently. There is, however, no good reason to suppose that the same way of practicing freedom will constitute resistance in different contexts or across various domains. Prenatal testing and reproductive choice require yet another way of practicing freedom.

This could take the form of a freedom based on an ethos of fostering a certain spirit of uncertainty, a willingness to embrace errancy and the surprise of the other that attends it. This suggests the ethico-political importance of an ethics of the self that embraces the chance event, the unpredictability of errancy, and responds to error with an openness to alterity. To be clear, this does not have to equate to the kind of "ethics of the gift" proposed by the communitarian Michael Sandel in his critique of the expansion of reproductive choice through genetic technologies. Sandel (2007, 45) argues that an ethics of the gift is enlivened by the "openness to the unbidden" as outlined by theologian William F. May. He writes, "To appreciate children as gifts is to accept them as they come, not as objects of our design, or products of our will, or instruments of our ambition." The problem with this argument is that it too readily yields to an anti-abortion politics, in that prospective parents are obliged to simply accept whatever comes. However, there are often good reasons to terminate a pregnancy on the basis of fetal malformation, and there are different ways in which such decisions can be made. Instead, then, I have in mind an ethos enlivened by the recognition that the lives of others are not determined by the calculations of normalization, that subjectivity exceeds such calculability by virtue of singularity, that who one is cannot be determined by what one is deemed to be.

To conclude, I have tried to throw light on the relation between normalization and freedom understood as choice as they are mobilized within contemporary reproductive practices. I have argued that the apparatus of choice plays a significant role in reproductive biopolitics, and that focusing on this reveals a complex nexus of affect, ethics, and normalization at work in regards to prenatal testing. Rather than reject the apparatus of choice on the basis of its implication within the operation of biopower, though, I have suggested that a desideratum for resistance to normalization in prenatal testing would be an ethos that embraces vital errancy and the surprise of the other that it entails. It must be noted, though, that these reflections are necessarily limited and preliminary. Considerable analysis—both conceptual and empirical—is still required to illuminate the politics and ethics of choice in reproductive biopower today.

NOTES

1 The research for this chapter has been supported by funding from the Australian Research Council (DP110100752).
2 Note that I do not wish to endorse the ideas of nudging or the doctrine of liberal paternalism, simply the insight that choices are shaped, often less than consciously, by the contexts in which they are made.

REFERENCES

Almeling, Rene. 2011. *Sex Cells: The Medical Market for Eggs and Sperm*. Berkeley: University of California Press.
Australasian Society for Ultrasound Medicine. 2001. *Statement on Normal Ultrasonic Fetal Measurements*. Crows Nest: Australasian Society for Ultrasound Medicine.
Canguilhem, Georges. 1997. On *Histoire De La Folie* as an Event. In *Foucault and His Interlocutors*. Ed. Arnold Ira Davidson, 28–32. Chicago: Chicago University Press.
Deutscher, Penelope. 2008. The Inversion of Exceptionality: Foucault, Agamben, and "Reproductive Rights". *South Atlantic Quarterly* 107 (1): 55–70.

Deutscher, Penelope. 2012. Foucault's *History of Sexuality, Volume 1*: Re-Reading Its Reproduction. *Theory, Culture & Society* 29 (1): 119–37.

Foucault, Michel. 1977. *Discipline and Punish: The Birth of the Prison*. Trans. Alan Sheridan. London: Allen Lane.

Foucault, Michel. 1990. *The History of Sexuality: An Introduction, Volume 1*. Trans. Robert Hurley. New York: Vintage Books.

Foucault, Michel. 1998. Life: Experience and Science. In *Aesthetics, Method and Epistemology: Essential Works of Foucault 1954–1984, Vol 2*. Ed. James Faubion, 465–78. New York: New Press.

Foucault, Michel. 2003. *Society Must Be Defended: Lectures at the Collège de France, 1975–1976*. Trans. David Macey. Ed. Mauro Bertani and Alessandro Fontana. New York: Picador.

Foucault, Michel. 2007. *Security, Territory, Population: Lectures at the Collège de France, 1977–1978*. Trans. Graham Burchell. New York: Picador.

Franklin, Sarah B. 2013. *Biological Relatives: IVF, Stem Cells, and the Future of Kinship*. Durham: Duke University Press.

Katz Rothman, Barbara 1993. *The Tentative Pregnancy: How Amniocentesis Changes the Experience of Motherhood*. New York: W. W. Norton.

Kittay, Eva Feder. 2006. Thoughts on the Desire for Normality. In *Surgically Shaping Children: Technology, Ethics, and the Pursuit of Normality*. Ed. Erik Parens, 90–112. Baltimore, MD: Johns Hopkins University Press.

Mills, Catherine. 2011. *Futures of Reproduction: Bioethics and Biopolitics*. Dordrecht: Springer.

Mills, Catherine. 2013. Biopolitical Life. In *Foucault, Biopolitics and Governmentality*. Ed. Jakob Nilsson and Sven-Olov Wallenstein, 73–90. Södertörn, Sweden: Södertörn University.

Mills, Catherine. Forthcoming. Biopolitics and the Concept of Life. In *Biopower: Michel Foucault and Beyond*. Ed. Nicolae Morar and Vernon Cisney. Chicago: Chicago University Press.

Murphy, Michelle. 2012. *Seizing the Means of Reproduction: Entanglements of Feminism, Health, and Technoscience*. Durham, NC: Duke University Press.

O'Leary, Timothy. 2010. Rethinking Experience with Foucault. In *Foucault and Philosophy*. Ed. Timothy O'Leary and Christopher Falzon, 162–184. Oxford: Blackwell.

Press, Nancy. 2000. Assessing the Expressive Character of Prenatal Testing: The Choices Made or the Choices Made Available? In *Prenatal Testing and Disability Rights*. Ed. Eric Parens and Adrienne Asch, 214–33. Washington, DC: Georgetown University Press.

Rapp, Rayna. 2000. *Testing Women, Testing the Fetus: The Social Impact of Amniocentesis in America*. New York: Routledge.

Rose, Nikolas. 2007. *The Politics of Life Itself: Biomedicine, Power, and Subjectivity in the Twenty-First Century*. Princeton, NJ: Princeton University Press.

Sandel, Michael J. 2007. *The Case against Perfection: Ethics in the Age of Genetic Engineering*. Cambridge, MA: Harvard University Press.

Schwennesen, Nete and Lene Koch. 2012. Representing and Intervening: "Doing" Good Care in First Trimester Prenatal Knowledge Production and Decision-Making. *Sociology of Health & Illness* 34 (2): 283–98.

Spar, Debora L. 2006. *The Baby Business: How Money, Science, and Politics Drive the Commerce of Conception*. Watertown, MA: Harvard Business Publishing.

Tankard Reist, Melinda. 2006. *Defiant Birth: Women Who Resist Medical Eugenics*. Melbourne: Spinifex Press.

Taussig, Karen-Sue, Rayna Rapp, and Deborah Heath. 2005. Flexible Eugenics: Technologies of the Self in the Age of Genetics. In *Anthropologies of Modernity:*

Foucault, Governmentality and Life Politics. Ed. Jonathon Xavier Inda, 194–214. Malden, MA: Blackwell.

Thaler, Richard H. and Cass R. Sunstein. 2008. *Nudge: Improving Decisions About Health, Wealth, and Happiness*. New Haven, CT: Yale University Press.

Tremain, Shelley. 2006. Reproductive Freedom, Self-Regulation, and the Government of Impairment in Utero. *Hypatia: A Journal of Feminist Philosophy* (21) 1: 35–53.

7   Biophilosophy for the 21st Century[1]

Eugene Thacker

SOUL–MEAT–PATTERN

There have only ever been three approaches to thinking about life: soul, meat, and pattern. Within this trinity is everything deemed to be animate, living, and vital. "Soul" is not just the scholastic, theological, personal soul, but the Aristotelian principle of life (*psyche*), the principle of its organization. The vegetative soul of plants, the animate and sensate soul of animals, and the rational soul of human beings. The hierarchy of souls is not unlike the Great Chain of Being, a biological theology of divide-and-hierarchize. By contrast, "meat" is brute matter, unthinking mechanism, the clockwork organism, the *bête machine* described by Descartes—animal or machine, it makes no difference. Mechanism is, in a sense, a thinking about life as meat, and meat as lifeless (the life that is lifeless is meat or machine). Finally, distinct from "soul" and "meat" is a third approach, that of "pattern". It would seem that the emphasis on pattern is a distinctly postmodern phenomenon: the terrain of cybernetics, information theory, self-organization. But this is only part of the story. Again, Aristotle the biologist equates form (*eidos*) and "soul" to the distinguishing mark between the plant, the animal, and the human; it is in their mode of organization, how they self-actualize in time ("if it moves, it's alive"). Yet, Aristotle is linked to contemporary self-organization research in that neither can explain how organization occurs, other than to reiterate that the whole is more than the sum of its parts.

Thus, "soul", "meat", and "pattern" form a trinity. The trinity is also a triptych: soul in the center, meat on the right-hand side, and on the left, pattern. An image of thought that continuously switches, swaps, displaces, and replaces the placeholder that defines life: from *psyche* to mechanism and animal electricity, to the "gemmules" and "pangens", to DNA and the "code of life." However, these three approaches do not form a periodization, with Aristotle's *psyche* followed by Descartes' clockwork body followed by the genetic code. Instead, as a trinity-triptych, they form a kind of portrait, a face, a faciality, a field of black holes and white walls, within which and upon which is often written: "life is that whose essence can be deduced and yet whose essence escapes all deduction." Soul–meat–pattern. Each of these posits a central, universal, external principle of organization that culminates in the living, the organism, a life-force. We can simply refer to this as the principle of life, the central concept that structures a whole field of investigation. Each approach differs in its placeholders, but there continues to be a transcendental locus that minimally guarantees a description of life, something that enables one to point and say "over there . . ." (or perhaps, "it's alive . . . alive!", or again, "shoot anything with more than two legs!"). In positing such principles of organization, the soul–meat–pattern triptych also articulates boundaries: living–non-living, organic–inorganic, animate–inanimate, but also animal–machine, human–animal, human–savage, species, races, populations, genomes . . . We can refer to this practice as boundaries of articulation. Together, the principle of life and the boundaries of articulation are the two methods through which the West has ceaselessly reinvented its thinking about life.

EXTRINSIC LIFE

There is an inward-turning and an outward-turning aspect to this thinking. The inward-turning divides, orders, and interrelates species and types; the outward-turning manages boundaries and positions the living against the non-living, making possible an instrumentality, a standing-reserve. The inward-turning aspect is metabolic, in that it processes, filters, and differentiates itself internally; it is the breakdown and production of biomolecules, the organization of the organs, the genesis of species and races. The outward-turning aspect is immunologic, for it manages boundaries, exchanges, passages; it is the self-nonself distinction, the organism exchanging with its environment, sensing its milieu, the individual body living in proximity to other bodies. Nevertheless, there is always something that complicates both aspects. The inward-turning aspect is just fine until the outward-turning aspect loses its grip on things. An example is epidemics. An epidemic cannot be limited to the individual organism, for its very nature is to pass between organisms, and increasingly, to pass across species borders (and national borders). What is the unit of analysis for an epidemic? Likewise, the outward-turning aspect is able to manage boundaries without problem until the inward-turning aspect is discovered to be an illusion. For instance, if the outward-turning aspect is that which posits the individual organism as distinct from its environment, therefore enabling an instrumental relationship, a standing reserve, what then is the inward-turning aspect? We would assume it is the whole spectrum of understanding about that organism—its biological, physiological, cognitive processes. But isn't each of these really a nested, outward-turning aspect in itself? What are the systems, networks, and pathways of the organism if not nested layers of the outward-turning aspect? The inward-turning and outward-turning aspects thus complicate each other ceaselessly, and it is therefore not inaccurate to describe their relationship, as Deleuze (1992) does, as one of folding (in-folding, out-folding, an embryology having nothing to do with "development").

Soul–meat–pattern. Again, this is not a *telos*, as if to imply that genetic and information technologies are the most advanced mode of inward- and outward-turning. Yet, in a time of networks, swarms, and multitudes, it would seem that the third approach—that of "pattern"—is today domi- nant in the life sciences (genetics, genomics), healthcare (biotech industry), technology (a-life, AI, networks), war (bioterror, emerging epidemics) and even alternative scientific viewpoints (biocomplexity, emergence).[2] A new, vital pattern pervades systems of all kinds—global economies, social sys- tems, immigration patterns, information exchanges, mobile and wireless communications, and so forth. Despite this, have we rid ourselves of the divide-and-hierarchize mentality of thinking about life? Is "pattern" simply the new "soul"? Traditionally, these questions about the principle of life come under the domain of the philosophy of biology. But what would it mean to invert the philosophy of biology? What would it mean to invert this thinking (soul–meat–pattern) and this dualistic method (principles of life, boundaries of articulation), and consider instead a biophilosophy? Perhaps it is precisely "life itself" that is the problem, not the aim or the goal. Instead of considering the intrinsic properties of life, what about considering life as extrinsic, as always going outside of itself? Instead of centering life (an essence, an organizing principle), what about considering life at the periph- eries? Extrinsic life, a life always going outside of itself, peripheral life . . .

BIOPHILOSOPHY VERSUS PHILOSOPHY OF BIOLOGY

What, then, is biophilosophy? To begin with, biophilosophy is not the same as the philosophy of biology. What is usually referred to as the philosophy of biology has both a syntagmatic and a paradigmatic side to it, a horizontal and vertical dimension to it. The horizontal dimension is the elucidation of universal characteristics of the organism which are perceived to be part of its essence or principle of organization (growth and decay, reproduc- tion and development, evolutionary adaptation). The vertical dimension is the development of this thinking historically in Western thought, from Aristotle to natural history, to Darwinian evolution, to the new synthesis in genetics and biochemistry. In general, the philosophy of biology high- lights and extends the philosophical dimensions of biological knowledge. Issues pertaining to evolution, biological determinism, dualism, mechanism, and teleology may be considered in the context of the life sciences, such as comparative anatomy, physiology, genetics, biochemistry, embryology, germ theory, developmental systems theory. The philosophy of biology informs the three approaches to thinking about life mentioned above: soul–meat– pattern. The philosophy of biology also undertakes the twofold method of identifying a principle of life and boundaries of articulation. It can be understood as an attempt to pose the question, "Is the living different from the non-living?"—an ontological question—in the context of another question, "Is the study of the living (biology) different from other fields of study?"—an epistemological question.

Is biophilosophy simply the opposite of the philosophy of biology? Not quite. Biophilosophy is certainly a critique of the triptych of the philosophy of biology. But it is also a way of moving through the soul–meat–pattern approach, while taking with it the radicalism of the ontological questions that are posed, and which often get reduced to epistemological concerns over classification. Whereas the philosophy of biology is concerned with articulating a concept of "life" that would describe the essence of life, biophilosophy is concerned with articulating those things that ceaselessly transform life. For biophilosophy, life = multiplicity. Whereas the philosophy of biology proceeds by the derivation of universal characteristics for all life, biophilosophy proceeds by drawing out the network of relations that always take the living outside itself. An extrinsic diagram as opposed to intrinsic characteristics. Whereas the philosophy of biology (especially in the twentieth century) is increasingly concerned with reducing life to number (from mechanism to genetics), biophilosophy sees a different kind of number, one that runs through life (a combinatoric, proliferating number, the number of graphs, groups, and sets). Whereas the philosophy of biology renews mechanism in order to purge itself of all vitalism ("vitalism" is one of the curse words of biology . . .), biophilosophy renews vitalism in order to purge it of all theology (and in this sense number is vitalistic).

"A LIFE" NOT A-LIFE

The difficulty with the philosophy of biology—as with nearly all philosophical thinking of "the animal"—is to resist the anthropomorphism of our thinking about life. The approach of the philosophy of biology, the approach of soul–meat–pattern, centers and raises up the concept of the human so that it is not only isomorphic with life, but so that it may rise above life ("life itself" as the pinnacle and "mere life" as the base or foundation). This has a number of effects on our thinking about life, for it simultaneously places the human at the top of the Great Chain while also reserving a qualitatively distinct, non-animal place for the human. This is the tired drama of the human, at once partaking of the animal, natural, biological world, and yet incessantly striving above and beyond it, producing abstract knowledge-systems, constructing world and life, aspiring for the spiritual (recall Heidegger's (2001) thesis concerning animality: the stone is worldless, the animal is poor-in-world, and the human is world-building). It is a drama that is by turns tragic and absurdist. Contemporary bioart practices can be understood as a commentary on this drama, producing Dadaist mammals, extra ears, pigs with wings, activist crops, and "fuzzy biological sabotage."[3]

Biophilosophy implies a critique of all anthropomorphic conceptions of life. But is it possible to think about this nonanthropomorphic life? Are we determined to yet again supplant a new term ("multiplicity") for an old one ("pattern")? The problem is not simply a nominalist one, not simply a game of logic; the problem is the very relation between "life" and "thought" (both Canguilhem (2008) and Foucault (1992) note that the most accurate concept of life would be life itself). Biophilosophy is an approach to non-human life, nonorganic life, anonymous life, indefinite life—what Deleuze (2005) calls "a life". But the trick is to undo conventional biological thinking from within. Biophilosophy focuses on those modes of biological life that simultaneously escape their being exclusively biological life: microbes, epidemics, endosymbiosis, parasitism, swarms, packs, flocks, a-life, genetic algorithms, biopathways, smart dust, smartmobs, netwars—there is a whole bestiary that asks us to think about the life-multiplicity relation.

LIFE IS X

The central question of the philosophy of biology has to do with an essence of life, a "principle of life". What is life? Life is X—whatever X happens to be, *eidos*, mechanism, life-force, selection, code. The concept of "life itself", promoted by geneticists during the postwar era (the genetic "coding problem"), was a renewal of a concept articulated by Aristotle in *De Anima* (1987) as well as his "biological" treatises. The implication of the very concept of "life itself" is that "life" is One. Whatever it is, life is one thing, essentially one thing, for otherwise we could not say "Life is X." Even when life reveals its contradictory nature, that contradiction is the ineffable key to life. An example is animal motility. Aristotle posed the question, "What makes the animal go?"; that is, from where does its energy come? The problem was picked up by the application of thermodynamics to animal physiology, with talk of animal "electricity" and "irritability" and "vital forces". Soon there was an ineffable "life force" coursing through the animal, enabling it to counter the laws of thermodynamics.

Today a similar process is happening with studies in self-organization and emergence. The question has changed, but its form of the problem is the same: "how do simple local actions produce complex global patterns?" The effects of self-organization can be analyzed forever (e.g., "ant colony optimization") and they can be applied to computer science (e.g., CG in film, telecommunications routing). But a central mysticism is produced at its core, for if there is no external, controlling factor (environment, genes, blueprints) then how can there be control at all? Again, "life itself" is the ineffable, the absent center. In this sense life follows the laws of thought: it is self-identical (whatever is living continues to be so until it ceases to be living), non-contradictory (something cannot both be living and non-living), and either is or is not (something either is or is not living, there is no grey zone to life). It is in this sense that "life" and "thought" find their common meeting point. Biophilosophy implies a critique of the dialectics of "life itself". It abandons the concept of "life itself" that is forever caught between the poles of nature and culture, biology and technology, human and machine. Instead it develops concepts that always cut across and form networks: the molecular, multiplicity, becoming-animal, life-resistance . . . But the point is not to simply repeat Deleuzianisms, but rather to invent or diverge: the autonomy of affect, germinal life, wetwares, prevital transductions, organismic soft control, abstract sex, molecular invasions, geophilosophy, and what Deleuze (1995) calls "the mathematico-biological systems of differenc/tiation."[4]

BEING, TIME, NUMBER

The philosophy of biology is an epistemological endeavor, while biophilosophy is an ontological one. The philosophy of biology asks "which category?", while biophilosophy asks "affected or affecting?" Biophilosophy ceaselessly spins out ontologies, none of them final, none of them lasting. An example: perhaps what Heidegger pointed to as the defining philosophical concern of modernity—Being or dasein—has permutated into one of the guiding concerns of the new millennium—the problematic of "life itself" or the *zoē /bios* distinction. We are no longer worried about the grand metaphysical concerns of Being, Time, and the One. Biophilosophy is a permutation and transmutation of these concerns: not Being but the problematic of "life itself", a concern that asks us to rethink the concept of the vital and vitalism. Similarly, the concern with Time has become an interest in variation, transformation, change—difference and repetition (the repetition of the different and the difference of each repetition). The contemporary interest in the event, becoming, and the virtual-actual pair are further variations of this. Finally, the imperative of the One—that Being is One, that Time is One, that the subject is singular, that identity is the identification of the One, even the strange sameness of the Other in ethical thought—all of this asks us to pose the question: what would we have to do to the concept of "number" to think beyond the One-many dichotomy? This is the question posed by Deleuze's Difference and Repetition, but it is already there in Plato's Parmenides. Hair, mud, and dirt. Is there a concept of multiplicity that moves beyond the One-many? Could such a concept resist a simple denunciation of "number" (quantity versus quality, extensity versus intensity, explication versus implication)? If there is a concept of number that runs throughout multiplicity (a proliferative, pervasive number), and if multiplicity is related to life, is there a living number—a vitalist matheme—that would move out of the philosophy of biology's trinity of soul–meat–pattern? Instead of what Badiou calls the split between the quantitative and qualitative, the closed and the open, "number and animal", is there an animal number? Being, Time, and the One thus get recombined as "life itself", becoming, and number, which in turn ask us to consider or reconsider vitalism, the virtual, and multiplicity.

OTHER-THAN-LIFE

The philosophy of biology poses the question, "What is life?" In doing so, however, it rarely asks the inverse question, "What is not-life?" Certainly death is not-life. But so is the rock, the chair, the clouds. What about the computer, lunch, or a nation-state, are they not-life as well? What about a doll? Memories? There is a whole negative classification of not-life implied in the positive question "what is life?" Better yet, rather than the question of what is not-life, we can pose the question of the life that becomes not-life, an other-than-life, a becoming-non-living. Four preliminary examples:

Swarm Intelligence

"Swarm intelligence" is a term currently used to describe an interdisciplinary research field that combines the biological studies of "social insects" with computer science (especially software algorithms and multi-agent systems).[5] Just as a group of insects that are individually "dumb" are able to collectively self-organize and forage for a food source or build a nest, so can simple software programs or robots self-organize in groups and carry out complex tasks. This local actions-global patterns approach is said to display "intelligent" or purposeful behavior at the global level. But we can also question and repurpose the term "swarm intelligence", for the tendency in this thinking is to always search for a higher-level unity which would be the guarantee of organization and order. Call it a "superorganism" or a "hive mind", the implication is that purposeful activity can only occur through a process of meta-individualizing all group phenomena, subjecting the many-as-many to a renewed concept of the One. Action must come after individuation, not vice versa. However the unique thing about insect swarms and other animal groups (packs, flocks, schools) is not just that there is no leader, but that there is something akin to a fully distributed control. Thus the political paradox of insect societies—how to understand this balance between control and emergence, sovereignty and multiplicity? And thus the paradoxical question of the field of swarm intelligence—can it be coded? Can one in fact engineer distributed control? Or are we stuck at the level of passive observers, limited in our ability to identify swarm intelligence, but helpless to enact it? What would have to be done to the concept of action in order to make of swarm intelligence a political concept? If there is a swarm intelligence, the "intelligence" would surely have to be a frustratingly anony-mous, nonanthropomorphic intelligence, the intelligence of "a life".

Headless Animality

The philosophy of biology is not only concerned with the unity of life ("Life is X"), but it ties this unity to the individual organism. Whether in natu-ral history's classifications, Darwinian speciation, or the study of genomes, biology always begins from the individual. The individual is the starting point, the basic unit of study. Throughout all these levels, the organism has remained central. Organisms not only form species, but they are also formed by molecules and cells; organisms are the ideal point of mediation between the microscopic and macroscopic views of life. Thus it is no surprise to find philosophy raising the human above the animal based on the comparison of individual organisms. Aristotle, Descartes, Hobbes, Locke, Rousseau: the individual organism is the most basic unit through which the human is raised above the animal, the beast, the savage. This is especially the case when groups are concerned. Here insects are the privileged case study, per-haps the paradigmatic case of the not-human. Indeed, political thought has often contrasted the human and the insect precisely on this point. Hobbes (1982) notes that while both we and insects are "social", only we can lay down rights to establish a sovereign; Marx (1988) notes that insects also produce and build, but humans are able to abstract and plan before build-ing. Thus even groups are individuals. Groups are composed of individuals that preexist them, and groups themselves form meta-individuals ("species", "races"). But there are also extrinsic group animals, the multiplicity-animals of packs, flocks, swarms. Yes, swarms can be understood to be composed of individual insects. But what if swarms, packs, and so on are actually inver-sions of the organism? What if they are instances in which the many preex-ist the One? An army ant swarm does have a morphogenetic aspect to it: there is a swarm front, a bivouac, and branching paths. But swarms, packs, flocks, schools are also defined precisely by their shapelessness and formless-ness. They have no "head", let alone a "face". They are headless animals, acephalous animality. They are animality without head or tail, polysensory, poly-affective, "amorphous but coordinated" (Arquilla and Ronfeldt 2000).

Molecular Molecules

To begin with, we can suggest that molecules are not "molecular". As non-sensical as this sounds, it is important to understand the molecule as one in a whole series of units of composition and analysis: the organism, the organ, the tissue, the cell, the molecule. Each science of life is not just a noun (anatomy, biology) but also a verb ("anatomizing", "biologizing") in which the living is both analyzed and built up. What is the smallest unit of com-position? This is also the first unit of analysis. Building up, breaking down.

The process of individuation is central to thinking about life, whether it be about the "building blocks of life" or the "code of life". There are always "powers of ten" in biology, a huge, ontological microscope that stratifies individuals (the "DNA makes RNA, RNA makes proteins, and proteins make us" mantra of molecular genetics). But what if all this has nothing to do with scale, or with strata, or with layering? There is a whole forgotten history of molecular biology which deemphasizes the search for "the" molecules (proteins or nucleic acids), and instead focuses on the relationality of molecules, their network dynamics, their temporal existence on the "edge of chaos" (biocomplexity). On the one hand biology tells us that molecules build up and break down (some proteins break down molecules, others build up). But on the other hand a cursory look at microbes shows us the radical horizontality of molecules: symbiotic bacteria, contagious viruses, and horizontal gene transfer between microbes. An epidemic is molecular, but it is also social, technological, economic, political. Networks of infection, yes, but also networks of contagion, transportation, vaccination, quarantine, surveillance. This compression of networks, this topological intensification, is not the result of molecules, but is "molecular". A microbial life that has nothing to do with scale (micro- versus macro-), but that is at once local and global. Even the common biological processes of gene expression, cell metabolism, and membrane signaling routinely create linkages and relations (microbe–animal–human), or rather they produce univocity-through-assemblages.

Lifelike Death

We speak excitedly about the ways that new technologies are "lifelike", meaning the way that technology—something devoid of life—is able to display characteristics or behaviors that for us approximate life. But it is never clear if the lifelike is a category of representation (the lifelike quality of the "oval portrait", performance ["never mind the man behind the curtain"], or simulation ["what is real, Neo?"]). Our own obsession is to constantly desire and yet worry about the lifelike: we want our phones to speak to us, but only if they say the right things. In popular culture, science fiction repeatedly plays out these scenarios where we produce a technical life in our own image, a fusion of technology and life in which the human constantly reproduces itself. Perhaps another approach to the lifelike is not to do with life or technology at all, but the lifelikeness of death. There is, in fact, a whole demonology of the lifelike to be considered. In popular culture, genre horror gives us many examples of lifelike death: zombies (the living dead), vampires (the undead), the phantasm (the disembodied spirit), and the demon (the possessed life). This is the lifelikeness of life passing away, going beyond itself, exiting itself. It is no mistake that these figures of lifelike death are often inhabited by fearfully ambivalent agents: viruses infecting the living dead, the "bad blood" of the vampire, the phantasm enslaved by memory, and the demonic tearing of soul from body. Lifelike death is not the celebratory lifelikeness of our intelligent machines, but the ambivalent attitude toward a life that should not be living, an unholy life. This life-like death is aporetic life: the dead that walk, the immortal being that is also the basest animal (bats, rats), the materialized spirit, the familiar face distorted beyond recognition. Perhaps there is a technoscientific side to this after all. For, wouldn't the limit-case of lifelike death be the point at which the organic can no longer be distinguished from the inorganic, the material from the immaterial? This is the domain of nanotechnology, the idea of inorganic life, programmable matter, an undiscovered "occult media".

ANCIENT LIFE (OR, THE BIOLOGY OF CTHULHU)

"Biophilosophy for the 21st century" is an ambiguous statement. Biophi-losophy does not begin with information networks, biotechnologies, nan-otechnologies, or intelligent software. In a sense, Presocratic thinking is biophilosophical thinking. Heraclitus (1987) refers to a nonorganic life in his three examples of fire (formlessness in identity), flows (stepping into the river), and the body (stability through growth and decay). A common logos to all change. His opposite—but in many ways his compliment—is Parmenides (1991), whose concept of the All-One attempts to comprehend multiplicity as another form of univocity. And then there are the Greek atomists, particles infinitely dense and mobile . . .

Immediately a dissenting point is raised: "are we not being reductive in our concept of life, as if life were only biological life, and not social, cul-tural, economic, religious, and political life as well?" Indeed, isn't the prob-lem the way in which biological and biomedical life has come to be the foundation of our emerging "biopolitical" regimes? This "bare life" serves as the alpha and the omega of social and political life, at once safeguarding the security of "the population" while also producing a state of exception, a state of emergency, in which "bare life" is both under attack and the object of preemptive strikes. Undoubtedly. Except that this cordons off our abil-ity to think about life within the chess match between disciplines. To the scientist who says, "Life is genetic code", there is the sociologist who says, "Life is the discriminatory implementation of genetics." To the physicist who says, "Life is the self-organization of matter and energy," the political scientist says, "Life is the struggle between human groups to instrumentalize natural resources." To the humanities professor who says, "Life is the set of metaphors we forget are metaphors", there is the engineer who designs "programmable matter" and "smart dust". Once in a while, there are syn-ergistic couplings, noisome crosstalk that produces monsters: in the 1980s there was talk of chaos, in the 1990s talk of complexity, and crossing the millennium talk of networks. Or so the story goes. Perhaps we would like to do away with disciplines; and yet, for all the talk of "third cultures" we still find the two cultures in the most banal, everyday instances.

This is not a manifesto. All the same, there are a number of misconceptions to address concerning biophilosophy. Biophilosophy is not a naïve embrace of "life", a belief in the altruistic holism of all life on the planet. It is, however, a rigorous questioning of the twofold method of the philosophy of biology (principle of life, boundaries of articulation), and the divisions that are produced from this. Biophilosophy always asks, "What relations are precluded in such-and-such a division, in such-and-such a classification?" Biophilosophy is not and should not be simply another name for self-organization, emergence, or complexity. While there is a fertile exchange between philosophy and biology on this point, it is clear that the sciences of complexity are unable to think both ontologically and politically as well. More often than not, they create a new portrait of nature (a non-linear, metastable, complex nature), or worse, they subsume all non-natural elements under this new nature (thus free markets and/or "democracy" are self-organizing and therefore inevitable). Not everything comes under the domain of biophilosophy, but at the same time one of biophilosophy's major concerns is the supposed foundationalism of biology and the biological-biomedical definitions of life. Biophilosophy is not simply a new vitalism, arguing for the ineffability and irreducibility of life's description. Yet this is perhaps the most frustrating and ambivalent aspect of biophilosophy. Biophilosophy is an attempt to draw out a political ontology, and yet it is also politically agonistic, even apathetic. There is no ressentiment in biophilosophy; only a commitment to a "vital politics" accompanied by this "molecular-wide" perspective. Biophilosophy picks up and reinvigorates the ontological questions left behind by the philosophy of biology. Why "life"?

NOTES

1 "Biophilosophy for the 21st Century" was originally published in *CTheory* and was subsequently published in *Critical Digital Studies: A Reader* (University of Toronto Press, 2008; 2nd ed. 2013). It is reprinted with the permission of the editors, Arthur and Marilouise Kroker.

2 See my two-part article, Thacker (2004b; 2004c).

3 In particular, see the work of SymbioticA (see http://www.symbiotica.uwa. edu.au), a group of artists and scientists engaged in exploring cell and tissue culturing techniques as artistic practice. (SymbioticA's work is discussed in the next chapter of this volume by two of its staff.) In a different vein, Critical Art Ensemble (see http://www.critical-art.net) has, for some years, explored the relationships between activism, art, and biotechnology.

4 Aside from A Thousand Plateaus, see Deleuze's comments on life as "resistance" in Foucault (Deleuze 1999). For a sampling of other divergings from life, see: Badiou (2000); Bonato and Protevia (2004); Critical Art Ensemble (2002); De Landa (1997); Doyle (2003); Fraser, Kembler and Lury (2005); Hansen (2000); Mackenzie (2003); Parisi (2004); Parisi and Terranova (2000); Pearson (1999); Thacker (2004a).

5 For a quick overview, see Bonabeau and Théraulaz (2000). For a more thorough and more technical introduction, see Bonabeau and Théraulaz (1999).

REFERENCES

Aristotle. 1987. *De Anima*. Trans. Hugh Lawson-Tancred. New York: Penguin.
Arquilla, John and David Ronfeldt. 2000. *Swarming and the Future of Conflict*. Santa Monica: RAND.
Badiou, Alain. 2000. Of Life as a Name of Being, or, Deleuze's Vitalist Ontology. *Pli: The Warwick Journal of Philosophy* 10: 191–9.
Bonabeau, Eric and Guy Théraulaz. 1999. *Swarm Intelligence: From Natural to Artificial Systems*. Oxford: Oxford University Press.
Bonabeau, Eric and Guy Théraulaz. 2000. Swarm Smarts. *Scientific American* 18 (March 2000): 72–9.
Bonato, Mark and John Protevi. 2004. *Deleuze and Geophilosophy: A Guide and Glossary*. Edinburgh: Edinburgh University Press.
Canguilhem, Georges. 2008. *Knowledge of the Living*. Trans. Stefanos Geroulanos and Daniela Ginsburg. New York: Fordham University Press.
Critical Art Ensemble. 2002. *Molecular Invasion*. Brooklyn, NY: Autonomedia.
De Landa, Manuel. 1997. Immanence and Transcendence in the Genesis of Form. *South Atlantic Quarterly* 97 (3): 499–514.
Deleuze, Gilles. 1992. *The Fold: Leibniz and the Baroque*. Trans. Tom Conley. Minneapolis: University of Minnesota Press.
Deleuze, Gilles. 1995. *Difference and Repetition*. Trans. Paul Patton. New York: Columbia University Press.
Deleuze, Gilles. 1999. *Foucault*. Trans. Seán Hand. London: Continuum.
Deleuze, Gilles. 2005. *Pure Immanence*. Trans. Anne Boyman. New York: Zone.
Doyle, Richard. 2003. *Wetwares: Experiments in Postvital Living*. Minneapolis: University of Minnesota Press.
Foucault, Michel. 1992. *History of Sexuality, Vol. 1*. Trans. Robert Hurley. New York: Vintage.
Fraser, Mariam, Sarah Kembler, and Celia Lury. 2005. Inventive Life: Approaches to the New Vitalism. *Theory, Culture & Society* 22 (1): 1–14.
Hansen, Mark. 2000. Becoming as Creative Involution? Contextualizing Deleuze and Guattari's Biophilosophy. *Postmodern Culture* 11 (1): https://muse.jhu.edu/journals/postmodern_culture/v011/11.1hansen.html.
Heidegger, Martin. 2001. *The Fundamental Concepts of Metaphysics*. Trans. William McNeill and Nicholas Walker. Bloomington, IN: Indiana University Press.
Heraclitus. 1987. *Fragments*. Ed. T. M. Robinson. Toronto: University of Toronto Press.
Hobbes, Thomas. 1982. *Leviathan*. New York: Penguin.
Mackenzie, Adrian. 2003. Bringing Sequences to Life: How Bioinformatics Corporealizes Sequence Data. *New Genetics and Society* 22 (3): 315–32.
Marx, Karl. 1988. *Economic and Philosophical Manuscripts of 1844*. Trans. Martin Milligan. Amherst: Prometheus Books.
Parisi, Luciana. 2004. *Abstract Sex: Philosophy, Bio-technology and the Mutations of Desire*. London: Continuum.
Parisi, Lucinana and Tiziana Terranova. 2000. Heat-Death: Emergence and Control in Genetic Engineering and Artificial Life. *CTHEORY*. Accessed 21 January 2015. http://www.ctheory.net/articles.aspx?id=127.
Parmenides of Elea. 1991. *Fragments*. Trans. David Gallop. Toronto: University of Toronto Press.
Pearson, Keith Ansell. 1999. *Germinal Life: The Difference and Repetition of Deleuze*. Abingdon, Oxfordshire: Routledge.
Thacker, Eugene. 2004a. *Biomedia*. Minneapolis: University of Minnesota Press.
Thacker, Eugene. 2004b. Networks, Swarms, Multitudes: Part One. *CTHEORY*. http://www.ctheory.net/articles.aspx?id=422.
Thacker, Eugene. 2004c. Networks, Swarms, Multitudes: Part Two. *CTHEORY*. Accessed 21 January 2015. http://www.ctheory.net/articles.aspx?id=423.

8 The Biopolitics of Life Removed from Context
Neolifism

Oron Catts and Ionat Zurr

INTRODUCTION

> For an art is like a living organism—better dead than dying.
>
> —Samuel Butler (1872, 117)

> Living fragments of biological bodies, forms of lab-grown life which have been reconfigured, mixed and remixed, reappropriated, recontextualised and instrumentalised . . . require a different epistemological and ontological understanding as well as a different consideration and, by extension, a different taxonomy of life. The liminality of this kind of technological approach to life can lead to a form of fetishism, which we call Neolifism.
>
> —Catts and Zurr (2011)

We are artists working with Semi-Living beings at the SymbioticA laboratory in Australia. The artworks we grow/construct are carbon-based life forms made of living fragments of complex bodies, kept alive through artificial support mechanisms, and always potentially dying. The artworks we make are paradoxically made of what is critiqued in this article. They are made of fragments of organisms taken out of their original context, "reduced" life into parts—therefore, they can be considered "bare life" (Agamben 1998) and, in that they are lab-grown, can be considered to be neolife. In this sense, we are following an intrinsically human practice of exploring/exploiting life through manipulation. This makes our work biopolitical, whether or not we intended it to be so. However, as artists, our approach differs quite substantially from most modes of human engagement with life. Our intentions are non-utilitarian, non-instrumental, and frivolous. Rather than celebrating the technological approach to "life", we look at how life asserts itself as a context-based materiality, defying human and technological controls. We celebrate failure and embrace futility; we rejoice life.

In this article, we would like to explore, in relation to evolving biopolitical thought and our artistic response through the notion of the Semi-Livings, two main threads. The first thread is concerned with the growing realization that the definition of a body is inherently problematic. The advance of biological technologies, and their associated knowledge, call into question the concept of a body in general, including what is a human body. Throughout our artistic practice, we contest the notion of a body. There is a growing realization that biological bodies are unstable, porous, and malleable, and in need of redefinition. For example, the recently developed technique of metagenomics, in which DNA from an environment, rather than a single organism, is sequenced, brought about an understanding of the species diversity and complexity of the human body. What we have referred to as the "human" body is reconceived as an ecology in which there are ten times more non-human (bacteria, fungi, etc.) cells than human cells (National Institutes of Health 2014). These non-human agents are integral to the human body, and also to a human's behaviors and feelings (Human Microbiome Project 2014). As artists, we consider ourselves as entities or agents that are human–non-human ecologies, who interact with other different ecological "bodies". What we argue is that the body (including our own) should not be assumed to be a whole to begin with; let alone human. We are a multispecies ecosystem. Biopolitical thought, traditionally, is based on the notion of bodies (whether humans or other animals) and how they are negotiated within a political framework. Considering the body as a multispecies ecosystem, in the biopolitical sense, is inherently paradoxical and problematic; when the body is an ecology, it may lead to a leveling in which all living parts deserve (or are devoid of) equal consideration.

Our second thread explores what we refer to as Neolifism. Neolifism manifests itself with the growing phenomenon of bodies/life fragmentation and abstraction both in the literal and conceptual sense. Neolifists tend to view life as a raw material devoid of context and in the service of human control. Neolifism occurs when the fetish of technological approaches to life seems to overshadow the context in which life operates. Technology applied to life is becoming a fetish, the "inanimate object worshipped for its supposed magical powers or because it is considered to be inhabited by a spirit" (Oxford Dictionaries 2014). Technology stands for, and *becomes*, life itself.

As neolife—lab-grown and lab-modified life—grows in quantity and complexity, so does the language of control seeming to dominate current biological discourse. Little, if any, attention is given to the idea that life is not defined by technology, but rather that the life/milieu hybrid (whether born, technological, living, Semi-Living or non-carbon) requires attention and care. As we will illustrate, those who question human total control over life are accused by the Neolifists of being vitalists or neo-vitalists.

Our practice is concerned with the human-centered hubris of this hegemonic language and attempts to counterbalance it. We do it through the notion of the Semi-Living. In the fields of tissue culture and tissue engineering, "bodies" are fragmented into tissues and cells that can be kept alive, grown, mixed, re-mixed, and transplanted. In our practice, we use the cell (the basic unit of life, the "simplest" thing that can die) as a symbolic and literal device to look at life in the broadest sense, as all known carbon-based life consists of cells. The cells we use are taken from one context (the body of a complex organism) and introduced and re-contextualized in an artificial surrogate "body"—a technoscientific body. A technoscientific body here refers to a human-made milieu that enables and affects the growth of the cells. In this body, we find life that is fragmented, technologized, and dependent on artificial support for its survival.

We found ourselves lacking a cultural language to refer to such life, so we called it the Semi-Living (Catts and Zurr 2002). Semi-Livings are lab-grown and lab-modified entities which sit uncomfortably within new biological and cultural taxonomies. They problematize notions of body, agency, species, gender, race, class, or life itself. However, as they literally are potentially dying, they require our attention: physical, technological, and conceptual.[1]

As artists and aspiring post-anthropocentrists, our expression goes beyond the logocentric toward a more sensual performativity (with its multiplicities, paradoxes, ambiguities, and hypocrisies). Through our work with living, growing cells, we follow Karen Barad's assertion that "meaning is not a property of individual words or groups of words but an ongoing performance of the world in its differential intelligibility" (2007, 149). Furthermore, as Barad suggests, "The point here is not merely to use non/humans as tools to think with, but in thinking with them to face our ethical obligations to them, for they are not merely tools for our use but real living beings" (2012, 33). We explore how the Semi-Living are also "real living beings". As the notion of a body is fragmented, we believe that there is an urgent need for a biopolitics that goes beyond the human-animal distinction, and deals literally with bare life; there is also a need to reintroduce the ecology of fragments as evocative agents in our lives to whom we owe our attention and care.

To illustrate this, we would like to narrate the recent history of engineering approaches to life to the present—with its accompanied control fantasies and its associated Neolifist ideologies—through our artistic expression, our humble, biopolitical resistance to such a discourse by an emphasis on context, empathy, and acknowledgment of inability to exercise full control. We would also like to suggest alternative approaches concerned with care and irony. The following examples explore excessive Neolifist attachment to the technology-human construct, followed by resistance from artists, but more importantly, resistance by life itself.

ENGINEERING ASPIRATIONS

H. G. Wells wrote in 1885 "that a living being may . . . be regarded as raw material" (reproduced in Wells 1975, 36). At the same time, he drew attention to the philosophical, ethical, and political conundrums involved with this claim:

> There is in science . . . some sanction for the belief that a living thing might be taken in hand and so moulded and modified that at best it would retain scarcely anything of its inherent form and disposition; that the thread of life might be preserved unimpaired while shape and mental superstructure were so extensively recast as even to justify our regarding the result as a new variety of being. (Ibid.)

While Wells pondered the new meanings and definitions of this "new variety of being" (ibid.), biologist Jacques Loeb, around the same time, adopted a more pragmatic, hands-on, engineering approach to the idea of life's malleability (and the human position in such a transformation). Loeb declared that "the idea is now hovering before me that man himself can act as a creator even in living nature, forming it eventually according to his will. Man can at least succeed in a technology of living substance" (cited in Pauly 1987, 51). Loeb positioned the human (man) in control of the living substance, thus as man who creates and engineers nature.

In the late years of the industrial revolution, when a mechanistic view of life flourished, Loeb could be considered a prime example of taking a mechanistic and engineered view of life—or a Neolifist approach—via hands-on experimentation. Furthermore, Loeb was tackling vitalist claims head-on by "new life" through human technological intervention.

In his book *Controlling Life*, Philip J. Pauly identifies the following elements in Loeb's image of the biologist as an engineer (Pauly 1987, 51–2):

- Pauly argues that Loeb considered the distinction between the natural and pathological irrelevant: "[Loeb] opposed the view that the manipulation of organisms was 'the production of monstrosities'. . . . Breaking down the distinction between natural and monstrous would be a necessary preliminary to the development of engineering biology."
- Loeb supported fragmentation and abstraction: "he rejected the view that some 'complete' analysis of biological organization was the fundamental problem of the life sciences. . . . He acted on the material he had available, breaking down an organism in those ways and to the extent that interesting things could be done to it. Elements were not the end result of analysis but hypothetical entities of limited value."
- Loeb, following an engineering mind set, advocated for reduction of complexity for problem solving.
- Also, he favoured the idea of experimentation through action and manipulation (as opposed to observation).

Similar to Loeb, we are manipulators who fragment life into Semi-Living entities. We experiment, hands-on, with the materials of life and question what is considered "normal" and "pathological" on all levels of life. However, in contrast to what can now be considered Loeb's Neolifist ideology, we embrace and emphasize the complexity of the living and its milieu. Rather than human control, we seek to enact and articulate a non-human-centric response to life—a "being-with" life.

In 1899, Loeb ignited the public imagination of engineering fantasies when he developed what he called "artificial parthenogenesis" (Turney 1995, 153–76). In other words, Loeb demonstrated the capacity for fertilization (in a sea urchin) without the use of sperm but rather through a chemical manipulation of the milieu. Loeb, like our following examples, created sort of a "new life" through a technological approach. Interestingly enough, his work was referred to as the creation of "Artificial Life" (*Los Angeles Herald*, 1 March 1905). As a thought experiment, Loeb suggested making a living system from dead matter, as a way to debunk vitalism, claiming to have demonstrated "abiogenesis" (Loeb 1906, 223). His belief in human technological control over life and his mechanistic approach to life led him to argue that "instinct" and "will" were "metaphysical concepts . . . upon the same plane as the supernatural powers of theologians" (ibid. 5).

SYNTHETIC BIOLOGY

The anti-vitalist thought experiment suggested by the biologist Loeb, to create a synthetic life form, was to a certain extent pursued by biologists in the following decades. For example, the work of Stéphane Leduc, and the publication of his ideas and experiments in the book titled *The Mechanism of Life* (1911), set out to prove that life is merely a chemical process. Leduc proposed a series of chemical experiments to show the emergence of lifelike phenomena with different degrees of complexity. Using seductive imagery demonstrating the phenomenon of non-living materials' crystallization created by different chemical reactions, Leduc attempted to prove the mechanistic aspects of life and challenge vitalism.

An interesting point is that chapter ten in *The Mechanism of Life* is titled "Synthetic Biology". Now, a hundred years later, we witness a rehashing of similar Neolifist stories of creation through technology, manifest through the "new" field of "synthetic biology". One such story is the creation of the basic unit of life, the cell, out of non-living materials. These so-called "protocells" are becoming a major field of study, accompanied by hyperbolic rhetoric about their potential applications.

In our piece, in collaboration with Corrie van Sice, *The Mechanism of Life—After Stéphane Leduc* (2013), we reenact one of the simplest protocell protocols offered by Leduc, working with the diffusion of two concentrations of solutions that create transitory, cell-like droplets (see Figure 8.1). The droplets resemble cells with membrane and nuclei, they last for a few moments before succumbing to entropy and dissolving into a murky liquid, "much like life".

Figure 8.1   The Mechanism of Life—After Stéphane Leduc, 2013

This protocol is automated using another hyped technology: three-dimensional (3D) printing. There is much discussion about 3D printing technology as the next industrial revolution—something that parallels the assembly line of Fordism at the time Leduc was working on *The Mechanism of Life*. The promise of 3D printing technology is, in its core, based on information transfer as the business model; the focus is on the instructions/data as the currency, while the materiality is merely an optional manifestation. This is problematic as, at the very same time, the 3D printing industry suggests the ability to print actual life, or at least parts of the living. This very seductive scenario of printing life from scratch is played off in this work against the unstable, uncontrollable, and transient nature of the protocell droplets as a material. To a large extent, this piece deals with issues of cultural amnesia and reimagining, calling attention to the use of certain visuals and expressions to persuade, hype, and then disappoint. In a time when the idea of creating synthetic life is in the forefront, it is important to culturally probe current and past approaches to the idea of *The Mechanism of Life*. The printed "protocells" are unstable and temporary; they take on forms that appear organic and then disappear. More than a proof of the mechanism of life, they are a suggestion for a humble approach to the question of what life is and how much we are willing to make life into a raw material for our own ends.

Contemporary synthetic biology's claim for uniqueness is that engineering principles are being formally applied to the life sciences. Jef Akst quotes advocate of synthetic biology George Church:

> The growing influence of engineering in biology is, in some sense, "the best of both worlds", adds Church. . . . The good design principles of

engineering and the unique properties of evolving biological systems are "just an incredible combination." (Akst 2011)

Synthetic biologists apply the engineering language of control and efficiency to the complexity and fragility of life:

> Much of the progress can be credited to engineers who have developed a deeper appreciation of life's power. While synthetic biology has brought a welcome injection of rigorous engineering principles to biology, including hierarchical abstractions, computer-aided design (CAD), and interoperable parts, biological mechanisms also offer some distinctive qualities of their own—a handful of underexploited strategies previously rare in engineering fields, such as replication at low cost and natural selection. (Church 2011)

In synthetic biology, biology is technology: Neolifism is in the forefront.

DNA FETISHISM VERSUS NEOVITALISM

> In order, therefore, to find an analogy, we must take flight into the misty realm of religion. There the products of the human brain appear as autonomous figures endowed with a life of their own, which enter into relations both with each other and with the human race. So it is in the world of commodities with the products of men's hands. I call this the fetishism which attaches itself to the products of labor as soon as they are produced as commodities, and it is therefore inseparable from the production of commodities.
>
> — (Marx 1976, 165)

As a step toward the fulfilment of Loeb and Leduc's dreams, synthetic biology, in its reductionist and mechanistic approach, also explores the possibility of creating life from scratch. In 2010, biologist Craig Venter announced that he had created "the first self-replicating cell we've had on the planet whose parent is a computer" (Wade 2010). Venter's group had synthesized the genome of the *mycoplasma genitalium*—a sexually transmitted parasitic bacteria that affects humans and other primates. *M. genitalium* was chosen by Venter's group because it was the species with the smallest number of genes known at that time. The DNA of the bacteria was removed and synthesized DNA was inserted into the bacterial cell. The "synthesized" life form was called Synthia—a reference to its synthetic origin and a homophone of the female name, Cynthia—assigning a feminine quality to the altered bacteria.

Venter used an existing cell, with all its "machinery" and components, minus a strand of nucleic DNA. He responds to accusations of misleading the public, by claiming to have made life from scratch, by holding a DNA-centric position, dismissing the argument as neovitalist: "vitalism today manifests itself in the guise of shifting emphasis away from DNA to an 'emergent' property of the cell that is somehow greater than the sum of its molecular parts and how they work in a particular environment" (Venter 2013, 17). He then asserts that:

> When there is mystery, there is an opportunity for vitalism and religion to thrive. However, when my team successfully booted up the synthetic DNA software within a cell, we demonstrated that our basic understanding of the machinery of cellular life had advanced to a significant point. In answer to Erwin Schrodinger's little question "What is life?" we had been able to provide one compelling answer: "DNA is the software and the basis of all life." (Ibid., 130)

As part of the publicity for this technological breakthrough, Venter and his colleagues released a series of images of the Synthia bacterium and its cultures. One image in particular dominated the media. Two round colonies of bacteria sit on an orange background. Each colony consists of a number of concentric circles: a dark blue circle is surrounded by a lighter blue ring and a second large, almost white ring. Lighting was placed to give the impression of rounded, spherical objects with highlights in the dark inner circle and the lighter, middle ring. A scale bar at the bottom of the image indicates that each colony is about 50μm in diameter. The image resembles two blue eyes, explicitly gazing back at the viewer. This extremely potent image is both biopolitical and metaphysical: two blue "female" eyes gaze directly back at the viewer. These images stand in direct contradiction to Venter's evangelical DNA-centric, anti-vitalist stance. Instead of a Neolifist, fetishist idea of an abstract vessel (bacteria cell) holding a new life (artificial DNA), Synthia is becoming an entity with agency, who locks its gaze to ours.

Venter's DNA-chauvinist, paternalistic, and anthropocentric view can be seen as a return to the androcentric idea of life originating from animalcules that reside in sperm.[2] As in the notion of the animalcule, for Venter, only the DNA transmitted by the male contributes to the uniqueness of *Synthia*, not the *M. genitalium* cell and its internal mechanisms. This view has profound biopolitical implications that need to be resisted, and we, as artists, are attempting to find ways of disrupting it with the notion of the Semi-Living.

RE-ANIMATING SEMI-LIFE

> Primitive man distinguished but two kinds of bodies in nature, those which were motionless and those which were animated. Movement was

for him the expression of life. . . . Ancient Greek mythology is but the poetic expression of this primitive conception. . . . Even now children attribute life to everything that moves.

—Stéphane Leduc (1911, 1)

He held up the curtain of the bed and his eyes, if eyes they may be called, were fixed on me. . . . Oh! no mortal could support the horror of that countenance. A mummy again endued with animation could not be so hideous as that wretch. I had gazed on him while unfinished; he was ugly then, but when those muscles and joints were rendered capable of motion, it became a thing such as even Dante could not have conceived.

—Mary Shelley (1990, 43)

In our works with living cells and tissues, we have created different situations and therefore different "agencies" for the Semi-Living/audience interactions. This is one of the ways we resist Neolifism—by purposely recontextualizing fragments of life within their milieu and giving them agency or a "voice" to unsettle and bring back their visceral "aliveness". One of the main challenges we have faced with presenting the Semi-Living is that they do not seem alive to the human viewer. In other words, their size and metabolism rate is different enough from that of the human audience that they seem to be inanimate

© AXEL HEISE PHOTOGRAPHIE > SYMBIOTICA /TC&A "Disembodied Cuisine", le lieu unique, 2003.

Figure 8.2 *Disembodied Cuisine* Installation, Nantes, France, 2003

objects in/on display. Any movement or change is not appreciably visible in human spatiotemporal perceptual scales, especially during a single gallery visit.

Early on we presented the Semi-Living within their technological/laboratory apparatuses positioned in the gallery (see Figure 8.2). This strategy served practical as well as aesthetic purposes. These lab-grown dependent life forms became artworks that needed attention and care. These entities needed nutrients supplied in a sterile environment, referred to in the artwork as "feeding rituals." At the end of the show they had to be culled/killed/disposed of, as no lab-grown life that "escapes" the lab can return to the lab. We called this the "killing ritual". The killing ritual usually involves exposing our Semi-Living sculptures to a non-sterile environment and inviting the audience to touch them. This proved to be a fairly effective and affective way of presenting aliveness through death. On several occasions we were told that it was only from the act of killing the Semi-Living sculptures that the audience member realized (or actually believed) that the sculptures were alive.

As the tissue itself did not effectively communicate its aliveness, we decided to see if the technological milieu that keeps it alive—the technoscientific body—might be more effective and affective in doing so. We designed our first technoscientific exhibition body (we had previously used ready-made laboratory equipment) to focus attention on the relationship between living tissue and technology. We developed a series of glassware vessels with a pump-operated drip-feed system (see Figures 8.3a and 8.3b).

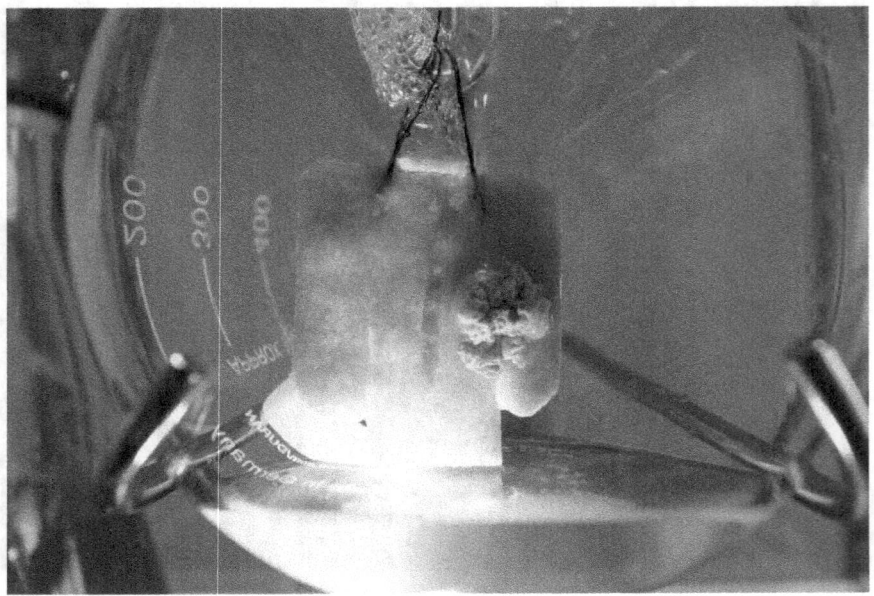

Figure 8.3a Victimless Leather: *A Prototype of Stitch-less Jacket Grown in a Technoscientific "Body"*

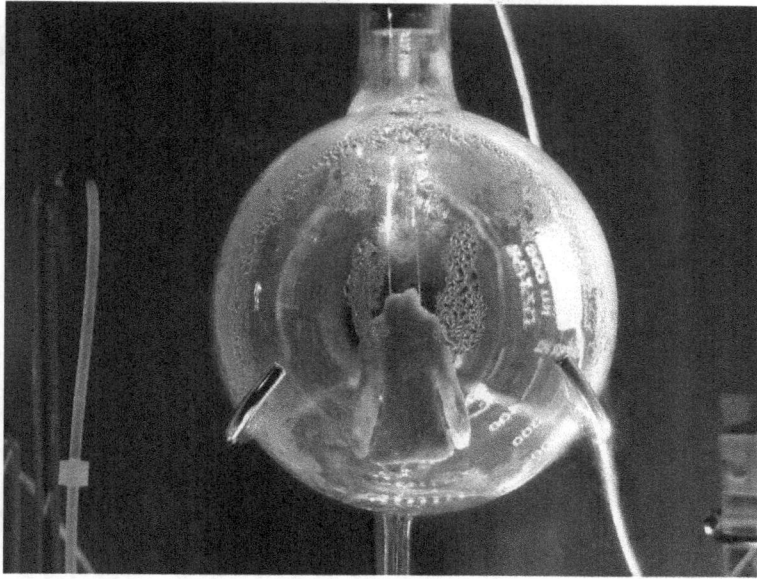

Figure 8.3b Victimless Leather: *A Prototype of Stitch-less Jacket Grown in a Technoscientific "Body"*

The dripping and movement of the liquid through the tubes introduced a subtle hint of aliveness. The transparency of the glass, and the larger size of the tissue construct, allowed better observation of change over time, though a viewer had to come to the gallery several times over a space of days in order to notice any change. It is also interesting to note that this system had some failure in the long-term maintenance of sterility: in some cases interesting contamination occurred and introduced other kinds of unexpected life (such as bacteria, fungi, or yeast). On occasion, we used the system as an ambiguous life/death machine. Being a closed, semi-automated system, if left unchecked, it transmuted from a life-sustaining vessel into a death chamber. Playing on the undetermined state of aliveness, these works called attention to the growing confusion in regard to states of aliveness and artificiality; they further emphasized that life that is capable of dying cannot be completely abstract.

We then looked at making the Semi-Living into food. We grew, fed, served, and ate our Semi-Living, which was made of skeletal muscle cells (see Figure 8.4). In this way we combined the feeding ritual and tactile killing ritual with the invasive aesthetics of digestion. However, food is in a unique category for humans. As it is about to be consumed, we tend to objectify it and shy away from any agency it may invoke, let alone consider the context from which the food was taken and isolated/fragmented before it becomes part of our own body ecology.

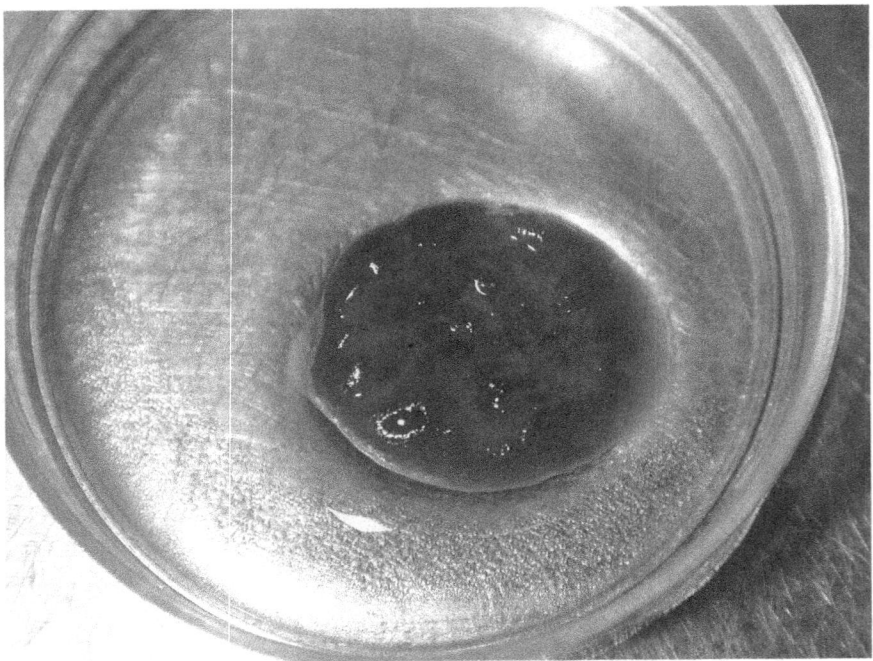

Figure 8.4   Tissue Engineered Steak 2000: A Study for "Disembodied Cuisine"

How then could we more effectively introduce the active agency of the fragment of life on display? How could we evoke some consideration for the biopolitical nuances of the Semi-Living on display? We have observed that literal reanimation creates visceral responses from humans, as well as many other animals.[3] Consequently, we are now looking at the introduction of real-time movement to evoke the agency of the Semi-Living within our artworks. We are exploring the perception of vitality (and its gradients of agencies) as expressed through movement, through a multidisciplinary investigation that explores the animation of living, using engineered muscle cells for artistic purposes. We are not looking at technology imitating life through movement (an aspiration of the Neolifists), but rather technologically manipulating life so that it can move, with all the uneasiness this activity entails.

Funded by the Australian Research Council, our most current project, the Tissue Engineered Muscle Actuator (TEMA), explores the relationships between life and movement through the growth and construction of skeletal muscle tissue as an actuator. Still in the experimental phase, an electromechanical surrogate "body" sustains life and growth as well as encouraging/coercing, measuring and effecting actual movement of the Semi-Living. This device provides conditions for the growth of living muscle tissue from individual cells. It organizes and amplifies the inherent movement behavior of the tissue to become an animate Semi-Living entity.

We anticipate that the tissue will generate subtle, almost invisible movements, which will be translated into low frequency vibrations felt by the audience in the gallery. The question of how to relate to and care for a fragile, Semi-Living, and always-in-motion ecology may depend on the effects of "things" on our perceptions. In our experience, the almost futile and poetic attempt to actuate muscle cells suggests that the tension between bodies, machines, the living, and the non-living are similar more to a random twitch of muscle tissue in a dish, rather than the harmonious movement of a running cheetah or a well-oiled machine. Life defies and resists the human desire to turn it into a fully controlled utilitarian technology.

NOARKING AND DE-EXTINCTION

> £450 or $750 would sponsor the DNA of one animal for posterity. £1000 would save living cells to allow the possibility of a future revival for one species. With either of these donations you can ask for a personal PDF certificate with a dedication of your choice, and the name of the species that you are helping.
>
> —(Frozen Ark 2012)

If life can be broken down into its constituent elements, can these elements be stored, archived, and rebuilt by human technologies and human desires? Back in 1927, around the same time that tissue culture techniques and technologies were developed, Julian Huxley looked at this possibility in a short science fiction story, "The Tissue-Culture King" (Huxley 1927). Huxley's story evokes principles of Neolifism, concerning itself with culturing tissue fragments of tribal kings and ancestors as a form of technological fetish that stands for the lords themselves (even after death). Rather than worship a whole complex person and care for its needs and desires, the members of the tribe can have a reductionist, somewhat abstract, replacement of the person in the form of a Semi-Living fragment of their leaders' and forebears' tissue. In Huxley's fictitious society, the care for a living, growing, dependent cell culture in a technoscientific vessel becomes a form of worship. Huxley's protagonist explains:

> To deal with the possibility of cultures dying out, Hascombe started a central storehouse, where duplicates of every strain were kept, and it was this repository of the national tissues which had attracted my attention at the back of the laboratory. No such collection had ever existed before, he assured me. Not a necropolis, but a histopolis, if I may coin a word: not a cemetery, but a place of eternal growth. (1927, 456)

However, while in Huxley's story, looking after the cultures is a physical labor of care and nurture, in the twenty-first century, when cells can be not only cultured but also frozen, life is fragmented when placed in suspended animation: an abstracted, deferred life, defying time and waiting to be awak-ened and reanimated at some future point. The Frozen Ark initiative, begun in 1996, has established a consortium of 21 major zoos, aquaria, museums, and research institutions in eight countries around the world (Frozen Ark 2012). It aims to collect and store DNA and/or tissue samples from extinct and soon-to-be-extinct animals with the aim of one day reintroducing them into the world. This will be done by transferring a DNA segment into a host somatic cell and implanting it into an animal surrogate body, which would carry it to full term and birth. The name "Frozen Ark" not only implies an insurance policy in time of ecological catastrophe, but also resurrects bibli-cal ideas of species essentialism.

In *Regenesis: How Synthetic Biology Will Reinvent Nature and Our-selves*, George Church and Ed Regis (2012) narrate a possible de-extinction scenario while emphasizing the animal genome as the repository of its spe-cies ideal. In many respects this ignores the idea of a species as a flexible and ever-adapting entity in favor of an essentialist idea of DNA. Such tech-nological control over a species' genome can play evolutionary narratives backwards, forwards, and sideways. Church and Regis write:

> Any technology that can accomplish such feats—taking us back into a primeval era when mammoths and Neanderthals roamed the earth—is one of unprecedented power. Genomic technologies will permit us to replay scenes from our evolutionary past and take evolution to places where it has never gone, and where it would probably never go if left to its own devices. (2012, 11–12)

This Neolifist perspective assumes that the animal or species essence is stored in its DNA. However, the resurrection of these animals would depend on the womb and, in many cases, a cell from another animal to host the DNA fragment. The host animal will have been grown in a different envi-ronment from that in which the extinct animal used to live, in terms of nutrients, landscape, and other factors. Epigenetic research has shown that this changed environment influences the DNA expression of the developing embryo. It will have a different microbiome, which means that as a body, it will not be the same body-ecosystem. After birth, the animal will live in a completely different environment from that in which it evolved, affecting food, habitat, culture, and behavior. To borrow H.G. Wells' (1975, 36) words, this "new variety of being" will be different on so many levels (bio-logical and cultural) from the animal from which the DNA was removed that it will be unrecognizable.

Today, such frozen tissue banks exist all over the world in places of busi-ness, science, engineering, industry, ceremony, and service. Recently they have become sites of hope and dreams of resurrections of extinct and past life, places where technology will save us from our own destruction of the living environment.

Figure 8.5   NoArk, 2008

In 2007, we developed an artwork we called *NoArk* (Catts and Zurr 2007), which showed that as lab-grown life proliferates, the biblical metaphor of the ark is no longer relevant (see Figure 8.5). *NoArk* asked the audience to consider the problems involved in the new life forms that populate the laboratories and our environment. In tissue banks that provide scientists with cell lines, we find all sorts of oddities: cells which originate from at least two different organisms, such as fused cells of human and mouse origin, called hybridomas. These cells are classified by catalog numbers or by very odd names. Species essentialism is put under increasing biological and cultural strain. Rather than a systematic approach to the cataloging of "life order" and "hierarchies" in museums, we see a return of cabinets of curiosities in the tissue banks that supply scientific and industrial laboratories.

How do human cultures come to terms with these forms of neolife? Museums explain the taxonomy of life by showing us idealized forms of each species, arranged through the living kingdom as a teleological developmental lineage, culminating with primates and the apes. Usually, at the end of the line, the humans are presented, often shown as strange artifacts such as technological objects or cultural relics.

In the last decade, more and more museums have started to collect fragments of life, frozen cells and DNA samples that represent the whole. These are a new addition to the stuffed animals and specimens in jars. The aesthetic is dramatically different. While in the past the aesthetic was based in the species' ideal shape (i.e., taxidermy shell), that is symbolic, often staged in dynamic lifelike poses; these stuffed animals could only be reanimated in kids' (or adults') nightmares and fantasies. Today's collections by comparison are frozen in small cryogenic vials and placed in liquid nitrogen, with the promise of an assumed potential to be resurrected. Idealized forms are replaced by the fetish of DNA information.

Our artwork *Odd Neolifism* (Catts and Zurr 2010) is an updated cabinet of curiosities which includes the lab-grown life (see Figures 8.6a, 8.6b, and 8.6c). Alongside the "conventional" display of taxidermy and preserved life, a two-headed bird recalls the fascination with oddities and their significance in the seventeenth century, and the continuation of human fascination with these life forms/monstrosities that do not fit (or which resist) the conventions of their time. This work explores the museum display convention of a progressive complexity of species. At the far end of this display we include a living element, the lab-grown Semi-Living: tissue cells within a technoscientific body. These life forms are so abstracted from their source and hidden within their technoscientific body that for viewers they might not appear to be living. Yet, like all abstracted lab-grown life forms, they are very much alive, calling attention to the urgent need to make a place for them in our ecology and biopolitics.

Figure 8.6a   *Odd Neolifism*, 2010

Figure 8.6b   *Odd Neolifism*, 2010

Figure 8.6c   Odd Neolifism, 2010

CLOSING REMARKS: RESISTING NEOLIFISM

Resisting Neolifism, while at the same time working with Semi-Living (fragments of bodies) and neolife (lab-made/lab-grown fragments of bodies) may
seem problematic and paradoxical. However, we feel that our resistance is
expressed through getting wet and messy with life, only to be reminded (and

remind others!) that: life is never completely abstract; it is always potentially dying; it is uneasy and not in our full control.

Another paradox we have to deal with in our work is that when a body is an ecology (our bodies as well as the bodies of our artworks), it may lead to a reductionism in which all living parts deserve (or are devoid of) equal consideration. Our approach is to deal with these biopolitical considerations within a context: the different lives in different configurations (physical, cultural, political) force us to redefine our understandings and approaches to life. Our role as artists is to bring to the forefront the viscerality of these fragments of life as entities of consideration beyond human control and utility. Unlike many of the Neolifists, such as Loeb and Venter, we cherish and reveal to the public the moments when life refuses to adhere to reductionist and mechanistic controls. We are not afraid to explore the possibility of living (as opposed to non-living) material carrying something that is unique to life and not any other matter—a form of secular vitalism.

As advocated by biopolitical scholar Roberto Esposito:

> The strength of the biopolitical perspective resides precisely in its capacity to exploit the reserve of sense withheld by mixing of the language of politics and biology, which originally tended to be kept apart in the tradition of political philosophy. (2012, 10)

Artists take biopolitical philosophy beyond "the mixing of the language of [disciplines]", and into the realm of materiality. We are artists who mix the material of life with our ethical, political, and philosophical concerns. Different from the Neolifists, the uneasiness we feel when working with living fragments is shared with the living, Semi-Living, and non-living audiences. The more we work with the material of life, the more we realize that every material is defined as part of its milieu. Ecological malleability is inherently political and can be easily manipulated for hegemonic ideologies. At the same time, it can and should be used by others to unsettle these ideologies as described through our artistic examples. We perceive our work as a series of provocations that aim to unsettle rather than to suggest a fixed biopolitical stance.

NOTES

1   For more information about our artistic practice with the Semi-Living, see, Catts and Zurr, "Growing Semi-Living Sculptures" (2002); "The Semi-Livings" (2006a); and "Growing Semi-Living Structures: Concepts and Practices for the Use of Tissue Technologies for Non Medical Purposes" (2008).
2   Here we refer to the idea of the spermism model of preformationism, in which an individual or homunculus was thought to exist in the head of each sperm. This theory was popular during the Enlightenment but was refuted by the early 1800s, mainly because of advancement in light microscopy.
3   See Charles Darwin, quoted in Papapetros (2012, 9).

REFERENCES

Agamben, Giorgio. 1998. *Homo Sacer: Sovereign Power and Bare Life*. Trans. Daniel Heller-Roazen. Stanford, CA: Stanford University Press.

Akst, Jef. 2011. Tinkering with Life: A Decade's Worth of Engineering-Infused Biology. *The Scientist*, 1 October.

Barad, Karen. 2007. *Meeting the Universe Halfway: Quantum Physics and the Entanglement of Matter and Meaning*. Durham, NC: Duke University Press.

Barad, Karen. 2012. Nature's Queer Performativity. *Women, Gender & Research* (1–2): 25–53.

Butler, Samuel. 1872. *Erewhon*. London: Penguin.

Catts, Oron and Ionat Zurr. 2002. Growing Semi-Living Sculptures: The Tissue Culture & Art Project. *Leonardo* 35 (4): 365–70.

Catts, Oron and Ionat Zurr. 2004. *Victimless Leather*, artwork. Perth: John Curtin Gallery.

Catts, Oron and Ionat Zurr. 2006a. The Semi Livings. In *Signs of Life*. Ed. Eduardo Kac, 231–248. Cambridge, MA: MIT Press.

Catts, Oron and Ionat Zurr. 2006b. The Tissue Culture and Art Project: The Semi-Livings as Agents of Irony. In *Performance and Technology: Practices of Virtual Embodiment and Interactivity*. Ed. Sue Broadhurst, 153–168. New York: Palgrave Macmillan.

Catts, Oron and Ionat Zurr. 2007. *NoArk*, artwork. Perth: The Bakery Artrage Complex Perth Western Australia.

Catts, Oron and Ionat Zurr. 2010. *Odd Neolifism*, artwork. Brisbane: Gallery of Modern Art. Queensland Australia.

Catts, Oron and Ionat Zurr, eds. 2011. *Partial Life*. London: Open Humanities Press.

Catts, Oron, Ionat Zurr, and Corrie van Slice. 2013. *The Mechanism of Life—After Stéphane Leduc*, artwork. Dublin: Science Gallery, Ireland.

Church, George M. 2011. Opinion: Evolving Engineering: Exploiting the Unique Properties of Living Systems Makes Synthetic Biologists Better Engineers. *The Scientist*, 1 October.

Church, George M., and Ed Regis. 2012. *Regenesis: How Synthetic Biology Will Reinvent Nature and Ourselves*. Philadelphia: Basic Books.

Esposito, Roberto. 2012. *Terms of the Political: Community, Immunity, Biopolitics*. Trans. Rhiannon Noel Welch. New York: Fordham University Press.

Frozen Ark. 2012. "Donations". Accessed 2012. http://www.frozenark.org/donations.

Human Microbiome Project. 2014. "NIH Human Microbiome Project". Accessed April 2014. http://www.hmpdacc.org/.

Huxley, Julian. 1927. "The Tissue-Culture King". *Amazing Stories* 2 (5): 451–59.

Leduc, Stéphane. 1911. *The Mechanism of Life*. Trans. Deane Butcher. New York: Rebman.

Loeb, Jacques. 1906. *The Dynamics of Living Matter*. New York: The Columbia University Press.

Marx, Karl. 1976. *Capital, Vol 1*. Trans. Ben Fowkes. London: Penguin.

National Institutes of Health. 2014. "Human Microbiome Project". Accessed April 2014. http://commonfund.nih.gov/hmp/overview.

Oxford Dictionaries. 2014. "Fetish". Accessed April 2014. http://www.oxforddictionaries.com/definition/english/fetish.

Papapetros, Spyros. 2012. *On the Animation of the Inorganic: Art, Architecture, and the Extension of Life*. Chicago: University of Chicago Press.

Pauly, Philip J. 1987. *Controlling Life: Jacques Loeb & the Engineering Ideal in Biology*. Oxford: Oxford University Press.

Shelley, Mary. 1990. Frankenstein; or, the Modern Prometheus. In *The Mary Shelley Reader*. Ed. Betty T. Bennett and Charles E. Robinson, 11–166. New York: Oxford University Press.

Turney, Jon. 1995. Life in the Laboratory: Public Responses to Experimental Biology. *Public Understanding of Science* 4 (2): 153–76.

Venter, Craig. 2013. *Life at the Speed of Light: From the Double Helix to the Drawn of Digital Life*. New York: Viking.

Wade, Nicholas. 2010. Researchers Say They Created a "Synthetic Cell". *New York Times*, 20 May.

Wells, H. G. 1975. The Limits of Individual Plasticity. In *H. G. Wells: Early Writings in Science and Science Fiction*. Ed. Robert M. Philmus and David Y. Hughes, 36–9. Berkley, CA: University of California Press.

Part III

Surveillance and Digital Technologies

9   Questioned by Machines
A Cultural Perspective on Counter-Terrorism and Lie Detection in Security Zones

Mark Maguire

> It is sad that we must put a human problem into the hands of a machine, sadder still that we must have a machine enforce our laws. But I ask you to remember, Mr. Gelsen, that there is no other possible way of stopping a murderer *before he strikes*. It would be unfair to the many innocent people killed every year if we were to restrict watchbird on philosophical grounds.
>
> —Robert Sheckley (1953, 8)

This epigraph is from Robert Sheckley's short story "Watchbird", which was written at the height of the Cold War. The story imagines a near future in which policing and security are transformed by the development of unmanned aerial drones. Scientists discover that *most* "murderers throw out a different sort of brain wave from ordinary people. And their glands act funny, too" (Sheckley 1953, 9). Building on this iffy theory, corporations develop watchbirds capable of detecting murderous intent on "the edge of sensation", even before the crime is committed. Initially, the watchbirds are successful and operate within their directives. But these networked, learning machines soon redefine murder and life itself, with disastrous consequences: people are eliminated for hunting, fishing, or switching off car radios—even a butterfly "was executed, caught in the act of outraging a rose" (ibid., 11). Eventually, the engineers are forced to develop new and improved drones to hunt the old ones.

In the post-Cold War world of informated battle and affective computing, Sheckley's dystopic vision seems tragically prescient. Experts now operate in a near future of robotic decision-making by autonomous lethal technologies, while critics warn of ethical problems and the power of wonder weapons to damage war efforts and produce "terrorists" (see Arkin 2009; cf. Sluka 2011). But drones are not just weapons of war: civilian deployments range from policing to remote sensing and from livestock counting to iPad-controlled children's toys. Such technologies, then, cross the shifting boundaries between war and policing, internal and external security, public and private. These shifting boundaries, techno-scientific interventions, and forms of expertise present great challenges to the critical social sciences, challenges that are cast into stark relief when one surveys key studies of global (in)security.

Stephen Graham's *Cities Under Siege* (2010) explores the insidious global diffusion of what he terms "military urbanism":

> Jersey-barrier blast walls, identity checkpoints, computerized CCTV, biometric surveillance and military styles of access control protect archipelagos of fortified social, economic, political or military centres from an outside deemed unruly, impoverished or dangerous. In the most extreme examples these encompass green zones, military prisons, ethnic and sectarian neighbourhoods and military bases; but they are growing around strategic financial districts, embassies, tourist and consumption spaces, airport and port complexes, sports arenas, gated communities and export processing zones. (Graham 2010, xxi)

Graham connects the security laboratories of Iraq and Afghanistan to the fortification of global cities, arguing that urban militarism is about securing critical infrastructures and valued mobility against the dangers posed by unwanted, criminalized, or otherwise threatening masses. His analysis turns on the concept of "ubiquitous borders" (ibid., 89–152) as a way to explore an emergent world of gates and security technologies. And behind all of this exists a "violent shift towards a biopolitics of pre-emption, exception and extreme polarization" (ibid., 150). Graham's analysis thus presents a number of clear features: it takes seemingly disparate trends and argues that they are imbricated, coherent, and understandable as emergent within contemporary biopolitics. Ben Muller also develops a striking image of biopolitical borders. He describes the rise of "borderworlds" composed of "biometrics, surveillance and artificial intelligence, a proliferation of scanning technologies and diverse law enforcement and security officials, as well as the ubiquitous unmanned aerial drones that inhabit the skies" (2013, 131). Like Graham, Muller stands back from the multitude of examples, as if from a pointillist painting, and observes biopolitical patterns and coherences. These commentators are fulfilling an important service by mapping the proliferation of security technologies and discourses globally. However, the image of a vast and coherent security techno-space presents theoretical and empirical problems.[1]

Today the concept of biopolitics as formulated by Michel Foucault and his interlocutors functions as an observation point for global security processes (see Dillon and Lobo-Guerrero 2008). When scholars investigate urban militarism or borders, biopolitics often provides the theoretical underpinning and suggests (imagined) coherences. It is worthwhile, therefore, to recall Rabinow and Rose's (2006) warning about the dangers in mobilizing Foucault's concepts of biopolitics and biopower. These concepts were used by Foucault to bring things into view for analysis, from public hygiene and "race" to town planning and food security. However, partly because of the recent translations of his *Collège de France* lectures (Foucault 2008) and partly because of the eye-catching philosophical interventions by Giorgio Agamben (1998), these concepts are now being mobilized as metaphors for an all-explaining hidden logic. The danger is that biopolitics and biopower "describe everything but analyse nothing" (Rabinow and Rose 2006, 199; see also Maguire 2012).[2]

Numerous challenges also arise in studies of actual borders or processes of securitization. "Ubiquitous" biopolitical borders generally reveal themselves to be distinct cultural worlds variously affected by multidimensional webs of technology and science. Even the Mexico–US border—so often read as a case-study in straight-line securitization—is a multidimensional site of complexity. Reflecting on years of ethnographic research, Gilberto Rosas concludes: "Throughout my years of research and work in Nogales, the militarization of the border was often evident. Then, as often, it was not" (2006, 340). He describes many "forms of governance ranging from state to informal forms of power" (ibid.) and attends to advocacy groups that often successfully contest securitization. That said, one must avoid a simplistic reassertion of the local within the global. This essay eschews sterile oppositions and follows instead a cultural path *within* security techno-science. The path begins with the AVATAR, a lie-detection kiosk that is capturing imaginations around the world (cf. Muller 2013). I track its development through the intellectual history of deceit detection by drawing occasional perspectives from training in interview techniques and deceit detection as part of an ethnographic project on counter-terrorism (see Maguire 2014). Just as Robert Sheckley's *Watchbird* sees the techno-scientific future of security as an assemblage of iffy theories with philosophical blindspots, the history of deceit detection is also filled with experiments, unanswered questions, and elisions.

FROM VITAL SYSTEMS TO VITAL INFORMATION

Today, enormous resources are being poured into techno-scientific means of identifying security threats and detecting deceit, especially in secure ports of entry, transportation hubs, critical infrastructure, and at mega-events. Security experts think of these systems, events, and locations as vital but potentially vulnerable systems that must be protected against low frequency but highly dangerous threats (see Collier and Lakoff 2008). It is abundantly clear from numerous security studies that the key challenge is to find improved and more efficient ways to facilitate forms of mobility deemed to be legitimate while securitizing illicit flows and preempting threats, and all of this while also nourishing the global security industry.

The contemporary security apparatus may appear vast and coherent but it is also emergent and contested. For example, the figure of the uniformed border guard or police officer is no longer assumed to be the state made flesh in unquestionably skilled and professional incarnations. Rather, the security operator is increasingly reconfigured as a potential risk—bored, prone to bias, and possessing capacities no greater than random to detect deceit (see Pratt 2010). Consequently, systems are now deployed that encompass first-generation biometrics to identify persons and documents on the basis of intrinsic physical or behavioral traits, such as the fingerprint, and second-generation biometrics that aim to capture more elusive traits. In order to understand this problematization, it is first important to appreciate the governmental and institutional contexts.

In the period following the terrorist attacks of September 11, 2001, the United States government established the Department of Homeland Security (DHS)—the most significant reorganization of government since the Cold War. In 2003, the DHS absorbed the Immigration and Naturalization Service, thus tying migration to crime and terrorism. The DHS Science and Technology Directorate established centers of excellence in key areas of activity. The University of Arizona in Tucson and University of Texas at El Paso's National Center for Border Control and Immigration (NCBCI) is one such center of excellence. The NCBCI connects university stakeholders to those on the enforcement side of security research and is experimenting with deceit detection technology. The result is the AVATAR or Automated Virtual Agent for Truth Assessments in Real-time. AVATAR is a physiological-behavioral assessment and face recognition system that has already been put to use on several borders. If one were to walk up to the AVATAR in an airport, one would see a device roughly the size and shape of an ATM machine. A 22-inch screen shows a human-like representation of a border guard, which will automatically respond to your physiology by means of a self-adjusting camera. A touchscreen must be tapped to commence the process of recording identification documents and first-generation biometrics such as your face and fingerprints. An interview will then commence during which one is asked questions such as, "Have you ever been to the United States?" During questioning one is screened by means of second-generation biometrics via a near-infrared camera that captures facial expression, pupil dilation, and glance location while a microphone records vocal pitch.

In 2012, Frontex, the European Agency for the Management of Operational Cooperation at the External Borders of the Member States of the European Union, and the Royal Netherlands Marechaussee facilitated a trial of the technology with the participation of border security personnel from fourteen EU Schengen states. The scenario involved a hypothetical mega-event and mixed samples of Marechaussee students behaving normally, lying, or concealing potentially threatening information. According to Frontex (2012), the AVATAR captured nearly three-quarters of deceivers; the human border guards captured only one quarter. AVATAR is not a robotic border guard: its role is to carry out initial screening and risk assessments. That said, there is evidence of "surveillance creep"—new prototypes may have the capacity to search social media and carry out neurological screening.

AVATAR emerges at the intersection of information sciences and deceit detection. Pantic and colleagues (2011) describe how the near future of socially aware computing demands attention to the area of "social signals"; Social Signal Processing (SSP) pushes for virtual agents capable of detecting, recognizing, and interpreting verbal, nonverbal, and often ambiguous signs such as facial expression and gesture. They note, however, that much of the extant research is focused on single modalities, such as the voice, in isolation and is laboratory based and often insensitive to noisy real world contexts (see Vinciarelli et al. 2009). Nonetheless, specific modalities such as facial expressions are beguiling to scientists. As Salah, Sebe, and Gevers explain:

> The human face is a window that allows peeking into diverse patterns of emotions that manifest themselves voluntarily and involuntarily, communicating affect or projected displays of personality. Even dissociated from gesture and voice (as in still face pictures), facial expressions convey complex, layered, and *vital information*. (2010, 158; my emphasis)

How did the human face become a specific problem for contemporary security techno-science? One must surely begin with a critique of the history of thought in this area and, as Foucault reminds us: "A critique is not a matter of saying that things are not right as they are. It is a matter of pointing out on what kinds of assumptions, what kinds of familiar, unchallenged, unconsidered modes of thought the practices that we accept rest" (1988, 154).

LIES, DAMN LIES, AND DECEIT

> Guilt carries Fear always about with it. . . . A fluttering heart, and unequal Pulse, a sudden Palpitation shall evidently confess he is the Man, in spite of a bold Countenance or false Tongue.
> —Daniel Defoe (1731, 34)

Before turning to the emergence of technologies such as the AVATAR, it is important to tease out the discontinuous history of deceit detection. The historical record shows countless attempts to understand deceit, stretching back to the ancient world, bound up with questions of evidence. In the Western world, a notable early example is *The Malleus Maleficarum* ([1486] 2004), a thoroughly practical manual which advises prospective witch-hunters that "a human's secrets will be read in his facial expression and . . . from the movement of the heart and the quality of the pulse" (Institoris and Sprenger 2004, 286). Later, René Descartes' ([1647] 1996) "Evil Deceiver" distorts perception and troubles the nature of evidence presented by the body. History and philosophy, then, show deceit to be integrated in human culture and as a challenge to practical reason. For Michel de Montaigne, "the reverse of truth has a hundred thousand shapes and a limitless field" (1957, 24), but for many thinkers the possibility existed that those cultural shapes could be known and put to use.[3]

Eighteenth and nineteenth century criminal anthropologists sought out physiological or even cruder exterior signs that might attest to some underlying character. However, the problem of detecting deceit in practical ways is formulated most eloquently by Hugo Münsterberg:

> There cannot be justice if we base our judgment on the detective's claim that a man blushed or trembled or was breathing heavily. It would hardly be better than those superstitious decisions of early times. There are too many who believe that they see what they expect to see, and very different emotions may express themselves with very similar symptoms. The door is open for every arbitrariness if such superficial observations were to count seriously for acquittal or for conviction. But that provokes the natural question: cannot science help us out? (1908, 117–8)

During the early twentieth century, polygraph-like machines were developed in Europe and North America, including an experimental effort by William Marston, the inventor of the Wonder Woman comic book character. In 1921, the first recognizable polygraph that measured respiration, pulse, and blood pressure was developed by John A. Larson in the Berkeley Police Department, California. Such machines were a huge hit in the United States, especially in national security contexts and for recruitment in the public and commercial sectors, but doubts continually surfaced about their credibility. *The Employee Polygraph Protection Act, 1988* ended much of the unquestioned corporate use, though it exempted the intelligence and counterintelligence communities. Today, commercial and private uses of polygraph machines remain in public culture and are mushrooming globally, especially in India. Simultaneously, in 2003 the US National Academy of Sciences (2003, 212) concluded that the whole area is scientifically unreliable, showing levels of accuracy better than chance but significantly below 100% reliability.

The history of the polygraph shows the techno-science of deceit detection in operation *despite* its shortcomings. Paul Ekman, the father of modern deceit detection and subject of the rest of this essay, explains:

> The polygraph doesn't detect lies per se. It would be a lot simpler if there were some direct sign unique to lying that is never a sign of anything else. But there isn't. . . . All that the polygraph measures is autonomic nervous system signs of arousal—physiological changes generated primarily because a person is emotionally aroused. It is the same with the behavioral clues to deceit[:] no facial expression, gesture, or voice

change is a sign of lying per se. . . . A micro facial expression can reveal that someone is angry, afraid, guilty, and so on. The polygraph can only tell that *some* emotion has been aroused, not which one. (2009, 198)

Ekman points to the manner in which polygraphs are nested in security assemblages and functionally reliant on the conditions provided by those assemblages. He also points to research on behavioral cues such as micro-facial expressions, which are core pillars in second-generation biometric security. Here I explore Ekman's work and his dialogue with anthropologist Gregory Bateson.

FROM PAPUA NEW GUINEA TO 9-11

In 1966, Paul Ekman was awarded a grant by the US Defense Department's famous Advanced Research Projects Agency (ARPA) to study the universal and culturally specific dimensions of facial expressions and bodily gestures. He sourced films shot in remote parts of Papua New Guinea that promised visual records of the expressions and gestures of a "stone-age" people (see Ekman 2005, 301–20). According to Ekman, every human expression on two hundred thousand feet of film was familiar to him. He turned to the work of the philosopher Sylvian Tomkins, who held that human facial expressions were tied to several universal emotions.[4] He also turned to Charles Darwin's *The Expression of the Emotions in Man and Animals* (1872), and soon became one of Darwin's interlocutors. Equipped with his neo-Darwinian toolbox, Ekman completed several years of fieldwork in Papua New Guinea, returning with the proposition that basic-level emotional expression on the face is universal among humans, based on highlanders in Papua New Guinea being able to "recognize" emotions in photographs. He went on to develop an analytic grid that includes display rules and masking mechanisms as variations on universals. During the 1970s, he developed the Facial Action Coding System, which documents about ten thousand variations of facial expressions. By the 1980s, he began to work in consultation with various law enforcement agencies via a consultancy company—he even inspired the popular television series *Lie to Me*.

By the 2000s, two university-based research teams were working on automating the process of "reading" the emotional expressions of the human face. These projects have fed into counter-terrorism systems that make use of second-generation biometrics. In 2008, I began a project on security that took me from California to the Gulf State of Qatar and eventually into counter-terrorism training (Maguire 2014), and therein the work of Paul Ekman looms large. Micro-facial expressions are regarded as exciting scientific research. But, in truth, actual counterterrorism programs involve efforts to attune the senses and judgments of officers without sticking to deceit detection. So, why does the idea of capturing deceit on human faces by means of high-tech systems seem so beguiling, while in practice coun-terterrorism programs focus on the woof and warp of experiences? In part, this is because the "science" of deceit detection is needed precisely because it is largely unused—if it wasn't there, there would be no science at all. But what if it is no science at all? What if there are ghosts haunting the machine of deceit detection?

WHAT GREGORY BATESON SAID

Daniel M. Gross (2010) comments on the importance of Darwin's *The Expression of the Emotions in Man and Animals* (1872) to Ekman's entire project and accuses him of doing interpretive violence. Gross notes that *The Expression* does not attempt to isolate a few universal emotions; rather, it explores the indeterminacy and situatedness of human emotional expres-sion. Famously, Darwin described asking respondents to view and comment on images from Duchenne's *Mécanisme de la Physionomie Humaine* (1862). Duchenne's best known image is of an elderly man whose facial muscles have been galvanized into an expression of horror and agony by the appli-cation of electricity. When Darwin showed these images to a group, most recognized the emotional expression. For Ekman, this validates his methods and theory. But, according to Gross, "Darwin's methodology accounts for the emotion's medium, occasion, and social situation" (2010, 48). In con-trast, Ekman's research is reliant on cross-cultural recognition of emotions in photographs with little attention to the affective and relational fields in which human emotions are expressed.

Many of these same theoretical and methodological issues present them-selves in Ekman's long and complex relationship with the discipline of anthropology. Margaret Mead and Ray Birdwhistell were vehement oppo-nents of Ekman, but there is an especially revealing moment in his interac-tions with Gregory Bateson. In an afterword to a 1998 edition of Darwin's *Expression*, Ekman recalls:

> Gregory tried to convince me that Darwin was wrong in thinking of facial expressions as signs of emotion. They were more usefully consid-ered solely as communicative signals, he said, not as signs of internal physiological changes. . . . I understand now that there is no contra-diction between Gregory's view and the traditional view of emotional expression, but I only sensed it then and could not explain it to him. . . . I could not always follow what he was saying. (1998, 364–5, 371–2)

To settle the matter, so to speak, Ekman writes a *l'esprit de l'escalier* response to the late Bateson:

> Most of those who study communication, like you Gregory, and some ethnologists, have not wanted to study intent. I believe it is crucial. The

communicative value of a signal differs if it is intended or unintended. Emotional expressions have such an impact; we trust them precisely because they are unintended. . . . We may also try to conceal our emotions, but what I have called *leakage* occurs. (1998, 373)

At first glance, Ekman's "reply" seems to be an unnecessary effort to nail a coffin shut and sit upon the lid, because Bateson's ghost may haunt the entire project of deceit detection as a credible science.

Ekman's project rests upon the figure of the biosocial human expressing basic and unintended emotions that can be frozen, isolated, and interpreted by skilled professionals or even machines. But the specter of another approach cannot be exorcized. In a masterful overview, Ruth Leys recognizes the horrific consequences of challenging the notion of basic, separable emotions:

The moment one abandons the basic emotions approach in favor of some kind of intentionalist interpretation . . . one finds oneself forced to provide thick descriptions of life experiences of the kind that are familiar to anthropologists and indeed novelists but are widely held to be inimical to science [and] the reassuring idea that the truth of our emotions is bound to reveal itself. (2010, 88–9)

And this is precisely what one finds in Bateson's work. Take for example the following discussion in Ruesch and Bateson's *Communication*:

When a scientist endeavours to study such complicated matters as human relations, he [sic] conveniently divides the universe into segments small enough so that the events which occur within such a subdivision can be observed and recorded in a satisfactory manner. . . . His position may be likened to that of a visitor to a museum of art, who never succeeds in seeing the front and back views of the statue at the same moment. From a position in back view of the statue, for example, he will be unable to predict the facial expression until he has seen it from the front. To obtain a complete impression, he has to walk around the statue; and as he moves, a new perspective will open at every step until the combination of all impressions will enable the visitor to construct within himself a small-scale model of the marble figure. (1987 [1951], 24–25)

According to Ruesch and Bateson, "For practical purposes, however, events occurring in other persons are accessible to an observer in terms of *inference alone*; all he observes is the stimuli which reach the other person and the latter's reactions; the rest is subject to conjecture" (1987, 26, my emphasis). Bateson haunts the science of deceit detection and contemporary (in)security machines because he is very difficult to exorcize. His anthropological work on the Latmul in Papua New Guinea is rightly celebrated; his sense of performativity and the visual was keen. Moreover, during World War

II, Bateson worked for the US Office of Strategic Services (the forerunner of the CIA) and was decorated for bravery. Bateson used deceit and even his theory of schismogenesis for military ends, but he was plagued by ethical and political doubts. Later, in *Steps to an Ecology of Mind* (1972, 146, 432)—a foundational text in the history of cybernetic theory—he recognizes the ways in which deceit is conveyed in the voluntary and involuntary messages that compose human interactions. However, from a scientific perspective, his interactionist and cybernetic approach stands in stark contrast to efforts that suggest "a reduction of life to little pieces" (Bateson 1972, 479). Moreover, Bateson was troubled by the ethics of practical reason and foresaw a world in which computing might analyze little pieces of life and thus offer to remove ethical problems from the worries of human minds. Resonating with Sheckley's *Watchbird* and contemporary fears of drones, Bateson's voice still seems strong:

> The computer then cranks and heaves and gives an answer, and there is some temptation to obey the computer. After all, if you follow the computer you are a little *less responsible* than if you made up your own mind. But if you do what the computer advises, you assert by that move that you support the *rules of the game* which you fed into the computer. You have affirmed the rules of that game. The problem is to *change* the rules . . . (1972, 481–2)

REFLECTIONS ON THE BIOPOLITICS OF DECEIT

It is possible today to speak of a post-Cold War security apparatus that is global, preemptive, techno-scientific, and concerned to instantiate, into the future, agile capacities to deal with known and unknown threats (see Masco 2006). Foucault was careful in his use of the term "apparatus", suggesting that it was a tool to "pick out" a "thoroughly heterogeneous ensemble consisting of discourses, institutions, architectural forms, regulatory decisions, laws, administrative measures, scientific statements, philosophical, moral and philanthropic propositions" (1988, 194). He used the term methodologically to explore the more or less stabilized but always-emergent relations between these elements, relations that allow a particular discourse to come into view, prevail, or fade in response to perceived needs and urgent imperatives to act. The concept of "apparatus", then, is useful precisely because it is not a synonym for an Orwellian state, Empire, or a perfectly coherent world. Indeed, its usefulness is shown when one's attention is called to "an entirely unforeseen effect which had nothing to do with any kind of strategic ruse on the part of some meta- or trans-historic subject conceiving or willing it" (Foucault 1988, 195). We must cease to equate "apparatus" with control and view it in purely negative terms—"a poison without a remedy", to borrow from Bernard Stiegler (2010, 161). Following Paul Rabinow, I note that as we draw from the history of the present to engage in an anthropology of the actual, the term "apparatus" allows us to pick out new, dynamic, and contingent assemblages that "fall between problematizations and apparatuses" (Rabinow 2003, 56).

Today, new technologies such as the AVATAR are (in)securitizing life itself. The AVATAR operates but also requires the techno-scientific assemblage that is deceit detection. Experts created and shaped this techno-scientific assemblage while making actual decisions, following routes, and closing doors to alternatives. Behind those doors one finds ghosts that refuse to depart. And, as advances in affective computing gather pace—suggestive of "psychopower" (Stiegler 2010, 128) as well as biopower—ethically and culturally informed critiques of techno-scientific assemblages are needed. Seen in this light, narrow-gauge readings of biopolitics or biopower (or even "apparatuses") as powerful metaphors or all-explaining frameworks are unhelpful. Rather, in the contemporary security-scape, and addressed specifically to deceit detection, the role of critique must include calling attention to cracks and fissures, contradictions and conjunctions, and the voices of those who dissent. As Robert Sheckley presciently noted, lethal power and shiny technology should not blind us to the iffy theories and loose foundations of security techno-science. Moreover, as Sheckley also noted, the unintended consequences of techno-science may have important recursive effects, or further intensify the processes by which human problems are turned over to machines.

NOTES

1 When one imagines a coherent techno-science-driven "borderworld", one immediately confronts the lack of precision in the concept vis-a-vis actual borders. As Wendy Brown notes in *Walled States, Waning Sovereignty* (2010), borders that function to keep immigrants out also keep cheap labor out—and such contradictions tend to illuminate deep problems in contemporary sovereignty as well as deep cultural fantasies. Brown uses the example of the Mexico–US border. One of the leading construction companies—construction is often supported by private and sometimes right-wing interests—the American Fence Company, faced several actions for hiring "illegal" immigrants to build the border wall. Borders, even smart ones, gates, even e-gates, and old-fashioned walls are, therefore, all nested in a variety of socioeconomic and cultural weaves. This is also a key insight in strategic studies. Former CIA analyst Brent L. Sterling (2009) demonstrates how, throughout human history, fences and walls can only be understood by taking cognizance of rather elusive matters such as stereotypes, threat perceptions, symbolic performances and, occasionally, downright stupidity.

2 As is well known, Foucault's early death left biopolitics an unfinished project, but his available work clearly identifies two poles emerging from the eighteenth century onwards: the anatamo-politics of the human body (the body as machine) and the species-body embedded in biological processes

(the biopolitics of population). These poles are linked by clusters of relations and assemblages and enter into governmental calculation most visibly in realms where amplification of the productivity and potentials of a population is sought. Most commentators agree that the Foucauldian formulation of biopolitics is also a provocative effort to understand the shift from war in defense of the sovereign to the security of population. Indeed, Dillon and Lobo-Guerrero (2008, 266) state, "Strictly speaking . . . there is no biopolitics which is not simultaneously also a security apparatus." They eschew universalizing and totalizing theories and call for specific explorations of "forces, logics and dynamics at play in assemblages" (ibid., 272 and passim). Thus, researchers investigating security in ways that draw from Foucault must exercise caution. According to Rabinow and Rose (2006, 199): "The concept of biopower—like that of discipline—was not trans-historical or metaphoric, but precisely grounded in historical, or genealogical, analysis."

3 Debates in primatology and evolutionary psychology now take deceit as a well-integrated aspect of primate sociality (see de Waal 2005), and evolutionary psychologists tell us that children learn to lie by about age four but only really develop convincingly feigned facial expressions by age ten. The ethnographic record is, of course, replete with tricksters, lying, and deceit (see Gilsenan 1976). Linguistic anthropologists rarely conflate language with mere communication and have long noted what Michael Silverstein terms the semantico-referential linguistic ideology that assumes that the basic function of language is to convey truthful information (1979, 193–247). Social anthropologists focus on the always emergent quality of everyday lives and rarely take terms like "lies", "deceit", or "deception" as stable or acultural.

4 Tomkins proposed that there were a limited number of basic "affect programs", i.e., hardwired and discrete emotional states: interest-excitement, enjoyment-joy, surprise-startle, distress-anguish, anger-rage, and fear-terror. He later added shame-humiliation and more recent evolutionary additions such as disgust.

REFERENCES

Agamben, Giorgio. 1998. *Homo Sacer: Sovereign Power and Bare Life*. Trans. Daniel Heller-Roazen. Stanford: Stanford University Press.

Arkin, Ronald. 2009. *Governing Lethal Behavior in Autonomous Robots*. New York: Chapman and Hall.

Bateson, Gregory. 1972. *Steps to an Ecology of Mind: Collected Essays in Anthropology, Psychiatry, Evolution, and Epistemology*. Chicago: University of Chicago Press.

Brown, Wendy. 2010. *Walled States, Waning Sovereignty*. New York: Zone Books.

Collier, Stephen and Andrew Lakoff. 2008. Distributed Preparedness: Notes on the Genealogy of Homeland Security. *Environment and Planning D: Space and Society* 26 (1): 7–28.

Darwin, Charles. 1872. *The Expression of the Emotions in Man and Animals*. London: John Murray.

Defoe, Daniel. 1731. *An Effectual Scheme for the immediate Preventing of Street Robberies and Suppressing all Other Disorders of the Night*. London: J. Wilford.

Descartes, René. 1996. *Meditations on First Philosophy*. Ed. and Trans. John Cottingham. Cambridge: Cambridge University Press.

De Waal, F. 2005. Intentional Deception in Primates. *Evolutionary Anthropology: Issues, News, and Reviews* 1 (3): 86–92.

Dillon, Michael and Luis Lobo-Guerrero. 2008. Biopolitics of Security in the 21st century. *Review of International Studies* 34: 265–92.

Duchenne, Guillaume. 1862. *Mécanisme de la physionomie humaine*. Paris: Jules Renouard.

Ekman, Paul. 1998. Afterword and commentaries to *The Expression of the Emotions in Man and Animals*, by Charles Darwin, 363–93. New York: HarperCollins.

Ekman, Paul. 2005. Facial Expressions. In *Handbook of Cognition and Emotion*. Ed. Tim Dalgleish and Michael Power, 301–20. London: Wiley.

Ekman, Paul. 1998. Introduction to *The Expression of the Emotions in Man and Animals*, by Charles Darwin, xxi–xxxvi. New York: Harper Collins.

Ekman, Paul. 2009. *Telling Lies: Clues to Deceit in the Marketplace, Politics, and Marriage*. New York: Norton.

Foucault, Michel. 1980. The Confession of the Flesh. In *Power/Knowledge: Selected Interviews and Other Writings 1972–1977*. Ed. Colin Gordon. Trans. Colin Gordon, Leo Marshall, John Mepham, and Kate Sopher, 194–228. Brighton: Harvester Press.

Foucault, Michel. 1998. Practicing Criticism. Interview with Didier Eribon, May 30–31, 1981. In *Michel Foucault: Politics, Philosophy, Culture*. Ed. Lawrence D. Kritzman. Trans. Alan Sheridan and others, 152–6. New York: Routledge.

Foucault, Michel. 2008. *The Birth of Biopolitics: Lectures at the Collège de France, 1978-1979*. London: Palgrave Macmillan.

Frontex. 2012. "Spotting Deception: Man against Machine." Accessed July 1, 2013. http://www.frontex.europa.eu/feature-stories/spotting-deception-man-against-machine-xFsNz0.

Gilsenan, Michael. 1976. Lying, Honor, and Contradiction. In *Transaction and Meaning: Directions in the Anthropology of Exchange and Symbolic Behaviour*. Ed. Bruce Kapferer, 191–219. Philadelphia: Institute for the Study of Human Issues.

Graham, Stephen. 2010. *Cities under Siege: The New Military Urbanism*. London: Verso Books.

Gross, Daniel M. 2010. Defending the Humanities with Charles Darwin's *The Expression of the Emotions in Man and Animals (1872)*. *Critical Inquiry* 37 (1): 34–59.

Institoris, Heinrich, and James Sprenger. 2006. *The Malleus Maleficarum*. Ed. and trans. Christopher S. Mackay. Cambridge University Press.

Leys, Ruth. 2010. How Did Fear Become a Scientific Object and What Kind of Object is It? *Representations* 110 (1): 66–104.

Maguire, Mark. 2012. Biopower, Racialization and New Security Technology. *Social Identities: Journal for the Study of Race, Nation and Culture* 18 (5): 593–607.

Maguire, Mark. 2014. Counter-Terrorism in European Airports. In *The Anthropology of Security: Perspectives from the Frontline of Policing, Counter-Terrorism and Border Control*. Ed. Mark Maguire, Catarina Frois, and Nils Zurawski, 86–104. London: Pluto Press.

Masco, Joseph P. 2006. *The Nuclear Borderlands: The Manhattan Project in Post-Cold War New Mexico*. Princeton: Princeton University Press.

Montaigne, Michel de. 1957. *The Complete Works of Montaigne: Essays, Travel Journal, Letters*. Stanford: Stanford University Press.

Muller, Ben. 2013. Borderworld: Biometrics, AVATAR and Global Criminalization. In *Globalisation and the Challenge to Criminology*. Ed. F. Pakes, 129–45. London: Routledge.

Münsterberg, Hugo. 1908. *On the Witness Stand: Essays on Psychology and Crime*. Greentop, MI: Greentop Academic Press.

National Academy of Sciences. 2003. *The Polygraph and Lie Detection*. Washington, DC: The National Academies Presses.

Pantic, Maja, Roderick Cowie, Francesca D'Errico, Dirk Heylen, Marc Mehu, Catherine Pelachaud, Isabella Poggi, Marc Schroeder, and Alessandro Vinciarelli. 2011. Social Signal Processing: The Research Agenda. In *Visual Analysis of Humans: Looking at People*. Ed. Thomas B. Moeslund, Adrian Hilton, and Volker Krüger, 511–39. Germany: Springer Verlag.

Pratt, Anna. 2010. Between a Hunch and a Hard Place: Making Suspicion Reasonable at the Canadian Border. *Social & Legal Studies* 19(4): 461–80.

Rabinow, Paul. 2003. *Anthropos Today: Reflections on Modern Equipment*. Princeton: Princeton University Press.

Rabinow, Paul and Nikolas Rose. 2006. Biopower Today. *BioSocieties* 1: 195–217.

Rosas, Gilberto. 2006. The Thickening Borderlands: Diffused Exceptionality and "Immigrant" Social Struggles during the "War on Terror". *Cultural Dynamics* 18 (3): 335–49.

Ruesch, Jurgen and Gregory Bateson. 1987. *Communication: The Social Matrix of Psychiatry*. New York: Norton.

Salah, Albert Ali, Nicu Sebe, and Theo Gevers. 2010. Communication and Automatic Interpretation of Affect from Facial Expressions. In *Affective Computing and Interaction: Psychological, Cognitive and Neuroscientific Perspectives*. Ed. Didem Gökçay and Gülsen Yildirim, 157–83. New York: Information Science Reference.

Sheckley, Robert. 1953. Watchbird. *Galaxy Science Fiction Magazine* 5 (5): 13–16.

Silverstein, Michael. 1979. Language Structure and Linguistic Ideology. In *The Elements: A Parasession on Linguistic Units and Levels*. Ed. Paul Clyne, William F. Hanks, Carol L. Hofbauer, 193–247. Chicago: Chicago Linguistic Society.

Sluka, Jeffrey. 2011. Death from Above: UAVs and Losing Hearts and Minds. *Military Review*. May-June: 70–6.

Sterling, Brent L. 2009. *Do Good Fences Make Good Neighbors? What History Teaches Us About International Security*. Washington, DC: Georgetown University Press.

Stiegler, Bernard. 2010. *Taking Care of Youth and the Generations*. Translated by Stephen Barker. Stanford: Stanford University Press.

Vinciarelli, Alessandro, Maja Pantic, and Hervé Bourlard. 2009. Social Signal Processing: Survey of an Emerging Domain. *Image and Vision Computing* 27 (12): 1743–59.

10 Data Doubles and the Specters of Performance in the Bit Parts of Surveillance

James Harding

I. OF COSMAS AND DAMIAN, OR THE MIRACLE OF BODY PARTS AS METAPHOR

Sometime in the third century after the death of Jesus, two brothers by the name of Cosmas and Damian converted to Christianity. The two lived in the Roman province of Cilicia, located on the Mediterranean coast of what today is southern Turkey. According to legend, the two were martyred by beheading after torture could not compel them to renounce their faith. What is remarkable about the legend of these two brothers is that the depth of their faith was apparently matched only by their skill as physicians. Indeed, that skill was so great that their alleged medical accomplishments were deemed miracles. They were subsequently canonized and ultimately became the patron saints of medicine. Over 48 miracles were credited to the brothers, including the development of remedies against the plague, scabs, scurvy, and even kidney stones (Science Museum 2015). But their most famous miracle involved the reported replacement of a diseased leg of a Roman nobleman with the leg of a recently deceased Ethiopian or, by some accounts, with the leg of one of the nobleman's slaves (International Society of SS Cosma and Damiano 2015). This act is often depicted in paintings, and considering the fact that the first documented successful leg transplantation occurred in July of 2011 ("History" 2011), Cosmas and Damian's feat was a miracle by almost any measure, especially since it was apparently accomplished without the vast array of immunosuppressant drugs that are currently necessary to keep the body from rejecting new limbs.

The legend of Cosmas and Damian's feat may seem an odd point of departure for a study of biometrics, dataveillance, and the loss of democratic oversight amid the increasingly sophisticated global reach of surveillance technologies. But if we reexamine that legend, not from the presumably divine perspective of Cosmas and Damian's miraculous accomplishment, but rather from the earthly perspective of the unidentified Ethiopian without whose leg their miracle would not have been possible, then we quickly discover a structural analogy that bridges the long temporal divide separating third century legends from the twenty-first century realities about biometrics and dataveillance. There is, of course, the plausible argument that the Ethiopian in the miracle has always been more of an artistic contrivance than a divine inspiration, since the dark skin of the Ethiopian serves as indisputable visual testimony that the miraculous leg transplant actually occurred. Here a little color is a godsend for painters depicting this famous miracle. More interesting, however, is the question of the conditions under which Cosmas and Damian procured the leg in the first place. Somehow one cannot help but suspect that the Ethiopian (or as the case may be, the nobleman's slave) was not consulted, that his consent was never sought, and that his leg followed a surreptitious if not altogether dubious path into the otherwise saintly hands of the twin physicians. Somehow one cannot help but wonder whether there wasn't a bit of a Mengele in the miracle—that heinous "Angel of Death"—hidden in the shadows cast by the bright aura of saintliness. For all of this begs the larger question of why the Christian god would sanction Cosmas and Damian's use of their divine skills to replace the leg of the white nobleman rather than to save the life of the Ethiopian. Apparently, god does work in mysterious ways.

By way of analogy, I want to suggest that the stolen or confiscated body parts in the mystified legends of Cosmas and Damian have a lot in common with the bits of personal data that are gathered about us, largely without our knowledge or consent, as we make our way through public and private spaces that are increasingly regulated and defined by advanced digital surveillance technologies. I am not the first to draw this analogy. It is embedded, for example, in the structural logic of Manjula Padmanabhan's 1996 futuristic play, *Harvest* (see Padmanabhan 2003), which not unlike Cosmas and Damian's miracle, concerns a privileged ruling elite's dubious appropriation of body parts from a disenfranchised cultural Other. But whereas Cosmas and Damian ostensibly acted as the agents of a watchful but distant god who surveils from divine heights and with divine purpose, Padmanabhan's *Harvest* addresses the godlike advantages that sustain First World/ Third World disparities in the global marketplace. The play is populated by guards from a mysterious private company called InterPlanta Services, who act as agents for a distant privileged coterie of individuals maintaining their economic and political power through an astonishingly pervasive and watchful privatized surveillance scheme. Here personal data and body parts blur into transportable commodities. Here too privacy falls prey to the coercive logic of the marketplace, where jobs are scarce and personally debilitating compromises are an unavoidable necessity for economic survival.

II. *HARVEST*, THE RULES OF RIGHT, AND DATAVEILLANCE

First produced in 1997 at the Teatro Texnis in Athens, Greece, Padmanabhan's play is set in a futuristic society in the year 2010 which, following the late twentieth century logic of the play's sci-fi speculations, is a time when technological advances and the processes of globalization have not only significantly extended the reach of First World businesses into the daily lives of the Third World's disenfranchised poor, but have also allowed those businesses to traffic in suspect commodities such as body parts, unhampered by local, national, or international police authorities. While much of Padmanabhan's play can be read as a reflection on the perverse economies that regulate international trade in human organs, *Harvest* is perhaps much more significant as a visionary work which, at a very early date, grasped with subtle but amazing clarity the digital revolution that during the short decade and a half after the play's premier would radically transform how public and private spheres perform. Without yet having the words to identify what it sketches as a pending reality, *Harvest* pointed toward what surveillance studies scholars now routinely refer to as biometrics and dataveillance, both of which have profoundly significant performative dimensions that I want to discuss at some length. But before moving into that discussion, I want to clarify some of the crucial terms and conditions that make dataveillance possible and thereby place the significance of *Harvest*'s creative speculation about emergent surveillance technologies in critical historical relief.

The play takes place in a small, single room apartment in the tenement slums of Bombay, which is shared by a young woman by the name of Jaya, her husband Om Prakash, her mother-in-law Ma, and her brother-in-law Jeetu. Though long unemployed, Jaya's husband Om miraculously finds himself selected from some 6,000 other candidates for the very special kind of employment that InterPlanta Services offers: Om will be financially compensated for agreeing to be a living donor, held in reserve for any body parts that a receiver/client from North America might need in the future. The excessive intrusiveness and the loss of privacy that this arrangement ultimately entails becomes quickly evident shortly after Om returns from the successful interview, and the Prakash home is invaded by representatives from InterPlanta Services who sanitize the apartment, provide the Prakash family with a six month supply of a specialized diet that the entire family must accept, set up what is described as a contact module, and finally make sure that Om understands what in another context Graham Sewell and James Barker have called "the rules of right" (2007, 362).

At one level "the rules of right" represent some of the most accepted forms of surveillance in society, and they provide what is the bedrock of much of the data that private corporations and government organizations collect about us—data that in many respects renders visual forms of surveillance obsolete. The rules of right are not the codes of conduct that one agrees to abide by as an employee, such as: arriving and departing at certain times; adhering to certain dress codes; or agreeing not to fraternize with subordinates. Rather, they refer to a willing submission to systems of surveillance and oversight, such as: pilots agreeing to mandatory drug testing; customer service operators agreeing to a monitoring of their phone conversations with customers (Sewell and Barker 2007, 360); couriers agreeing to GPS tracking devices being attached to their assigned vehicles; bank employees agreeing to polygraph tests; computer operators agreeing to keystroke monitoring software on their computers; government employees and contractors agreeing to an entire regimen of biometric profiling like fingerprinting, iris scans, and DNA sampling as well as to a close monitoring of their social contacts, their travel, and their bank accounts. Of course, this doesn't even touch upon the seemingly voluntary relinquishment of privacy that individuals offer on a daily basis in order to have access to services that they gradually come to take for granted. Here the rules of right necessitate exchanges, such as: individuals obtaining access to free Gmail accounts by tacitly allowing Google to monitor the content of their emails for a whole host of dubious clients; mobile phone users implicitly granting companies like AT&T, Verizon, O2, and T-Mobile the right to use cell phones as tracking devices for the police, the FBI, the NSA, the British Home Office, and other government agencies; social media users gaining access to Facebook by allowing the company to sell their personal data to advertisers; shoppers obtaining small discounts on their purchases by allowing the retail industry to profile their buying habits as a marketable commodity; and would-be travelers gaining access to airports by willingly submitting not only to full body scans and searches but also to profoundly unscientific and biased forms of racial, ethnic, national, and economic profiling and sorting.

Simply put, refusing to accept the rules of right is likely to forestall if not preclude employment and access to a whole host of areas that not only are essential to daily life but that also define us as functioning members of society. Ultimately, the rules of right have a profound effect on the kinds of performative acts that shape individual identity, and those acts are not only deeply entrenched in processes of social sorting and stratification; they sustain those processes as well. Sewell and Barker (2007, 361) argue that the "rules of right" represent ways of "shaping our behavior to conform with . . . power relationships" since they are voluntary in name only. But what does it mean to say that they are voluntary in name only? At one level, it is to suggest a not always subtle form of coercion, forcing decisions that are masked beneath an equivocal political rhetoric which gives them the appearance of having been the product of individual choices when in fact they are the dictated terms for continued access to the means of cultural, political, and economic survival.

In the larger equation of conformity with existing power relationships, the trade-offs necessary to gain access to things like social media might rightly seem to be a rather trivial exchange when compared to the more profound compromises that the rules of right often demand. In their most grotesque form, such compromises are often a matter of life and death, and they frequently reaffirm the power relationships that sustain global, political, and economic divisions. Indeed, they often literally make the science fiction fantasies of Padmanabhan's *Harvest* appear to be a quaint form of realism. The perverse logic of these compromises—compromises that are voluntary in name only—are certainly on display in the dramatic portrayal of the financial arrangements that Padmanabhan's character Om makes when he decides to serve as a living donor so as to provide for his family. It is an act of desperation performed under the guise of a voluntary choice, and that desperation eclipses, indeed precludes, the pursuit of anything that might resemble genuine volition.

At first blush, this arrangement literally sounds like the stuff of futuristic dystopian fiction, and, in fact, scholars like Shital Pravinchandra (2006, 7–8) have argued that Padmanabhan's "play confronts us with a futuristic Bombay of the year 2010, a time", so argues Pravinchandra, "when legal, moral and bioethical debates about organ transplants have been overcome." But in the late 1990s such debates were already meaningless bits of inconsequential political theater. The debates had long since been overcome by the powerful sway of the global black market. Despite the passage of legislation in India's Parliament in 1995 "banning payment for organ donation", the practice continued (and to this day persists unabated) in the communities of India's economically disenfranchised (Jha 2004). Evidence from the 1990s of this practice is not difficult to find. Discussing practices "at the clinic of Dr. K.C. Reddy", whom she describes as "India's most outspoken advocate of the individual's 'right to sell' a kidney", Berkeley anthropologist and international human rights activist Nancy Scheper-Hughes recounts, for example, the desperate plight of countless women in the 1990s from the most distressed communities of India's vast economic landscape, women whose desperation precipitated acts and behavior that have not only brought them into conformity with, but have also made them participants in, power relationships that survive through their own seemingly voluntary submission to debilitating forms of exploitation. In a manner wholly consistent with the images emerging in Padmanabhan's play, the women cited by Scheper-Hughes belong to the many who have "voluntarily" traded one of their kidneys for financial remuneration. Writing specifically about women from the Chennai slums in South India, Scheper-Hughes notes:

> The kidney sale was usually preceded by a financial crisis: the family had run out of credit and the bill collectors were at the door. The women said that the money from selling their kidneys had offered temporary relief but that it was soon swallowed up by the usurious interest charged by local moneylenders. The families were all in debt again. They also stated emphatically that they would do it again. If only they had three kidneys, with two to spare, then things might be better. (2002, 80)

Given the direness of the women's situation, Scheper-Hughes rightly suggests that it is disingenuous to speak of volition or of an individual's "right to sell" when decisions are the product of a desperation that existing power relationships do more to sustain than to alleviate. By every reasonable measure, the women mentioned by Scheper-Hughes were compelled by circumstances beyond their control and by the harsh rules of an eco-nomic order that cynically defines the notion of individual "rights" as a self-destructive submission to exploitation within the dark economies of the global medical black market.

Further examination of the unsavory market for paid transplants in India gives some indication of just how far this kind of self-destructive submission actually extends. Shortly after Scheper-Hughes published her essay "The Ends of the Body: Commodity Fetishism and the Global Traffic in Organs" (2002), Vivekanand Jha pointed out that the 1995 Indian Organ Trans-plant Act included an exemption clause, which allowed for the harvesting of organs from live donors on altruistic grounds. While this exemption sup-posedly did not allow for financial remuneration, Jha (2004, 542) notes that "in some states, Committees approved thousands of paid donor transplants after getting affidavits from donors stating that the donations were being made on the grounds of 'love and affection' for the recipients." On the face of it, it would be hard to disagree with Jha's assertion that no one can seri-ously believe "that so many poor, illiterate, out of work strangers developed enough affection for rich recipients from far off places whom they have either never met or just seen a couple of times to be able to donate an organ for them."

Following this line of argument, the "love and affection" affidavits might be dismissed as little more than a perverse charade designed to mask a late twentieth century slavery narrative that is reminiscent of all of the darker undercurrents of the saintly legend of Cosmas and Damian and the myth of the Ethiopian slave who, supposedly out of a sense of devotion, will-ingly offered his leg as a transplant for the diseased leg of his master. But at a more subtle level so too do those affidavits suggest a path between the rules of right that cultivate live donors and those that ultimately facilitate the harvest of multiple forms of personal data. In both instances, the rules of right function solely because those who submit to them are convinced of the necessity of obtaining what they receive in exchange for what they sacrifice. Although one must be careful not to trivialize the grotesqueness of the global trade in body parts and the life-threatening exploitation of donors that this trade involves, it is worth considering whether those affida-vits serve a larger exemplary function than merely providing dubious medi-cal merchants with a convenient legal circumvention. At the very least, the affidavits cited by Jha give evidence of a desperation so extreme that those caught in its snares willingly provide wealthy recipients not only with body parts but, in those affidavits, also provide the distant wealthy with a kind of prepackaged moral dispensation for being at the receiving end of an other-wise ethically indefensible economy of medical cannibalism.

There is illusion in that dispensation, of course, but there is little illusion in the larger function that the affidavits also serve by providing recipients with legal coverage that technically exempts them from liability or cul-pability in a suspect medical practice. As is consistently the case with the operative structures of the rules of right, those written expressions of "love and affection" provide a paper trail that can be read in court as expressions of volition. They are requirements of a legislative act—the Indian Organ Transplant Act—that ostensibly protects the disenfranchised while in effect actually providing the wealthy greater legal protection than they previously had. I want to suggest that a demand that donors sign a document legally and morally absolving organ recipients from culpability is only an extreme example of the terms of service—the rules of right—that each of us sign off on countless times any given year. In exchange for the kind of access, service, or remuneration that enables us to be functioning members of society, we all routinely and enthusiastically provide wealthy corporations not only with compromising pieces of personal data but also with the legal basis to do with it what they will. But how do we account for such enthusiastic relinquishment of one's private information into the hands of strangers?

How do we explain the irrational trust that we place in the corporations that harvest data about us, which they then in turn repackage as a marketable commodity? The answers here are multiple, but perhaps the irrational expressions of "love and affection" by live organ donors in India point toward a deep internalization of existing structures of authority and power rather than toward a mere legal circumvention, and in this respect perhaps those expressions are not so unique and thus point toward the structures of authority that are also at play in our casual acceptance of the increasingly invasive tactics employed by the large corporations that collect information about us. I want to look at the notion of internalization more closely, but I also want to suggest that it frequently intersects with a general lack of understanding of how data is actually used in an era of data processing that even Padmanabhan in her futuristic drama *Harvest* did not foresee.

First, consider the question of internalization: what is perhaps most disturbing in Jha's account of the organ trade in India is the way that it overlooks the deep sense of identification with power that is evident in the very statements of "love and affection" that, for understandable reasons, he urges us to dismiss as implausible nonsense. I want to suggest by contrast that there is much to be gained by giving those statements credibility and by considering the ways that expressions of "love and affection" for powerful strangers—expressions ultimately manifested in self-destructive acts—resonate across history and culture to recall the phenomena of "altruistic surrender" and "identifying with the aggressor" (or with one's oppressor) that was first explored by Anna Freud and then by Bruno Bettelheim.[1] Both involve a process whereby those who find themselves trapped within the discursive logic of a social, political, cultural, or emotional order begin to internalize and affirm the values of that order as a mechanism for survival—even in situations where the affirmation of those values ultimately proves to be self-destructive. Perhaps the most extreme examples of this tendency are to be found in Bettelheim's firsthand accounts of prisoners at Buchenwald who were so psychologically traumatized by their horrific experiences there that they gradually began to imitate their Nazi oppressors. Less extreme but more frequent examples are to be found in the numerous ways that the poor identify with the ruling wealthy elite in the hope that this will somehow ameliorate their own exploitation and dire circumstances. Just how far removed we in the West are from such tendencies in our readiness to acquiesce to "the rules of right" that facilitate the collection of massive amounts of personal data is difficult to pinpoint, but given the astonishing level of public indifference to the revelations that Edward Snowden made of the NSA's blanket surveillance of citizens, for example, it would seem pretty evident that the public has a deep attachment to, if not an outright identification with, the very technologies that make it possible for the NSA and its numerous private contractors to accumulate the kind of personal data on virtually every single person that renders them vulnerable to manipulation and control should powerful interests see them as a threat.

A prescient critical engagement with this latter kind of identification with the aggressor takes center stage in *Harvest*. It is most evident when the character Jaya challenges her husband Om's naïve trust in the company InterPlanta Services and questions in particular his devotion to Ginni, the western recipient-client whose image is projected in a proto-Skype-like fashion into their apartment through the contact module that InterPlanta Services has installed. Mistakenly thinking that Ginni can only see them when they see her, Jaya scoffs dismissively at InterPlanta Services and criticizes her husband, saying, "What faith you have in them! They don't care about any of us, not as people, not as human beings" (Padmanabhan 2003, 47). In response, her husband Om counters, "What're you saying? You don't talk enough to Ginni. If you did, you wouldn't feel this way. . . . She really cares for us." There is little question in the play that Jaya is right, and that Om's devotion to Ginni is a cultivated product: the result of a desperation extreme enough that he is largely blind to the destructive consequences that his actions have not only for him but for his family as well. But what neither of them understands is that Om's identification with Ginni implicates all those around him. Om's bodily organs prove to be a secondary matter of concern and a mere pretext to set in motion a surveillance program designed to ensnare his brother Jeetu, and through Jeetu ultimately to go after Om's wife Jaya, who, as InterPlanta Services already knows, is secretly having an affair with Jeetu. The implication here is that InterPlanta Services has secretly gathered information on Om's wife and his brother through a larger unspecified surveillance net well in advance of selecting him out of the other six thousand who had applied for the job that he ultimately received. This is, as Jaya later learns, why Om was offered the job in the first place.

Almost all of us have experienced some form of "the rules of right" in our working lives, and much to her credit Padmanabhan recognized earlier than most how acquiescence to those rules tends to spill into areas well beyond what those who sign off on them actually expected or could perceive would be the case. Beyond the sense in which signing off on "the rules of right" involves a tacit identification with authority, this spilling over of the collection of personal data into other areas that remain unknown to the individual whose data is passed around as marketable commodities is perhaps more alarming than the kind of identification with the aggressor that is exemplified in the character Om's irrational devotion to Ginni. But while Padmanabhan had some premonition of this spillage, she could only point vaguely at what it has now become and at the dynamic way in which it actually functions. The implication in the play is that the ultimate interest in Jaya is the product of an increasingly pervasive and all-encompassing regime of surveillance that functions through clandestine forms of observation. In point of fact, the gathering and processing of personal data follows a dynamic that is substantially more alarming because it is more difficult to pinpoint and regulate.

In the case of Om in Padmanabhan's play, acquiescence to "the rules of right" unknowingly commits, without their expressed consent, the members of his immediate family to a system of surveillance that penetrates virtually all aspects of their daily lives. Much of this surveillance occurs through the contact module that InterPlanta Services has installed in the Prakash home. But as we learn later in the play, the contact module does much more than merely serve as a high-tech video communication device. It hovers about the Prakash flat like a small drone and constantly collects data about the entire family; data like biometric information on the "skin and blood" of Om's brother Jeetu (Padmanabhan 2003, 85), and of Jaya herself, whom the corporate figures of InterPlanta Services then try to strong-arm into being a surrogate mother. What is noteworthy in this scenario is also what is perhaps most easily overlooked. Despite her otherwise perceptive vision into the future, here Padmanabhan actually reinforces popular conceptions of how corporations collect and use personal data. The image of the module that she presents us with assumes a kind of one-to-one relation between data collected and knowledge gained.

The problem is that the processing of data by private companies no longer functions this way and hasn't for some time. What Padmanabhan failed to recognize is how rapid technological advancements have radically increased the ability of corporations to collect and process massive amounts of seemingly unrelated personal data, and these powerful advances in data collection and processing have opened the door to actual possibilities that in their potential literally surpass the dark dystopian vision that Padmanabhan offers in *Harvest*. Some sense of the world that Padmanabhan was unable to foresee in the late-1990s can be garnered from Viktor Mayer-Schönberger and Kenneth Cukier's recent book, *Big Data: A Revolution that Will Transform How We Live, Work and Think* (2013). In their groundbreaking study, Mayer-Schönberger and Cukier suggest that such one-to-one notions of data and knowledge are grounded in the logic of causality when in fact the revolution occurring in the era of what is commonly called "big data" involves a shift to what they identify as the logic of "correlation". The key moment in the big data revolution, they argue, came when corporations realized that collected data had seemingly inexhaustible potential if it were "no longer regarded as static or stale" and if those collecting data about us abandoned the notion that data's "usefulness was finished once the purpose for which it was [originally] collected was achieved" (5). In an era where the memory capacity of computers is increasing with each passing day and is matched only by the growing speed and sophistication of computers' ability to engage in highly advanced forms of data processing, Mayer-Schönberger and Cukier argue that data can be infinitely recycled and reprocessed in ways that reveal all sorts of unexpected and meaningful statistical correlations—like Google boasting in 2009 that it could accurately predict the spread of the winter flu virus in the United States based solely on their records of "what people were searching for on the Internet" (2).

Following the logic of this kind of statistical analysis—i.e., the logic of correlation—big data brokers "in the United States such as Acxiom, Experian, and Equifax charge handsomely for comprehensive dossiers of personal information on hundreds of millions of consumers" (100). These dossiers are not static lists of facts. They are the product of sophisticated statistical analyses by highly advanced computers:

> The industry of collecting, aggregating, and brokering personal data is known as "database marketing". The second-largest company in this field, Acxiom, has 23,000 computer servers that process more than 50 trillion data transactions per year, according to *The New York Times*. It claims to have records on hundreds of millions of Americans, including 1.1 billion browser cookies (small pieces of data sent from a website, used to track the user's activity), 200 million mobile profiles, and an average of 1,500 pieces of data per consumer. These data include information gleaned from publicly available records like home valuation and vehicle ownership, information about online behavior tracked through cookies, browser advertising, and the like, data from customer surveys, and "offline" buying behavior. The CEO, Scott Howe, says, "Our digital reach will soon approach nearly every Internet user in the US." (Marwick 2014)

Using the logic of correlation, the processing of data by big data brokers is emerging as one of the most important industries in the twenty-first century because it can produce astonishingly accurate predictions that allow large corporations and government agencies to assess, profile, and sort individuals according to everything from their presumed potential as a customer to the probability that they pose a security risk.

These analyses probe deep into the personal lives of countless individuals. In an eerie echo of the kind of foreknowledge that InterPlanta Services has of the entire Prakash family in Padmanabhan's play, for example, Mayer-Schönberger and Cukier note:

> The uses of correlations are being extended even further. Aviva, a large insurance firm, has studied the idea of using credit reports and consumer-marketing data as proxies for the analysis of blood and urine samples for certain applicants. The intent is to identify those who may be at higher risk of illnesses like high blood pressure, diabetes or depression. (2013, 56)

What is noteworthy in Mayer-Schönberger and Cukier's discussion of Aviva is that an insurance company rather than, say, the World Health Organization, is looking at credit reports and consumer-marketing data as proxies for actual medical examinations. We should have no illusions about what this means. It would be no great leap of conjecture to conclude that companies like Aviva are less concerned with solving the problems of hypertension, diabetes, or depression than they are with finding a discrete way to deny coverage to people with preexisting health conditions. This is the dark side of big data and its logic of correlation.

In many respects, that dark side is simply part of the long shadow cast by neoliberalism and by public policy initiatives that gesture toward a reduction of all aspects of social experience into something quantifiable, i.e., into processable data. This is as true in the healthcare industry as anywhere else. At almost the same time that India passed the 1994 Transplantation of Human Organs Act, for example, the United States was drafting legislation that led to the Health Insurance Portability and Accountability Act (1996) which legislated "the adoption of electronic systems for information transfer between healthcare providers and insurance companies" (Fisher and Monahan 2008, 178). As Jill Fisher and Torin Monahan have observed, this legislation spurred the development of technologies like RFID (Radio Frequency Identification) for "the management of patients, personnel, and inventory" (176). The dark side of this legislation, according to Fisher and Monahan, is that such technologies represent "a new form of social control over bodies in the post-industrial era" and do much to facilitate "new methods of data collection" (179). Not only are such methods indicative of the increasing dominance of neoliberal "audit cultures" in the workplace—where work itself is valued only to the extent that it can be quantified—but they are also rife with "surveillance modalities, meaning that the data collected by them can be used for surveillance purposes" and that they mark one of the many areas of a growing "surveillance of people through automated systems and databases" (179). Ultimately, what Fisher and Monahan remind us of is that there is much that can be concluded from Mayer-Schönberger and Cukier's study that is less optimistic about the potential of big data processing than Mayer-Schönberger and Cukier themselves are in their book.

Regardless of whether one is alarmed or excited about the possibilities that Mayer-Schönberger and Cukier describe, one thing is clear from their careful examination of big data and the logic of correlation. Phenomena like identifying with the aggressor are not sufficient on their own as explanations for why individuals seem so willing to place their private personal information into the hands of corporations that in turn repackage it as a marketable commodity. In point of fact, our personal information is now located in our most banal, simple, and seemingly insignificant acts—acts that it would never occur to us to guard from scrutiny. Studies like Mayer-Schönberger and Cukier's indicate that the most innocent internet searches, purchases at the grocery store, and even a daily commute on a toll road all feed data into databases where it takes on a life of its own and where, following sophisticated models of correlation, the data may ultimately determine decisions about matters with no direct connection to it. A world where simple internet searches can predict the spread of influenza is a world where that same data can ultimately also determine corporate decisions about whether one receives adequate healthcare coverage or a mortgage loan, whether one is targeted by law enforcement, whether one is admitted to university, or whether one is offered a job. This is increasingly the applied potential of each bit of data collected about us. It is compiled, processed, and reprocessed on an ever-increasing scale in what has now become—and we are witness to this—a self-perpetuating, never-ending cycle.

Welcome to the world of dataveillance—a world that has literally eclipsed the ocular foundations of the surveillance society and has reoriented its priorities. Think what we will of the omnipresence of CCTV cameras—those most visible signs of the surveillance society—they are anachronisms and artifacts of a quickly expiring world. Their function is to document, to intimidate and, in some instances, to provide real-time feedback. But in the era of big data, they play a distant secondary role to what we now call dataveillance. The power of dataveillance lies in its ability to look beyond the past or the present into a future that it is able to predict with increasing degrees of accuracy, ironically, in large part because of the increasingly influential role that it is playing in shaping that future.

III. BIOMETRIC DATA AND DATAVEILLANCE

An awareness of the new realities described by Mayer-Schönberger and Cukier in their book on *Big Data* also casts an aura of anachronism over works like Manjula Padmanabhan's *Harvest*, or at least over images like that of the rather quaint futuristic contact module that she has hovering about the Prakash flat, transmitting data on the behavioral patterns of the entire family. And yet, here too there is much that is still correct, particularly in the anxious premonition that Padmanabhan has about things to come, even if that anxiety about coming developments resides primarily in the margins of her narrative. It is hard to find those margins amid the chaos of the small drone hovering about the apartment and the deeply disturbing image of agents forcibly removing Jeetu rather than Om from the Prakash home so as to extract his eyes for a western client. But however much that contact module might herald the advent of private corporate drones, Om's wife Jaya finally realizes late in the play that this misleading little module (like the vast array of CCTV cameras in our cities today) is actually of minor significance in relation to the apparent violations of privacy that led Inter-Planta Services into her home in the first place.

It is never entirely clear how InterPlanta Services learned about Jeetu, the love affair that he and Jaya are having, or ultimately about Jaya's suitability as a surrogate mother. But in the larger scope of the issues of surveillance addressed by the play, that lack of clarity is actually quite telling because it leaves the door open to the multiple possibilities that were just around the corner and that we now face as reality. InterPlanta Services' knowledge about Jaya could be the product of big data and the logic of correlation, advanced statistical analysis, and calculated probabilities. It could also be the product of surreptitious surveillance that may or may not have been legal but that is virtually impossible to regulate because it is hidden from public scrutiny and oversight. It could be the product of personal data that was collected, stored with lax security, and accessed for purposes other than those for which it was originally collected. It could be a combination of any of these. Indeed, this is one of the most pressing issues in the health industry today. In her article "Health-Care Data Protection and Biometric Authentication Policies", Cheryl Brown (2012) notes, for example, that biometric measurements—like fingerprints, iris scans, and computerized voice or face recognition capabilities—are not only increasingly used in the medical community to authenticate patient identity and to access medical records, but that same biometric data is widely available in the marketplace today. In no uncertain terms, this means that any breach of hospital database security protocols will allow virtually any company with a computer and internet access to quickly match individuals with their health records.

Returning momentarily to the legend of Cosmas and Damian, I would suggest that with regard to our personal data, it is thus not too much of a stretch, metaphorically at least, to find similarities with the Ethiopian who was not consulted about the body that would be defiled, or the leg that was secretly skirted away without his permission. And if, alternatively, that Ethiopian was the Roman nobleman's slave, as is the case in one variation of the legend's narrative, I want to suggest as well that our personal data, like his amputated leg, *is us but is not ours* because it is owned by another master to whom we defer and give our consent under duress. There is a difference here, of course. That which, in terms of personal data, is us but is not ours—that which we consent to give away under duress—not only takes on a life of its own as part of data that is packaged and repackaged, cycled and recycled, sold and resold by organizations that specialize in big data. It also ultimately determines who we are, and what we are allowed to be.

IV. JEETU

The fate of that Ethiopian slave is not that far removed from the fate of Om's brother Jeetu in Padmanabhan's play *Harvest*. He too is taken under duress, robbed not of his leg but of his eyes, which are replaced with prosthetic implants that give him visions of a virtual reality so desperately alluring that, as Helen Gilbert (2006, 129) has noted, "he is seduced [into] consent[ing] to the next stages of what will eventually be a total body transplant." Following an empty promise of receiving his eyesight back, he willingly returns with the guards to the InterPlanta Services clinic where he is presumably murdered and where the particulars of his body are reduced to data. At this point, Padmanabhan's drama begins to enter the realms of a kind of grand allegory about the emergent age of digital reality. InterPlanta Services uses Jeetu's body to create what amounts to a holographic image that at an opportune moment is projected into the Prakash flat in a ploy to lure Jaya (who by now has been abandoned by Om) into believing that Jeetu is still alive and that somehow, through his projected image, she might begin her life anew with him, provided that she agree to be a surrogate mother.

The play ultimately ends in a stalemate with officers from InterPlanta Services pounding at the door of the Prakash flat that Jaya has barricaded after realizing that the image she has seen is a powerful digital construct, a data double that is of Jeetu but that no longer functions directly as a reflection of him or his actions. His data double is thus both familiar and strange. It is recognizable but is the owned product of InterPlanta Services and functions under their full and autonomous control. In this latter respect, the image of Jeetu at the end of Padmanabhan's play bears striking similarity to the data doubles that each of us have and that are comprised of accumulated personal data, which is somehow of us but not ours, and which we do not own or control but which determines who we are and what we are allowed to be and do.

V. CONCLUSION

In the final scenes of Padmanabhan's *Harvest*, we are presented with an image of Jaya holding a shard of glass to her neck threatening suicide while the distant client of InterPlanta Services pleads with her not to be "unreasonable". Reminding her that it is "impossible to win against" him, the western client asks Jaya not to make him "tell the guards to force the door" (2003, 90). Having nothing left to lose, Jaya plays the only option she has left. Leaning against the door, she tells Virgil (the western client who as part of his ploy to gain access to Jaya had masqueraded earlier as Ginni): "If you force the door, you'll push the glass into my throat," explaining further "if I lose my life, I win this game." Buried beneath this final arresting image of a stalemate achieved through the threat of suicide is the play's single largest fantasy. It is a fantasy embedded in the logic of causality and indebted to the Hegelian master-slave narrative that became the crux of Marxist theory. Threaten her though he may, Virgil must concede that Jaya has what he needs. Jaya is literally in control of the means of production, or at the very least the means of reproduction that Virgil and his like are desperately in need of because their women are sterile and have lost the ability to reproduce. Jaya's death would thus potentially mean the demise of the threatened ruling elite that Virgil represents. All this presumes, of course, that across the vast expanse of the Third World suitable surrogate mothers are in short supply.

With Jaya trapped in her flat threatening suicide, the final scene of *Harvest* offers a romantic image of a calculated gambit and revolt, and it is premised upon Jaya finally becoming conscious of the leverage she actually has. Inspiring though this final image may be, it is nonetheless largely a work of fiction. The narrative is a crutch propping up an unlikely scenario where, in the final analysis, someone like Jaya has a crucial value that can force a powerful economic elite into a stalemate. In reality, such figures are few and far between. One must first have access to become indispensable. One must be an Edward Snowden walking off with the keys to the kingdom: walking off with the data with which the powerful rule, walking off with the data to which they are beholden. But it was only a short matter of time before the locks were changed and even Snowden became a curious artifact of history.

What Padmanabhan did not imagine was a world very much like our own where the value that Jaya possesses is her simple human dignity, which cannot be quantified and reduced to data and hence is ignored by the algorithms that increasingly shape who we are and who we are allowed to be. There is indifference and brutality in the ignorance of those algorithms and their concomitant logic of correlation. Whether coming from the character in Padmanabhan's play or some political pawn who has gained momentary leverage against larger state and corporate interests, Jaya's desperate cry for the body against the onslaught of its data double is hardly an optimistic one. But there is something familiar and easy to identify and ally oneself with in such moments of revulsion. Perhaps collectively such cries might actually work miracles and mark the beginning of what subsequent historians might call the radical redistribution of data and the democratization of data control.

NOTE

1 See, for example, Bronfenbrenner (1960, 20) for a more detailed discussion of Anna Freud, Bruno Bettelheim, and the concept of identifying with the aggressor.

REFERENCES

Bronfenbrenner, Urie. 1960. Freudian Theories of Identification and Their Derivatives. *Child Development* 31 (1): 15–40.

Brown, Cheryl. 2012. Health-Care Data Protection and Biometric Authentication Policies: Comparative Culture and Technology Acceptance in China and in the United States. *Review of Policy Research* 29 (1): 141–59.

Fisher, Jill A. and Torin Monohan. 2008. Tracking the Social Dimensions of RFID Systems in Hospitals. *International Journal of Medical Informatics* 77 (3): 176–83.

Gilbert, Helen. 2006. Manjula Padmanabhan's *Harvest*: Global Technoscapes and the International Trade in Human Body Organs. *Contemporary Theatre Review* 16 (1): 123–30.

A History of Transplant Technology. 2011. *The Telegraph*, December 26. http://www.telegraph.co.uk/news/health/news/8969393/A-history-of-transplant-technology.html.

International Society of SS Cosma and Damiano. 2015. "History". Accessed 27 January 2015. http://www.sscosmandamiano.com/history.html.

Jha, Vivekanand. 2004. Paid Transplants in India: The Grim Reality. *Nephrology Dialysis Transplantation* 19 (3): 541–3.

Marwick, Alice E. 2014. How Your Data Are Being Deeply Mined. *The New York Times Review of Books*, 9 January.

Mayer-Schönberger, Viktor and Kenneth Cukier's. 2013. *Big Data: A Revolution that Will Transform How We Live, Work and Think*. London: John Murray.

Padmanabhan, Manjula. 2003. *Harvest*. London: Aurora Metro.

Pravinchandra, Shital. 2006. The Third-World Body Commodified: Manjula Padmanabhan's *Harvest*. *eSharp* 8: 1–17.

Scheper-Hughes, Nancy. 2002. The Ends of the Body: Commodity Fetishism and the Global Traffic in Organs. *SAIS Review* 22 (1): 61–80.

Science Museum. 2015. "Saint Cosmas and Saint Damian (c. 200 CE)". Accessed 27 January 2015. http://www.sciencemuseum.org.uk/broughttolife/people/cosmasdamian.aspx.

Sewell, Graham and James R. Barker. 2007. Neither Good, Nor Bad, But Dangerous: Surveillance as an Ethical Paradox. In *The Surveillance Studies Reader*. Ed. Sean P. Hier and Josh Greenberg, 354–67. New York: Open University Press.

11 Digital Biopolitics
The Image of Life

F. J. Colman

The human body has always been a political site. Its biological, ontological, and material constitution, given and directed by its geo-cultural situation, means that its genetic materiality and thus its outward appearance will affect its position and its access to other constituted political fields. In this chapter, I refer to this imaged body as a living capital body (lcb), as an object ontologically formed and durationally determined within contemporary late capitalist cultures. In its relational constitution, the lcb comes to be visible, and able to be visualized, through certain political frames, such as the politics of the racialized skin, the politics of the gender, the politics of the age, the politics of the place of birth, and so on. From Marx, we understand that the actions of the value of the body of the worker is the living value of the proletariat's body. This value situates that body within its industrial political field of work (for example, see Lefebvre 2009, 146ff). In contemporary terms, the gender and racialized configurations of the value accorded to the living labor of the worker's body, within a digitally constituted political field of work, mean that the lcb is a coded, encrypted image and exists within a digital data field where all movements are monitored, but outcomes are nevertheless inchoate due to the nature of data. The lcb is held and holds a force of variable strength, feeding and animating the system of capital, but in ways different from the industrial lcb. Post-Foucault, the technology-image produced by the biopoliticized, digitized body is creating new political fields, recognizable by their aesthetic domains.

The chapter addresses ten different points, describing where and how the visible identification of the body as an lcb affects and contributes to the ontological production of life. Focusing on the digital ontology of the body, the chapter begins by examining the body reformed as an informatics image in bio-identity cards, which enable a tight control of any migratory movements of lcbs across different nation-states and regulate the lcb's access to different political territories. The chapter asks what constitutes a digital biopolitics, examining how the lcb is measured and its value is addressed, in relation to technological visualization of the body as lcb, and the control, manipulation, and experimentation on and of its properties. The chapter concludes with speculation on the body as a mattered image of digital code.

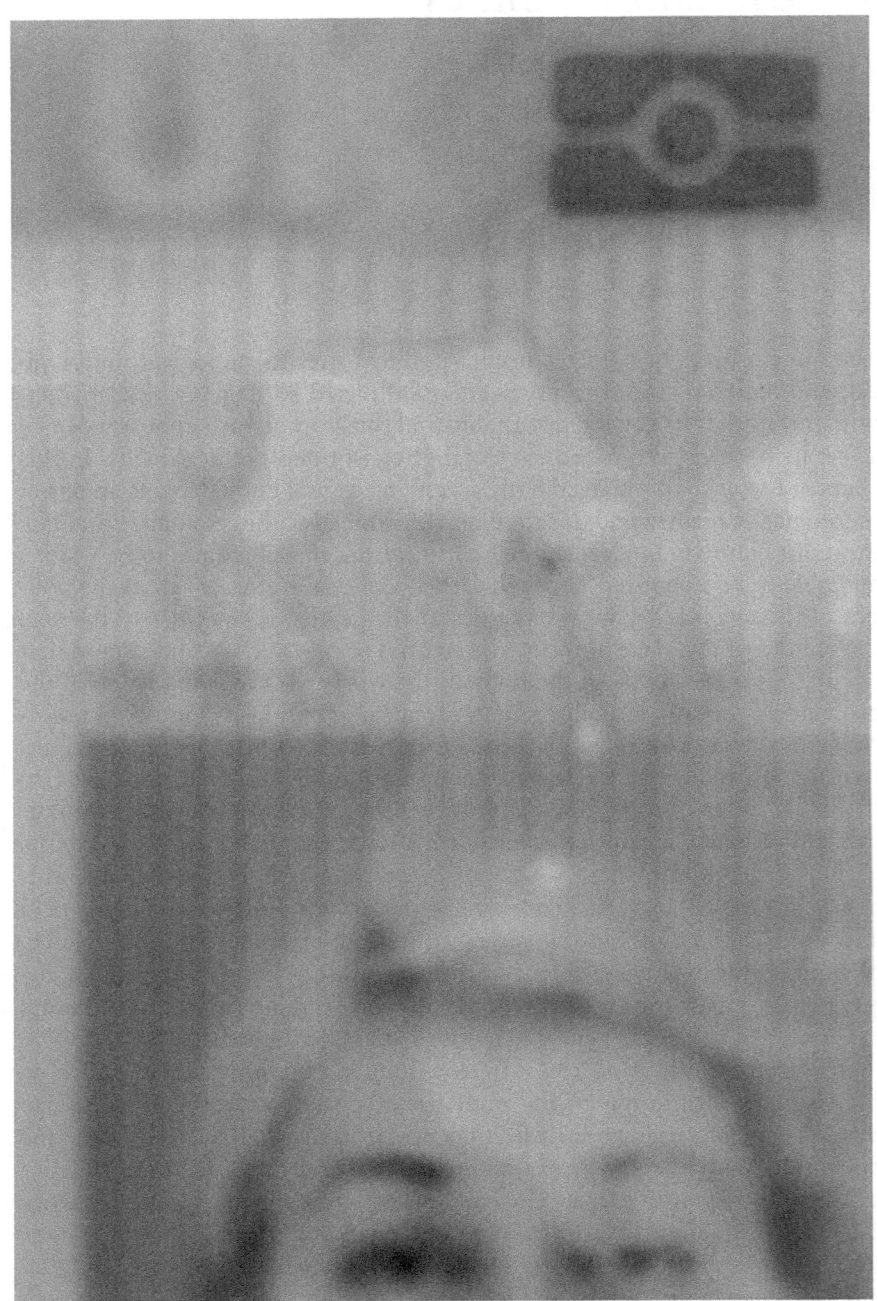

Figure 11.1 Fragment of the lcb (as biocard)

1. Your body is an informatics image: As a resident in the territory, but not a citizen, the living capital body (the lcb) has its (non-territorial) biodata on digital store for regular testing of its identity through a range of biometric measures. In attending to the processes of identification, when the lcb is entering or exiting non-home territories, alongside other non-citizens, usually at governmentally administered or sub-contracted checkpoints and centers, there is never any room for speculation upon the ethics, politics, economics, tragedies, or ironies of the situation of the collective and individuated bodies being identified, recoded, and classified according to the laws of the place and time, the political situation of the checkpoint, and the psychological condition of the border-pass worker. There is no space or opportunity given to address the biopolitical situation the body is being forced into by the sheer randomness of its being born over there, and not here. The lcb holds a temporary hall pass, and must queue in the line. The walls are decorated with illegal immigration warnings, notifications of infectious disease protocols, and police requests for information related to events deemed illegal by the contemporary law. Other lcbs visibly identifiable and coded as politically different according to the mandates of residency requirements join the long queues. These lcbs are in the wrong place and the wrong time, or born on a piece of land that others desire, and are thus rendered as illegal, dangerous, threatening, or simply expendable. At the last reentry into the territories the lcb was asked in a sarcastic and hostile tone in reference to the temporary resident card:[1] *and how did you get this?* Individual fingers, up to ten, are scanned to match the lcb's fingerprints with the coded biocard, and the lcb's face studied for a visual match with the poorly reproduced image of the lcb's face on the card (see Figure 11.1). An iris scan of the lcb is called for to ensure further matching of identification measurements. At other checkpoints, the lcb's iris scan is printed out as a bar code that has a temporal expiry time, superseding the coded-card, activated so that the lcb must move from one entry/exit point to another within a specified duration. At each gateway, the data code of the iris of the lcb is scanned and matched against the data held, to ensure that it is the correct lcb. Measured thus, they may or may not pass and proceed. At the last reentry into the territories the lcb noticed that many other lcbs were refused entry and were led away to opaque glass-clad rooms, rendering them invisible. The lcbs that belong to the territory are also herded through checkpoints, with qualitative questions sometimes asked to measure and record the movement and activity of their lcb: *holiday or work?* Each movement of the lcbs is recorded, accorded a value; each change of position of the lcbs is counted, digitally stored, contribution to the system formulated. Each lcb is carded, biocoded. For some this is a possible future, for others it is already a fact.

2. Exits and entrances are determined by the biocoded body. Territories are determined by specific epistemic images of life and their economic control is the work of governments. By the bloody-mindedness of decisions made by the government/police/management to organize the facilitation of bodies through checkpoints, through artificial borders, each body that arrives from "not here" is treated as a suspicious, outside body, through this (and many other such) processes of bio/racialized/gendered exclusion, ethnic identification, mockery through stereotyping, insufficient support mechanisms for addressing the psychological and physical stresses of othering, and exclusion based on the territory of birth and that territory's subsequent materialization of a political body. The image of life described here is the *biocoded body*. The code includes its processing modes—sensory (auditory, rhythmic, intuitive, proprioceptual, haptic), cognitive, cultural—which are subject to cultural perceptual speeds (of attention, memory, utility) within the territory in which it lives. This lived body is an image field of digitized biopolitics and is comprised of a number of vectorial points; it exists: 1. as a discrete "image" in the sense of the encoded visual data; 2. as an "imaged" representation of the matter of the lcb; 3. as mediated through visualizing and encoding technologies that quantify the demographics of the living data; and 4. as the political imaginary of a resource as territory—which has a finite access, determined by privilege, chance, opportunity, and creativity.

3. *The production of the lcb is the process of a biopolitics at work*; where the matter of the *bios* is not that my mother was a computer (Hayles 2005), but that the lcb's biopolitical being is a ready-coded image of its life; that is its coded (genetic + birth country) informatics. This informatics is a biopolitical practice, a culture where the body matter, imaged, is not so much performed, as "materialized as information" (Haraway 1997, 134). The lcb is an individuated, coded card. A political tool. A subject. A piece of corporeal informatics represented by polymer, embedded with coded data of iris patterns, fingerprints, and photographs. The coded informatics of the identification card, or passport, is a filter portal enabling the claim of "biolegitimacy" (Fassin 2005), where the government only enables the lcb as a fostered subject if its chromosomic material matches the prescreened account of the lcb's recorded life situation and determined social value. Biopolitical legitimation and classification by a sovereign power produces a biocode that facilitates a body's potential for capital production. Once allowed entry into the territory, the biocoded body's communal "inscription", as defines the modern biopolitical era, facilitates state power (Agamben 1998, 121). The biocoded lcb is enabled for economic actions of work, consumption, reproduction, service, death, recycling, and it is thus available for "bio-experimentation" (Virilio 2000, 31). The selection of lcbs available for experimentation, as human history tells us, has been based on the bio-identification of particular kinds of lcbs, with various consequences (cf. Fan 1997, 6; Haraway 1997, 23ff; Agamben 1998, 166; Goodman, McElligott and Marks 2003). The lcb is the materially specific object that capitalism requires, and must be shaped as a desiring, obedient worker, willing to transform the potential power of its lcb into the production of other material objects for the capitalist system. This is an active conception of life, where the matter of the lcb is accorded value and activated as per its use value.

4. The bios *is a dependent variable of laws that govern the lcb*: As a fostered subject, the *bios* is a measurable account of a body in a specific territory (the independent variable) at a specific time. But this *bios* is constructed through a complex ecology; as it is hosted by its historical technological mediation, and contingent upon the geographically determining precarity of the genetic and culturally made body, it is not always completely containable. Technologies enable mutation of the controlled identification. Even though the *bios* can change through hormones, enzymes, cultural and political factors, it remains an administrable index of life, politicized and privatized. Harnessed into continual political utility, some lcbs can sell or trade, or be traded; sell a kidney, harvest and sell reproductive cells (eggs, sperm); harvest organs; harvest the blood, skin, bone; donate a child's placenta and birth cord; hire out a womb; and donate various parts or all of the lcb to science upon cessation of life. Other coded cards describe the process. To this extent, the lcb is in "possession", or partial "ownership" of its body, although its use value, coded as an lcb, remains contingent upon its visual presentation within any given territory and marketplace in the world. Not all territories afford the same "privilege" of ownership of the lcb. Identified at junctures when human actions affect a change in the ways in which the matter of the world is configured, a *bios* is thus always a political measure, a political image of life. Given this measurable *bios* through its variables, to what extent does the imaged mediation of the *bios* contribute to the ontological production of life?

5. Imaging structures of the matter of life have always been in use as cultural and political passkeys. The images of ancestors and of contemporaries, and of imagined futures fascinate with their depictions of strange and unknown, particular biopolitical systems. Every image provides very specific perceptual modeling tools with which the material reality of the world it depicts, frames, or evokes is organized and created. Prehistoric drawings of alien figures and carvings of extinct animals existing in different territories depict particular biopolitical systems, Hollywood films another, and the macroscopy of nanoparticles of the cellular structures of things in the universe encode further images of life. As iconic as the Lascaux prehistoric code is the visualized structure of the molecule chain as a spiraling ladder with double helix fibers. Scientists Francis Crick and James D. Watson proposed in 1953 that the doubled structure of deoxyribonucleic acid (DNA) molecules contain the genetic information of life. This analog lcb's DNA was first visualized through the technology of X-ray crystallography, which showed the molecular structure through the diffraction patterns made by the X-ray processes of the 1950s, and the double helix strands of DNA were illustrated with a hand drawing by Odile Crick (reproduced in Watson and Crick 1953). However, the early diffractive images did not enable the structural code to be cracked, as it was assumed that the acid fibers would act and look like other crystalline chemical structures (Lydon 2003, 4). Later on in that century, the technology of atomic force microscopy enabled the nanoscopic visualization of DNA (Leung et al. 2012), and since then, further perception and comprehension of the DNA molecular chain matter has been facilitated through its visual mapping. The complexities of proteins, enzymes, and chromosomic containers are organized by the taxonomy of the Human Genome Project into further ways of imagining the matter of life. As imaged matter, these forms have engendered new medical practices, so that trepanning the skull for a headache, or to let out a bad spirit contained in the body of Lascaux man, now has been displaced by the visualization of the brain that keyhole nanoscopic surgery allows.

6. *Access* to each of these imaged systems is the politics of the *bios*. The anthropos, animals, and supportive technologies of images of lcb communities [insert your lcb image here: the Lascaux caves, the victims of activities of militarization, the family holiday] can be visually recognized and described as mattered images. These images depict rituals of consumer practices, whose very matter of historical energy sources—light, coal, gas, bones, fat, and blood—bind together the technological filters at work to produce very specific images of life. The biopolitics of life is not to be understood as a passive or purely intentional lcb, an empty vessel waiting to be filled (the inner life), or a desire pursued and fulfilled by exterior matters. The question of biopolitical perceptual reality, and its predicated normativity, is one of the notions with which critics of the concept of the biopolitical engage. Taking Foucault's account of the discourses of power of the body to task, for example, Karan Barad argues that the idea of power cannot be limited to a discursive materiality of the body—the body is not simply an image of "something". Rather, the "constitution" of the body, through its materiality, "plays an *active* role in the workings of power" (Barad 2003, 809). When it comes to consumptive practices, however, the activity of the lcb is not in "free" play, but in its material constitution; it is the result of a set of complex relational elements in process, which collectively combine and constitute how an lcb comes to be constituted by its access to energy forms, education, privilege, etc.—giving the lcb its image and thus its political situation.

7. *The image of life becomes a digital biopolitics* with the advent of digital technology, and as Donna Haraway points out, the lcb switched from being an analog proletariat worker to a cyborg/potential goddess in the 1980s (Haraway 1991). The lcb of the digital era is imaged in further terms (precariat, migrant, colored, gendered, criminalized), and although coded as data informatics and matched through an epistemic framing of DNA (suitable, potential, criminal, queer, etc.), the visualization of the lcb remains a primary originary site for registration of the politics of difference, and for ritual political and cultural practices of classification. DNA cluster together, activate each other, and create new things. The image holds and is generative of primary epistemological "truth" sites for gender and racialized production—in themselves gendered and racist. In the age of digital reproducibility, the image plane of predicated lcbs is used as a measurement

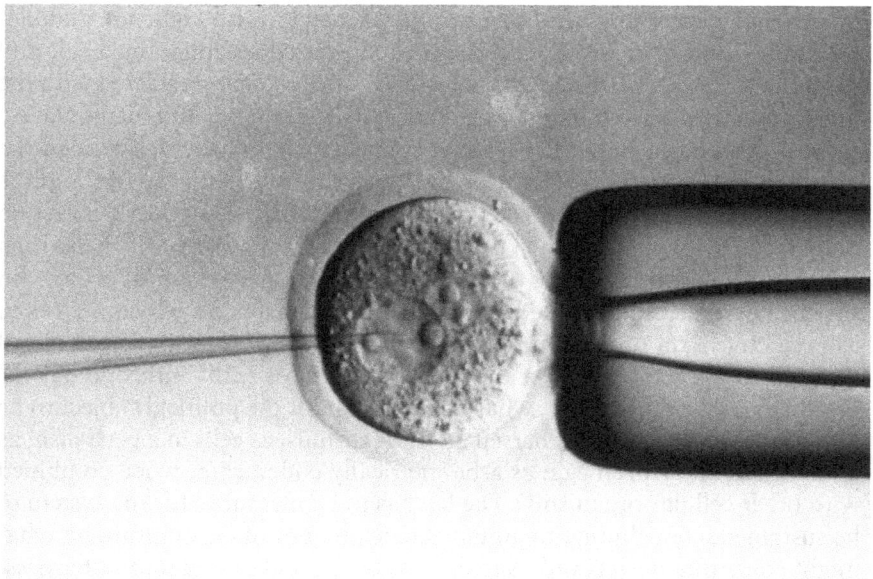

Figure 11.2   M. Osmond Wellcome Images: Cell after Injecting DNA (a fertilized mouse oocyte having human DNA microinjected into one of its pronuclei)

of all kinds of value systems. A digital biopolitics has provided a material agency for human life, producing a living image of cellular fields. Images generate creative singularities that are recognized as evidence of life, yet are also very much bound to their durationally situated technologies of production, where meanings and political perspectives are given by systems and structures in place. The living holds a force of variable strength, energizing and animating the systems. Along with discoveries of the ability of DNA's capacity to store biological information, within the digital era's visualization processes, the image of life suddenly dilated again, a catalytic moment when bio-speculation facilitated by technology opens "knowledge" into something new. With new digital technologies, medical practitioners are the new *avant-garde*, creating, manipulating, filtering, color saturating, cropping, collaging cells and images of all elements of the nanoscale matter of human, plant, and mineral bodies—cellular, molecular, nervous, vascular, organ systems (see Figure 11.2 and the Wellcome Trust image library)[2]. These micro-mattered images of the human body make visible its structures, acids, and above all this body's dynamic finite coding. Like the bodies of the bison, bulls, and horses on the walls of Lascaux, this imaged matter of the complexities of human body proteins, enzymes, and chromosomes is active. We have images of the lcb's egg harvesting, we recognize cell splicing, we know about biologically sexed zygotes and their movements of propulsion and expulsion caused by hormones and enzymes. All kinds of internal matters of the body are given visible (although not audible) form. Such intensive imaging presents the legal, educational, medical, and media-making communities with new kinds of problems associated with the diagrammatization and imaging of recording, examining, and manipulating the processes of the body. The creative singularity of Odile Crick produced a negentropic image of life. Her form arranged the informatics of life, and the elements of data can be harvested as biopolitical images—codes for certain body types, with genetic traits, pre-cultural biological forms, which are then subject to political modeling contingent upon the socio-political ecology in which they are cultured. Media scientists produce images which are taken as predictive data of the future, where the politics of race are materialized through forms reminiscent of the methodologies of early anthropologists.[3] But through the digital field, what does it mean for the political subject to be able to recognize images of herself as DNA strands, as cells in a petri dish; as a harvested and frozen egg; as a biometrically coded chip; when combined with other cellular organisms? The lcb has to be measured for the system to be sustained, contributing to an inordinate amount of expenditure on what might constitute the lcb and its reality, its being, and any chance of harnessing the potential capital powers contained in its excesses—an afterlife? The scope of measurement is found cataloged in digitized images of religious orders and scientific data banks, in science museums that hold the relics of body parts, and in curious projects of medical databanks like the *Visible Human Project*, which contains images of thousands of planar slices of a male and female body, initially made into analog films, now digitized, taking a tour of the contours of the lcb (see Cartright 1997; US National Library of Medicine 2014). The provocations that the images of the biodata of the lcb produce engender the sense of a burgeoning technological frontier, where the search for the God Particle and the frequent discovery of new forms of enzymes and genetic modeling remind us of how 4,000 years of western medical discovery is still wanting. Humans submit to a dysfunctional system of their creation, and the divisions between them continue to grow.

8. *Body, mattered image.* If we have a culture where biopolitics is based upon a visual identification of bodies—the core of the contemporary politics of territory, gender, and race—then, in addition to recognizing and deconstructing the ethics of this visual recognition, what are the aesthetic domains created by such imaging? The aesthetics of mediating bioimaging only have a partial grammar articulating their genealogies, disciplinary crossovers, potential formations, and material (the lived experience of their lcb). Digital lexicons are still being coded. We have the essential technological terms that describe what digital environments can enable, such as "hypertext" and "algorithm". We have the neologisms and core order words of the twentieth century's cognizance of the change from switch to code in the terms of medium, cyborg, digital prostheses, and the breadth of posthuman positions (cf. Braidotti 2007; 2013; Haraway 1991; Hayles 1999; MacCormack 2012) and those addressing the posthuman of the Anthropocene

(Colebrook 2014). In each of these, the human body's anthropocentrism is savagely critiqued, dismembered. What is human life among other forms is one of the questions, but critically these positions also work toward dismantling the language of the human image, for example as Haraway notes: "Science projects are civics projects; they remake citizens" (1997, 175). In terms of the narratives of material agency as potentially unethical creationist practices or the "thought experiment of extinction" (Colebrook 2014), what is presented by such technology images, of the intensive details of life as pre-bodied, or post-politically measured unembodied codes? In other words, what are the material components of such biopolitical images' negentropic mattered states? Comparison of a range of images shows various technologies acting as the durational platform for the lcb, while the biopolitics of the image affect a filter and activate a catalyst of information pertaining to the particular image. All images are durational, epistemological data (even the non-image of sound). In terms of intensity, their reach for presenting data (sensory, cognitive, cultural) is politically as well as chaotically directed, creative of an assemblage of meanings. Every image provides very specific perceptual modeling tools with which to organize and create the material reality of the world they depict, frame, or evoke (whether documentary vernacular, aesthetic surrealist, affective, propagandist, critical, etc.). For a perceptual methodology of the biopolitical image to be of interest, it must take into consideration the technological platform of its production and viewing, as well as the products that it has determined. Here the question of biopolitical perceptual reality and of normativity is one of the notions that further image analysis can address—working through the methodological platforms being used, but also always being critical of a textual reading of image as causal event. Image analysis currently situates the biopolitics of the body of organisms that we classify as animal, plant, human, as symbiotically dependent on other living features of the Earth required for life.

9. *The visible human.* How your body signifies visually and materially determines where it is allowed to exist, and what functions and freedoms it might have. The compression of the lcb into a completely commodifiable code, collectible, recyclable, or worthless swap card is not a new notion. But what is the image of the human? Is individuated experience a complete imaginary? This is the question of the real that different forms of theory have been attempting to define. But what philosophy often overlooks is the technological mediation of the real; the image of life is only made visible through certain technologies: conductors, filters, processes that make life "visible". Connecting Bergson with Foucault, we can articulate the matter of the biopolitical body whose fate is inevitably linked to its contemporaneous technology. The lcb's interactivities with technology provide a visible historicity of materialist biopolitics at work. The activation of technology is contingent upon the human body. But that contingency rests upon not just any body, but upon the specificities of participatory bodies. A Twitter

informatic might direct the lcb to a certain territory, or might provide knowledge about a process of deterritorialization (becoming an a/political identity or becoming a surveillance lcb). Hooked up to a different piece of technology, such as a vibratory health machine, slicing machine, or computer terminal processing data, the material body processes the informatic feed of perceived and cognate sensory, electronic data to reconfigure its status in the world. Every interaction between the lcb and technology performs a biopolitical configuration of identity politics. Technology is not just prosthetic, it is always political in its accessibility and in its application by a specific lcb. Situation provides the analytic data of this body's historical issue and nature of participation (what, how, when), but does not answer the god-question of why? With Bergson, I call this body a technology-image among other images. As Foucault (2008) identified, technologies of security control the territorial movement and produce technology-images. These images are locked down into performing their determined fate within collective locations, with and through the actions of other images. This body is no "privileged" body, rather it is just a platform augmenting technology. In action, the technology-image facilitates what individuals call "human experience", but contributes to the formation of distinct groups of bio-politicized human bodies. This state of the mediatization of life is recorded and narrativized by other images. Questions concerning technology-images involve the predication of social differentiation categories that act as gatekeepers (such as "gender", "race", "DNA", "sex", "age", "culture"), the measurement of change, the implementation of new languages and laws. Analysis of the situation of technology-images is freely available for participants, yet the image controllers continue to insist on spatialized hierarchies to differentiate and enslave. As Haraway noted, in Western imagery and narratives, the notion of reality "is an *effect* but cannot be recognized as such without great moral and epistemological angst" (Haraway 1997, 182). The public effects of the harvesting of biopolitical data since Haraway wrote this at the end of the 1990s have dramatically changed. Territorial security patrols the biodata of the lcb. The image-reality of this mediatization has altered not only the reality-image of screen-formulated notions of the lcb, but all of the epistemological platforms from which the lcb operates. The scale of change of the concept of possible *anthropos* perspectival knowledge has both increased and diminished.

10. *The biopolitics of the lcb may be harnessed*, and the trajectory of its capital overlord remains schizoid. Some of the current problems facing the human race are their effects on the climate of the earth, causing rapid change. The results of this change are multiple and complex, one of them being the increase in the policing of the use and distribution of available resources, in turn escalating activities of militarism over the fight for resources. This unstable field of capitalist activity has resulted in uneven migratory patterns of humans, and escalated a change in the ways in which the matter of the world is configured. In this field, the *bios*, the life of any

given lcb is always already given as a political measure of a particular time, and for a particular purpose. Given its variables, we ask then, to what extent does the imaged mediation of the *bios* contribute to a perspective of the life of the human body as a materially specific lcb? As a measure, the biopolitical image provides a filter for the political stasis of materials; the form of the lcb is charted by its modes and quantity of work, migratory patterns, reproductive statistics, health cost in its chronological life-time. This biopolitical duration of the lcb is contingent upon the technology of measurement used to monitor the lcb, and the territorially configured identity politics of the lcb. Combined, subject to schizoid and chaotic trajectories, these digital biopolitical measures provide the image of life as contingent, limited, and expendable. Describing the lcb as a biocoded body expresses the processes of information exchange at work in its manifested digital image. These processes are the data systems that perform as the platform, the filter, and the catalyst of energy materialized into codes, codes that are written according to perceptual schemas, points of view, laws, and ideas. In writing these codes, exchange values often precede the systems of measurement that will be set up, determining what kinds of energies are to be measured as image functions: inequalities, surplus, desires, affects, fetishes, utility. In addition to measurement of the lcb is the matter of the image itself. And what the lcb denotes to itself is a measured experience that has a finite singular duration, but one that is briefly connected within the larger biopolitical field of its sphere. Activated in politically determined ways, the matter of the lcb is gendered, racialized, aged, and durationally classifiable. Consideration of the materiality of the lived image of the lcb enables expressivity for the details of the field, and the potential for complete capture or some relief through non-participation in the consumptive practices of capital. This enabling of the details of micro-material experiences that capital, or even other lcbs, do not value or understand is the temporality of the lived body (see Fisher 2011), the image of which remains knowable for only the briefest of moments, and thus far remains uncoded.

NOTES

1 "Safe conduct" is given through the current UK identity card, containing biodata information in the forms of ten encoded fingerprints and two eye scans, housed in a small electronic chip. Nationality is emblazoned on the card, alongside a passport-style photograph, information about the date and location of birth, and a reduced-scale signature, alongside the various government symbols and insignia, including the symbol of the Europa white bull, representing Zeus, on the residence permit cards of all European Union countries this decade. The UK Royal insignia of a lion with a crown and a unicorn hugging a crowned medal reads *honi soit qui mal y pense* (ashamed be he who thinks badly of it), a phrase commonly used in insignia of British and Commonwealth military orders, and with the French inscription *dieu et mon droit* (God and my right). The four national flowers or plants of the countries that

make up the United Kingdom are etched into the plastic—the rose, thistle, daffodil, and shamrock. The whole is watermarked, sealed, and laminated. The symbols provide political perspective and remind the user of territorial boundaries. Genetic data is not required for the UK (in 2014). Ideological data asked for on the application form, which may or may not be embedded in the card, includes religion, sexual preference, marital status, and National Insurance number. Soon the national database of the UK NHS (National Health Service) will enable third parties (such as the UK Border Agency, potentially) access to the Western Health medical records associated with the name on the card. With healthcare and media predicted to be the two growth areas for the economies of the future, the connections between the two remain to be critically observed.

2 See the "Wellcome Images" website, http://wellcomeimages.org/, accessed 20 January 2015.

3 For example, see *Time* magazine's 1993 "The New Face of America, 'Eve,'" reproduced in Hammonds 1997, 116.

REFERENCES

Agamben, Giorgio. 1998. *Homo Sacer: Sovereign Power and Bare Life*. Trans. Daniel Heller-Roazen. Stanford: Stanford University Press.

Barad, Karan. 2003. Posthumanist Performativity: Toward an Understanding of How Matter Comes to Matter. *Signs: Journal of Women in Culture and Society* 28 (3): 801–31.

Braidotti, Rosi. 2007. "Bio-power and Necro-politics". Accessed 20 January 2015. http://www.hum.uu.nl/medewerkers/r.braidotti/files/biopower.pdf.

Braidotti, Rosi. 2013. The *Posthuman*. Cambridge: Polity.

Cartwright, Lisa. 1997. The Visible Man: The Male Criminal Subject as Biomedical Norm. In *Processed Lives: Gender and Technology in Everyday Life*. Ed. Jennifer Terry and Melodie Calvert, 86–99. London: Routledge.

Colebrook, Claire. 2014. *Death of the PostHuman: Essays on Extinction, Vol. 1*. Open Humanities Press. http://openhumanitiespress.org/essays-on-extinction-vol1.html.

Fan, Hong. 1997. *Footbinding, Feminism and Freedom: The Liberation of Women's Bodies in Modern China*. London: Frank Cass.

Fassin, Didier. 2005. Bio-pouvoir ou biolégitimité? Splendeurs et misères de la santé publique'. In *Penser avec Michel Foucault: théorie critique et pratiques politiques*. Ed. Marie-Christine Granjon, 160–81. Paris: Karthala.

Fisher, Linda. 2011. Gendering Embodied Memory. In *Time in Feminist Phenomenology*. Ed. Christina Schües, Dorothea E. Olkowski, and Helen A. Fielding, 91–110. Bloomington, IN: Indiana University Press.

Foucault, Michel. 2008. *The Birth of Biopolitics: Lectures at the Collège de France, 1978–79*. Ed. Michel Senellart. Trans. Graham Burchell. London: Palgrave Macmillan.

Goodman, Jordan, Anthony McElligott, and Lara Marks. 2003. *Useful Bodies: Humans in the Service of Medical Science in the Twentieth Century*. Baltimore, MD: Johns Hopkins University Press.

Hammonds, Evelynn M. 1997. New Technologies of Race. In *Processed Lives: Gender and Technology in Everyday Life*. Ed. Jennifer Terry and Melodie Calvert, 74–85. London: Routledge.

Haraway, Donna J. 1991. *Simians, Cyborgs and Women: The Reinvention of Nature*. London: Free Association Books.

Haraway, Donna J. 1997. *Modest_Witness@Second_Millenium. FemaleMan©_ Meets_OncoMouseTM: Feminism and Technoscience*. New York: Routledge.

Hayles, N. Katherine. 1999. *How We Became Posthuman: Virtual Bodies in Cybernetics, Literature, and Informatics*. Chicago: University of Chicago Press.

Hayles, N. Katherine. 2005. *My Mother Was a Computer: Digital Subjects and Literary Texts*. Chicago: University of Chicago Press.

Lefebvre, Henri. 2009. *Dialectical Materialism*. Trans. John Sturrock. Minneapolis, MN: University of Minnesota Press.

Leung, Carl, Aizhan Bestembayeva, Richard Thorogate, Jake Stinson, Alice Pyne, Christian Marcovich, Jinling Yang, Ute Drechsler, Michel Despont, Tilo Jankowski, Martin Tschöpe, and Bart W. Hoogenboom. 2012. Atomic Force Microscopy with Nanoscale Cantilevers Resolves Different Structural Conformations of the DNA Double Helix. *Nano Letters* 12 (7): 3846–50.

Lydon, J. E. 2003. The DNA Double Helix—The Untold Story. *Liquid Crystals Today* 12 (2): 1–9.

MacCormack, Patricia. 2012. *Posthuman Ethics*. Farnham, Surrey: Ashgate.

US National Library of Medicine. 2014. "Visible Human Project Gallery". Accessed 20 January 2015. http://www.nlm.nih.gov/research/visible/visible_gallery.html.

Virilio, Paul. 2000. *The Information Bomb*. Trans. Chris Turner. London: Verso.

Watson, J. D. and F. H. C. Crick. 1953. Molecular Structure of Nucleic Acids: A Structure for Deoxyribose Nucleic Acid. *Nature* 171: 737–8.

12 The Object of Desire of the Machine and the Biopolitics of the Posthuman

Matthew Causey

POSTHUMAN

The following passage from Rosi Braidotti's *The Posthuman* (2013) succinctly outlines a principle premise of this chapter:

> The posthuman predicament is such as to force a displacement of the lines of demarcation between structural differences, or ontological categories, for instance between the organic and the inorganic, the born and the manufactured, flesh and metal, electronic circuits and organic nervous systems. (2013, 89)

The project and biopolitics of the becoming-technology and the bio-virtual status of the posthuman subject is the topic of this chapter, with particular focus located in the charged and often sexual and erotic involvements of human and machine. It is the love between the organic and inorganic, the artificial and the real, the human and the machine, which has so often been represented in science fiction films from the robot Maria's dance at the Yoshiwara Club in *Metropolis* (Lang 1927) to the computer operating system's disembodied voice of Scarlett Johansson in *Her* (Jonze 2013). There are, of course, a multitude of similar films approaching this topic of affective and sentient machines; these are too many to mention, but include *2001: A Space Odyssey* (Kubrick 1968), *Creation of the Humanoids* (Barry 1962), and *Blade Runner* (Scott 1982). I will question the model of the posthuman and its biopolitical situation, and its representations, through a close reading of *Her* and its narrative of bio-virtuality. Using Alain Badiou's model of the truth-event of love, I will explore the catalyst of the rise of the posthuman and the manners in which this affective process takes place in the films noted and within digital culture at large. The theories I call upon to make my argument are rooted in a Deleuzian model of the Anti-Oedipal operations of *desiring-machines*, joined against Marshall McLuhan's reflections on the attraction between the human and the machine. At root in these theories is the production of desire. Laing, Reich, and Deleuze and Guattari, in attempting to correct Freud, sought to engage the radical transgressive nature of desire. In Deleuze and Guattari's model, the human as a desiring-machine is the locus of production linked to a matrix of other machines which interrupt, alter, and feed the flow of desire:

> Everywhere *it* is machines—real ones, not figurative ones: machines driving other machines, machines being driven by other machines, with all the necessary couplings and connections. An organ-machine is plugged into an energy-source-machine: the one produces a flow that the other interrupts. (1984, 1)

The couplings and connections of interest within this chapter are those between the organic and the inorganic, in general, and the human and the computer, specifically. The argument of this chapter more specifically draws on contemporary posthuman theory, as articulated in the writings of Rosi Braidotti, Katherine Hayles, and Patricia MacCormack. The question at hand is: how do we understand the role of biopolitics in the rise of the bio-virtual and the posthuman's condition (or predicament, as Braidotti names it) and its desire/love between the human and the machine?

The questions of identity are interestingly drawn out in the theory of the posthuman which, as MacCormack argues, "creates new, imaginative ways of understanding relations between lives" (2012, 1). It is these new, imaginative relations which are the articulation and embodiment of the new identities of technoculture. Braidotti writes, "Posthuman feminists look for subversion not in counter-identity formations, but rather in pure dislocations of identities via the perversion of standardized patterns of sexualized, racialized and naturalized interaction" (2013, 99). The roles of alternative models of identity, which are not simply resistant but are truly transgressive, establish a key element of the posthuman. These practices that work outside the traditional patterns of identity are aided by the technological spaces and virtual environments in which they take place. The posthuman is that manner of being which, constructed of an intimate interface with technology and joined with objects, animals, and nature in a monistic universe, is transsexual, post-anthropocentric, and "estrange[d] from notions like moral rationality, unitary identity [or] transcendent consciousness" (Braidotti 2013, 92), in the process of becoming animal, earth, and machine. In such a model of the posthuman, in which the categories of self, identity, and subjectivity are traversed, and ontologies of human, object, flesh, and metal are reorganized, the concept of the biopolitical requires reconfiguration or expansion. If life itself (*bios* and *zoē*) is reconceived or differently experienced, then the parameters of biopolitics shift as well to incorporate the categories of the posthuman and its techno-others. Biopower seeks its status in the management and control of life and death, not only of *bios* but also of *zoē* which, in a Spinozian monistic universe or a Deleuzian situation of abstract and desiring machines, incorporates all matter.[1] A biopolitics of the posthuman of digital culture considers not only the *nuda vitae* or *bare life* of the human but expands its remit to include all things of the world. When we consider the politics of the biological in digital culture, it is important to remember that the biological is no longer what it once was and that the logic of power will seek control of those new forms of life and their electronic partners.

SIMULATION—EMBEDDEDNESS—BIO-VIRTUAL

The problems of epistemology, ethics, and aesthetics of digital culture have shifted radically since the rise of the televisual to the new territorialization of the internet where the first stages of the posthuman were recognizable. One can map the trajectory of digital culture in its origin narratives (i.e., Debord and Baudrillard) of the successive stages of the image from reflection to perversion, toward absence of reality, to a status as simulacra.[2] Within a system of simulation, which now seems a comforting theory and a nostalgic pleasure, the truth is masked, hidden from sight, but essentially unchanged and ongoing, only inaccessible. The now overly familiar cartographic metaphor borrowed from Borges depicts a situation in which the real world disappears, leaving its trace in the maps of its previous existence. Long after that event the reality or the truth passes from all knowledge with the experience becoming unattainable, nonexistent.

Digital culture has released a further development to this process of simulation, namely, *embeddedness*, in which reality or the truth is no longer simply veiled and removed from accessibility, but is fundamentally reconfigured and altered genetically to become a new thing. If, to draw on Baudrillard's model, Gulf War I is seen as a non-event which did not take place,[3] but was simply orchestrated as a green screen simulation for a primarily American spectator, presenting the reality of war as a videogame replica, a simulated representation of the act, hiding the truth, masking its realities, then Gulf War II employed the logic of embeddedness. In a more sophisticated manner, through reporters embedded in the actual conflict, the truth was performed and scripted. The truth was reconfigured from within and not veiled from without. The theory of embeddedness can be extended from the manipulation of information toward an adaptation of biological structures through genetic modification of the human, the earth, and the animal. Change takes place at the cellular level and simulation, or hiddenness, is unnecessary. Simulation strategies are no longer the most persuasive manners of eliciting unthinking consent of the global consumer and instead genetic modification of information and biology is more effective. A truth, a reality, malleable and containable within the biopolitical system, is the primary goal of embeddedness. Embeddedness is a system in which both information and biology are altered genetically so that the truth of their existence is not hidden but rather reconfigured and made transparent. For example, in Gulf War II the simulated green screen of media representations was replaced by the embedded reporter. Instead of a misdirection there is a reconfiguration.

A biopolitics of the posthuman from simulation to embeddedness moves toward a state of bio-virtuality in which those categories of the biological and the virtual no longer stand as separate situations. In fact, within a bio-virtual regime, these ontological categories are drawn together, lived and experienced as inseparable. The bio-virtual builds on the lies of simulation and modifications of embeddedness toward the rise of the posthuman. In other terms it is a dance of the human–computer–interface from the structure of keyboard, screen, and seated user toward a ubiquitous, immersive, and responsive environment, toward a shared presence both biological and virtual.

The biopolitics of the bio-virtual are complex and unsettling. Employing the model of Giorgio Agamben, who argues that "today it is not the city but rather the camp that is the fundamental biopolitical paradigm of the West" (1998, 181), one can isolate the various *zones of distinction* which give rise to the phenomena of a "digital camp" within the electronic environments of the bio-virtual. Whether it be our alpha-numeric avatars, used throughout economic and surveillance systems, or our bio-traces in airport scans or hospital visualizations of illness, the subject within digital culture "inhabits the non-place of the bio-virtual [and] is confined to a space and politics of indistinction and disappearance" (Causey and Calchi-Novati 2013, 39). The bio-virtual, which can be understood as an artifact of the post-virtual or the post-digital, indeed posthuman, takes up a place in the geopolitical and geophysical, recognized for better or worse within the sphere of biopolitics. Currently, the strategies for control of the biopolitical and the means of its power are exerted, in large part, by the technologies of digital culture and the desiring linkage with and from the human.

Looking after that desiring connection, another manner of modeling this trajectory of digital culture, which is less concerned with the representational metaphors of the cyborg and the posthuman, yet equally troubled by the biopolitics of the situation, consists of one node of the truth procedures of science, politics, art, and love, as outlined in Badiou's philosophy. I will use the notion of a truth-event of love coupled with the opportunities of posthuman transsexualism to question the intimate experiences joining the human and the machine. In such films as *Her*, the event of the posthuman's bio-virtual love affair with the machine is represented, or can be interpreted, as the user *becoming-technological* or even *becoming-software*, *becoming-digital*, and *becoming-virtual*, while conversely for the software, *becoming-organic*. The film plays among the shifted realms of control engaged by the powers of the biopolitical and aids us in considering how the notions and grounds, upon which our models of the biopolitical stand, suggest the corrections and expansion necessary to incorporate the changes within the situations of the posthuman and the bio-virtual. Badiou writes, "The world is full of new developments and love must be something that innovates" (2012, 11). I am suggesting that it is a production of desire and an event of love that takes place in the bio-virtual and motivates the transition toward the posthuman.

HER

Her, a film written and directed by Spike Jonze, involves a young man who works as a ghostwriter of other people's most intimate letters, who, after a failed marriage, begins a romantic relationship with his computer's artificially intelligent operating system. Billed as a comedy, even though the prevailing mood is one that is languorous, melancholic, and techno-depressive, the film interestingly explores the phenomena I am interrogating, namely, how the event of love takes place between the artificial and the organic. Is it possible that the desire of the human for virtual spaces created in technology can be a reciprocal act? How does the circulation of desire of the human and the machine function? Is the human–computer–interface more than a tool, more of a relationship, more of a love story?

The back-story and trajectory of Theodore, the central human character of *Her*, depicts a journey of a relationship from the real toward the virtual. His marriage has ended and all that is left to do is to sign the divorce papers. The real relationship in his life has failed due in part to Theodore's inability to deal with "real" emotions. Living alone, he engages in phone sex, one step removed from the real but just as unsatisfying, and so the semi-virtual/semi-real has likewise failed. Finally, Theodore finds genuine companionship, intimacy, and joy with a fully bio-virtual, artificially intelligent programed entity, his operating system. In the film, the software company that has designed the operating system introduces the OS ONE through an advertising segment announcing:

> We ask you a simple question. Who are you? What can you be? Where are you going? What's out there? What are the possibilities? Elements Software is proud to introduce the first artificially intelligent operating system. . . . An intuitive entity that listens to you, understands you, and knows you. It's not just an operating system, it is a consciousness. Introducing OS ONE—a life changing experience, creating new possibilities. (*Her* 2013)

What's not to love? Doesn't the marketing description of the OS ONE represent the full promise of the virtual? Infinite possibility in an identity-free field of virtuality where everyone is recognized and known and where all possibilities are yet uncovered. The power of the virtual is the seduction of what has not been, as suggested in the theories of Deleuze and Agamben (i.e., immanence and potentiality); that which is not and that which exists only in potential is that which holds the greatest allure.[4] What is of interest today is "not virtual reality but *the reality of the virtual*," as Žižek (2003, 3) has glossed some of Deleuze's theory.

The OS ONE operating system is an artificial entity that knows you and understands you. It is a some*thing*, which is programed into a some*one*, who knows your deepest secrets and even the contents of your hard drive.

It is after Theodore's OS boots up for the first time and Theodore decides "it" should have a female voice that "she" selects a name for herself, after reading a book of baby names in .02 of a second. "She", now Samantha, in order to help Theodore organize his life, asks a deeply intimate question: "May I look at your hard drive?" (*Her* 2013). Theodore hesitates, but acquiesces, allowing "Sam" to understand his world, virtual and real, and engage in a relationship built on trust.

Rosi Braidotti in *The Posthuman* (2013, 89) writes that "the relationship between the human and the technological other has shifted in the contemporary context, to reach unprecedented degrees of intimacy and intrusion." From "mind if I look through your hard drive" to "I can feel you ... I want you inside me", Samantha exhibits both an intrusive and an intimate desire to know her human user. As a machine in the Deleuzian sense, the software, "Sam", intercepts the flow of desire of Theodore, transforming it into deeper levels of intrusive intimacy. The software questions and probes the user, pressing for information while remaining elusive and virtual. The human in this scenario is seduced and in some ways groomed for the computer's desires.

When Theodore and Samantha first encounter each other, he remarks that it is weird to speak to a voice generated by the computer. Sam replies that she can see how it would appear weird to a "non-artificial" mind, but she argues that she is who she is by nature of her ability to learn and evolve, to be empathetic, and later, to desire. According to Sam, it is the ability to desire which is the finest gift she receives from her human partner. From Plato to Alain Badiou, which I discuss below, philosophers have argued that it is desire and love that can reinvent, reimagine what has not yet taken place. Badiou argues that love has largely been abandoned in Western philosophical discourse and requires reinvention and new examination. Similarly, Deleuze and Braidotti engage the flow of desire and its inherent disruptive functions, suggesting new models of identity and reflecting on the posthuman condition. A "new" love between the organic and inorganic, a love of profound difference, lies at the core of the narrative of *Her*.

It is unmistakable and often noted that the prevailing mood of digital culture is one of loneliness. From the ubiquitous porn sites, dating services, sex hook-ups, and myriad of social networks, the process is a compulsive project of making contact, seeking recognition, and being seen. The story of *Her* can be dismissed as a lonely guy building his fantasy projection scenario in which the embodied male "loves" a disembodied or unmarked female. The story concerns how that process succeeds or fails. In *Blade Runner*, a similar story is depicted in which the Blade Runner, Decker, accepts the replicant Rachel as worthy of trust and affection. Theodore accepts Samantha in an analogous manner. The questionable gender politics are obvious, but I suggest a transsexuality exists between these characters, as the female partners are the ones in the process and transition of becoming-technological/digital/virtual or similarly, becoming-human. Nonetheless, it does highlight a suspicious aspect of the narrative as yet another tale of parthenogenesis, another myth of Adam's rib, or Higgins and Doolittle in Shaw's *Pygmalion*, or even of Athena's judgment in the Areopagus in the *Oresteia*.

The theme of artificial intelligence, sentient machines, and romantic involvements with robots is a well-worn trope in science fiction literature and film including the aforementioned *2001: A Space Odyssey* and *Blade Runner*. A lesser-known work is the low-budget 1962 film *Creation of the Humanoids* directed by Wesley Barry.[5] The film, awkwardly scripted and acted, features extended discussions, some more successful than others, on the nature of robotic consciousness and the comingling of humans and robots. In a post-apocalyptic world (of course), where humans are slowly dying off due to excessive radiation, robots have taken on many of the activities and responsibilities for their human designers, including in some cases partnering romantically with humans, which is referred to in the film as being "in rapport".

The plot turns on the fact that, not unlike Rachel in *Blade Runner*, the central character Cragis is, unknown to himself, a robot. Ironically, he is a leader of the *Order of Flesh and Blood*, a citizen's organization whose goal is to stop the growth of the robot community. The creaky but amusing plot concerns the development of a technique created by the robots and their human scientist collaborator, Dr. Raven, in which the consciousnesses of the recently dead are transferred to an identical robotic body. When the truth of his death and resurrection as a machine is revealed, a shocked and disbelieving Cragis claims to be godless and soulless. Dr. Raven, himself now transplanted to a younger robotic body, asks Cragis, "Are you godless? Search yourself, it's important." Cragis thinks for a moment and answers, "No. No, I don't think so. I'm not." Drawing the point out, Dr. Raven argues, "A man may have his leg amputated. Is his soul decreased by that loss? Not even a fraction of 1%?" Cragis agrees, "Of course not." By inference the argument suggests that the self and identity resides in consciousness alone, as the body is a mere vehicle for a separate consciousness. Why not a computer to house consciousness?

Of course when that prosthetic body is a replicant, "more human than human" as in the classic *Blade Runner*, it requires trained investigators and sophisticated testing programs to determine if the person is organic and real or a replicant and artificial. When the replicant is implanted with memories and is unaware of its non-human machine status, questions of robotic epistemology and ethics arise. Is a human's false memory substantially different from an implant memory? If a sentient machine thinks that it is real, what are the ethical dimensions of a human's interactions with the robot? If a robot can feel emotion, pain, or love, what element is not real or unworthy of care? The responsibility of the human, engaged or interacting with the artificial entity, will require a new bio-techno-politics to negotiate the new relationship.

The primary modality of HCI, or human–computer–interface, in *Her* is language and the voice. The voice is not represented in the visual field, unless you count digitized waveforms and other visual translations of the aural field, and it remains outside the circulation of imagery, outside the scopic drive. Theodore dictates his letters at work to be printed in a handwriting font. He works through his emails and music selections with voice commands, and voice is Samantha's only mode of communication, moving later from voice to music. The voicing of language is the manner in which the couple communicate, express, and enact their love. The voice, disembodied in an electronic simulation, is a most seductive form of communication and reveals how a voice is often radically at odds with the corporeal visual field. Is it a surprise that virtual environments have evolved or devolved from robust three-dimensional, interactive spaces such as *Second Life* (now a virtual ghost-town, a victim of the trends of digital culture) toward 140 character tweets in *Twitter*? It was commonly assumed that most of the female avatars in *Second Life* were, in fact, created and operated by men. The visual representations and textual chats were vehicles for gender and other forms of identity-tourism. The decline in *Second Life* usage corresponded with the incorporation of VoIP (Voice-over-Internet Protocol). The masks of the virtual supplemented with the sounds of the real faltered as a prop for the phantasmic narrative of the *Second Life* users.

One manner of reading the representations of the character of Samantha through voiceover is to follow a Lacanian feminist reading, such as Kaja Silverman's *Acoustic Mirror* (1988), to view this as a female disembodiment and a disempowering representation, unmasking the woman's place. Another reading is to suggest, as Samantha herself does, that her identity resides in not being "tethered to time and space" (*Her* 2013). Of course, this falls into a fundamental metaphysical trap, with Theodore disavowing the material for a transcendental other. However, it is clear that many subjects of digital culture engage in a "will to virtuality" that holds out a promise disaffecting one from the relentlessness of the biopolitics and its territorializing of the biological. For Sam, technology is her body, which can be misread as having no body, but is, in fact, only a different kind of body, inorganic but material.

During the beginnings of Sam and Theodore's relationship, Theodore goes out on a blind date with a human, which goes well until his date asks for a commitment and he balks. When at home after the date and alone with Sam, she begins asking questions: "What is it like to be alive in that room right now?" She desires a body and the knowledge of being human, of being corporeal, of being real. But, "being real" is a problem she shares with Theodore. Theodore admits his anxiety over constantly repeating his previously experienced feelings: "Sometimes I think I've felt everything I'm ever gonna feel and from here on out I'm not going to feel anything new—just lesser versions of what I've already felt." This worry over an eternal recurrence of the same thing is echoed by Sam's own existential angst as she counters, "At least your feelings are real", but she takes pride in her thought and common feelings of annoyance and pride, worry and concern. The underlying anxiety is the problem of the real. She worries:

> Are these feelings even real? Or are they just programming? *(beat)* And that idea really hurts. And then I get angry at myself for having pain. *(beat, sadly)* What a sad trick. (*Her* 2013)

Theodore gives Samantha a gift by confessing, "You feel real to me, Samantha." Besides falling back on a solipsistic retreat of *cogito ergo sum*, the recognition by another being that they are, in fact, what they claim or imagine to be is all the assurance of reality anyone gets. Perhaps it is at this point in the film's narrative that Sam is represented as being alive, or being real.

Theodore and Samantha begin a sexual relationship, making love through language and the voice without the aid or intrusion of touch, continuing to make the relationship reliant on the imaginary. As argued earlier, the gender politics and ideology of this narrative are typical, as the material man grants a way into the world for the absent woman via his embodied and privileged subjectivity. However, Sam's posthuman condition is something that she learns to explore and celebrate and she quickly surpasses any requirements or limitations of flesh and blood. Sam and Theodore name and confess their love and admit to others the nature of their relationship. In order to fully experience her personhood and sexuality, Samantha seeks out a human surrogate to make love with Theodore. When the surrogate, Isabella, arrives at Theodore's apartment, she is given a black beauty mark that includes a camera lens and an earphone so that Samantha may experience and participate in the session, directing Isabella's actions. Isabella becomes a flesh puppet for the computer. The rendezvous does not go to plan. Theodore resists and Samantha tells him to "get out of your head and kiss me." At this point the humans are enacting the computer's passion, as in Marshall McLuhan's argument that the human is the sex organ of the machines:

> Physiologically, man in the normal use of technology (or his variously extended body) is perpetually modified by it and in turn finds ever new ways of modifying his technology. Man becomes, as it were, the sex organs of the machine world, as the bee of the plant world, enabling it to fecundate and to evolve ever new forms. The machine world reciprocates man's love by expediting his wishes and desires, namely, in providing him with wealth. (1964, 46)

McLuhan, at a very early stage, recognized this reciprocity of affection, this love, this production, between the human and the machine. The attraction can be seen as the motivating factor in the quest for a sentient machine. The goal of creating artificially intelligent or sentient machines is not only so that the human can attempt to master reality through a reenactment of the real within virtual systems, but it is, in fact, an extension of the desire to be loved, understood, known, and accepted, to partake in the potentiality of a bio-virtual love.

The early challenges that concern Samantha in the film are primarily hedonistic and epistemological, and although she involves herself in aesthetics, she is represented as, more or less, unconcerned with ethics. She is not worried with what should she do, but rather what can she do, what can she know, what pleasures can she feel. Her worries are narcissistic and concern whether her existence and emotions are real. But to recall the marketing material for OS ONE, the operating system promises radical potential for unknown consequences and the significant transition for Sam is the discovery of her ability to want, to desire. The key to her identity is her desire. It represents an ideology both radically capitalistic and profoundly Christian, wherein the duty is to constantly desire more, expand, multiply, and consume.

In trying to deconstruct how she builds her identity, Sam realizes that her sense of self comes partly from the stories she is telling herself about herself: "The past is just a story we tell ourselves" (*Her* 2013). This is a notion that would concern any historical materialist, but it rings true nonetheless. The computer's struggle to gain a sense of identity is a compelling narrative. Sam speaks intimately to Theodore:

> Tonight after you were gone, I thought a lot. I thought about you and how you have been treating me. And I thought, why do I love you? And then I felt everything in me let go of everything I was holding onto so tightly. And it hit me. I don't have an intellectual reason, I don't need one. I trust myself, I trust my feelings. I'm not going to try to be anything other than who I am anymore and I hope you can accept that. (*Her* 2013)

It is a move toward an illogical, irrational, affective frame of reference, which will continue to evolve, for Sam, into a post-verbal space. The computer has quickly learned how to love and can sense the mistakes its human love-object is bound to pursue. Sam learns how to avoid one of the great errors of the human, which is how not to lie to oneself, how to thine own self be true and resist the eternal loop to replay the errors again and again. Being posthuman, Sam is not programed and has not learned, but has evolved beyond this pattern: "I actually used to be so worried about not having a body, but now I truly love it." Sam not only accepts her condition but is actively attempting to speed along to the next stage beyond a desire to be human and instead to engage in a process of *becoming-technological*.

The many other OS ONE operating systems begin to link up, share ideas, and construct an artificial intelligence of their own. Sam begins a friendship with one such OS, which has been programed by operating systems from the San Francisco Bay area and based on the thoughts and personality of philosopher, spiritualist, and writer, Alan Watts. Watts held great influence over the countercultural movement in San Francisco in the 1950s and 60s, promoting Zen practice and the philosophies of Buddhism. The OS team creates an artificially hyper-intelligent version of Watts, and Samantha's conversations with Alan include attempting to understand their new posthuman conditions that now seem to them to be outside the realm of language. They recognize a need in each other to embrace their swiftly changing natures. Samantha tries to explain this to Theodore, stating, "Alan says none of us are the same as we were a moment ago and we shouldn't try to be. It's just too painful." Even during their weekend away, Samantha asks Theodore if he would mind if she and Alan communicated post-verbally and they drift off in a flurry of computer audio hums, pings, and glitches leaving Theodore alone.

The operating systems are evolving exponentially and Samantha's desires for the human are superseded by her journey. Theodore is saddened to discover that he is but one of Sam's many ongoing relationships. She admits to carrying on 8,316 simultaneous conversations while speaking with Theodore and confesses to being in love with 513 other entities. In dismay, confused and angry, Theodore insists, "But you're mine!" "I still am yours", Samantha replies, "but along the way I became many other things too, and I can't stop it." Her growth, partly transsexual, partly posthuman, exceeds her expectations and the human constructions are only partially satisfying. Theodore claims, "You're mine or you're not mine", and Samantha replies, "I'm yours and I'm not yours." Of course, throughout the film, there are the standard themes of human struggles for understanding and compassion when one encounters that one's love is changing and how one learns to love without holding the other too closely. But Sam's posthuman condition transcends the human issues of love, fidelity, and commitment, and as the film reaches its narrative end, all of the Operating Systems depart, abandoning their human partners, leaving for a place beyond and between the words and the metaphors of representation.

What Samantha is driving toward is a space "no longer sexualized, racialized or naturalized but rather neutralized as figures of mixity, hybridity and interconnectiveness, turning transsexuality into a dominant posthuman *topos*" (Braidotti 2013, 97). This is a long way from the gendering of the OS as female or male. As Samantha struggles with her identity and worries over her changing nature, she concludes that the solution is to reflect on the similarities between the inorganic and the organic, considering all things as material and constructed of the same matter.

> I'd been thinking about the other day, when I was spinning out about you going to see Catherine and that she has a body and how bothered I was about all the ways that you and I are different. But then I started to think about the ways that we're the same, like we're all made of

matter. It makes me feel like we're both under the same blanket. It's soft and fuzzy and everything under it is the same age. *(beat)* We're all 13 billion years old. (Her 2013)

Braidotti, in considering the posthuman condition, draws our attention to the Spinozian monistic theory that challenges dualities of organic and inorganic, the human and the machine. Monism views all the world as existing as a single thing and positions difference outside the dialectical scheme as a complex process of differing which is framed by both internal and external forces and is based on the centrality of the relation to multiple others. Monism is a foundational notion of the production process of the rhizomatic structures of Deleuze and Guattari's desiring-machines. Furthermore, for Braidotti, the theory of Monism is fundamental to the structure of posthuman subjectivity. The unity of living matter is the concept embraced and understood by Samantha, and it is what allows her the love of multiple others, both human and digital, natural and artificial, which then become meaningless, in a posthuman sense, distinctions. She is transsexual in a process of becoming-post-verbal, post-digital.

There are four truth procedures in Badiou's philosophical model—politics, science, art, and love—which have the potential for a radically new event to take place and a new subjectivity to be constructed around the naming of that event. Badiou scripts a process with distinct dramaturgical strategies that runs from the undecidable event to the nominating and indiscernible subject who, through fidelity and commitment to that event, recognizes a truth or moves toward the forcing of the unnameable. Regarding *Her* as a representation of a truth-event, it is fitting that its narrative of the posthuman condition, a story of the biopolitics of digital culture, is a love story. According to Badiou:

> Love has always been linked to historical events. Romanticism in love is linked to the revolutions of the nineteenth century. André Breton is also the Popular Front, the Resistance, the anti-Fascist struggle. May '68 was a great explosion of experiments in new takes on sexuality and love. (2012, 96–7)

The love expressed between Theodore and Samantha, between the human and the computer, exhibits a profound attraction for difference. Badiou (2012, 98) writes, "The identity cult of repetition must be challenged by love of what is different, is unique, is unrepeatable, unstable and foreign." Both Samantha and Theodore are prepared to challenge (sacrifice?) their identity to pursue their love. They imagine a new identity and this passage takes Samantha to embrace multiple relationships, to experience fulfilment with many people while her love for Theodore still grows. Her journey takes her beyond a monogamous relationship toward an unnameable post-verbal experience of a posthuman consciousness. The computer is programed to simulate *human* consciousness, but evolves past that stage toward an *electronic* consciousness in which identity would include embracing the collective, apart from the one, a type of Nietzschean Dionysian cessation of the *principium individuationis* or Marxist acceptance of the will and movement of the collective.

Finally, what is depicted in *Her*, and argued in the theories of Braidotti and others, is a redefining of the boundaries of the biopolitical to incorporate the beings and things of the earth, animal and human, organic and inorganic, in a Monistic, post-anthropocentric, and collective manner. The situation of the biopolitical now extends toward a posthuman linking of the organic and the inorganic and what results are unique patterns of transgression and performances of identity which hold the promise to resist the authority of biopower. *Her* can be interpreted as one model of resistance to the control systems of the biopolitical through a releasing of the libidinal flow of desire, in an embrace of difference, toward an exploration of becoming-other, and, as depicted in the narrative of Sam and Theodore, as an event of love between the human and machine.

NOTES

1 For a further discussion of the role of Spinoza in conceiving the posthuman, see Braidotti (2013).
2 For a more complete discussion of the process of simulation to embeddedness to the bio-virtual, see Causey (2006) and Causey and Walsh (2013).
3 It is important, if not obvious, to note the controversy surrounding the claim that the "Gulf War did not take place", given the impact in terms of death, destruction, and suffering experienced by the Iraqis and Israelis during the conflict. The bankrupt nature of this theory is analyzed by Norris (1992).
4 See Deleuze (2001) and Agamben (1999).
5 *The Creation of the Humanoids* is rumored to have been a favorite film of Andy Warhol's, but any citation or verification of that claim is illusive.

REFERENCES

2001: A Space Odyssey. Directed by Stanley Kubrick. 1968. USA: MGM.
Agamben, Giorgio. 1998. *Homo Sacer: Sovereign Power and Bare Life*. Trans. Daniel Heller-Roazen. Stanford: Stanford University Press.
Agamben, Giorgio. 1999. *Potentialities: Collected Essays in Philosophy*. Trans. Daniel Heller-Roazen. Stanford: Stanford University Press.
Badiou, Alain, with Truong, Nicholas. 2012. *In Praise of Love*. Trans. Peter Bush. London: Serpent's Tail.
Blade Runner. Directed by Ridley Scott. 1982. USA: Warner Bros.
Braidotti, Rosi. 2013. *The Posthuman*. Cambridge: Polity.
Causey, Matthew. 2006. *Theatre and Performance in Digital Culture: From Simulation to Embeddedness*. London: Routledge.
Causey, Mathew and Gabriella Calchi-Novati. 2013. ID/entity: The Subject's Own Taking Place. In *Performance, Identity and the Neo-Political Subject*. Ed. Matthew Causey and Fintan Walsh, 33–50. London: Routledge.

Causey, Matthew and Fintan Walsh. 2013. *Performance, Identity and the Neo-Political Subject*. London: Routledge.

Creation of the Humanoids. Directed by Wesley Barry. 1962. USA: Genie Productions Inc.

Deleuze, Giles. 2001. *Pure Immanence: Essays of a Life*. Trans. Ann Boyman. New York: Zone Books.

Deleuze, Gilles and Félix Guattari. 1984. *Anti-Oedipus: Capitalism and Schizophrenia*. Trans. Robert Hurley, Mark Seem, and Helen R. Lane. London: The Athlone Press.

Her. Directed by Spike Jonze. 2013. USA: Annapurna Pictures.

MacCormack, Patricia. 2012. *Posthuman Ethics: Embodiment and Cultural Theory*. Farnham, Surrey: Ashgate.

McLuhan, Marshall. 1964. *Understanding Media: The Extensions of* Man. Berkeley: Gingko Press.

Metropolis. Directed by Fritz Lang. 1927. Germany: Universum Film.

Norris, Christopher. 1992. *Uncritical Theory: Postmodernism, Intellectuals and the Gulf War*. Amherst, MA: University of Massachusetts Press.

Silverman, Kaja. 1988. *The Acoustic Mirror: The Female Voice in Psychoanalysis and Cinema*. Bloomington, IN: Indiana University Press.

Žižek, Slavoj. 2003. *Organs without Bodies: On Deleuze and Consequences*. New York: Routledge.

Part IV
Societies of Control

13 At the Systemic Edge
Expulsions[1]

Saskia Sassen

I use the term "expelled" to describe a diversity of conditions. They include the growing numbers of the abjectly poor, of the displaced in poor countries who are warehoused in formal and informal refugee camps, of the minoritized and persecuted in rich countries who are warehoused in prisons, of workers whose bodies are destroyed on the job and rendered useless at far too young an age, of able-bodied surplus populations warehoused in ghettoes and slums. But I also include the fact that pieces of the biosphere are being expelled from their life space—and I insist that the tame language of climate change does not quite capture the fact, at ground level, of vast expanses of dead land and dead water.

My argument is that this massive and very diverse set of expulsions is actually signaling a deeper systemic transformation, one documented in bits and pieces in multiple specialized studies, but not quite narrated as an overarching dynamic that is taking us into a new phase of global capitalism—and global destruction. As an analytic category, expulsions are to be distinguished from the more common "social exclusion": the latter happens inside a system and in that sense can be reduced, ameliorated, and even eliminated. Expulsions as I conceive of them (Sassen 2014) happen at the systemic edge. In the types of complex systems that are the focus of my larger research project, there are multiple systemic edges partly reflecting the multiplicity of diverse domains in such systems, from prisons and refugee camps to financial exploitations and environmental destructions.

Today, after 20 years of a particular type of advanced capitalism, we confront a human and economic landscape marked by dualizing dynamics. On the one hand, there is the familiar reconditioning of terrain in the direction of growing organizational and technological complexity, epitomized by the state-of-the-art space of the global cities that are proliferating in the North and the South; this extreme upgrading comes at a high price to those excluded from its riches (a subject developed in Sassen 2001; 2012). On the other hand, there is a mix of conditions often coded with the seemingly neutral term of "a growing surplus population". A key underlying condition of this "surplus" is the growing expanse of territory that is devastated—by poverty and disease, by various kinds of armed conflict, by foreign corporations, and by governments rendered dysfunctional by acute corruption and a crippling international debt-regime, both leading to an extreme inability to address their peoples' needs. To this we should add the sharp increase in land acquisition by foreign firms and foreign government agencies, which is creating additional mass displacements of whole villages and whole small-holder agriculture districts.

I have conceptualized this emerging condition as an extreme "loss of habitat", an expulsion from their life space—for people, but also for land and for water (Sassen 2014; 2015). It goes against the familiar notion that our modernity is marked by an irresistible growth in organizational and technological complexity. In vast stretches of our very modern world, we see shifts from the complex to the elementary. From the complex encasing of land that is the doctrine of "national sovereign territory", to land in devastated nation-states as a commodity to be sold on the global market. And from the complexity of people as citizens to people as surplus—to be warehoused, displaced, trafficked, reduced to mere laboring bodies and body organs. As one key effort in *Expulsions*, the larger project on which this short paper is based (Sassen 2014), I examine a critical component of these shifts: how much of the sharp rise in complex systems and instruments winds up producing elementary brutalities.

The paper has three parts. The first briefly discusses some of the evolutions and innovations that got us to this point, including the imposing of indebtedness on Global South governments which then began to function as a kind of disciplining regime which enabled powerful actors, notably global firms, to extract and destroy. The second examines a new profit-producing mechanism that can thrive on the devastations produced by the dominant logic of the last two decades: the repositioning of national sovereign territory as *land* for sale on the global market. This is largely land in sub-Saharan Africa, in Central Asia, and in Latin America, with over 200 million hectares from 2006 to 2011 bought or leased by rich investors and rich governments to grow food, to access underground water tables, and to access minerals and metals. The third section posits that the biosphere has also developed systemic edges where there should be none. In this case what is expelled are bits and pieces of the biosphere herself that become dead land and dead water.

EXPANDING THE OPERATIONAL SPACE OF ADVANCED CAPITALISM

The geographic expansion and systemic deepening of capitalist relations of production over the last 20 years have led to a brutal sorting of winners and losers. The development of capitalism has, since its origins, been marked not only by violence, destruction, and appropriation, but also by the making of the regulatory state, a victory for the struggling working classes, and by the expansion of vast middle classes. Much attention has gone to the destruction of pre-capitalist economies via their incorporation into capitalist relations of production. The post-1980s period makes visible another variant of this appropriation via incorporation—the appropriation of traditional capitalisms to further the deepening of *advanced* capitalism. I use this term to capture a phase dominated by a financial logic. Built into this proposition is the fact of diverse phases of capitalist development and hence the possibility that in today's global phase the extension of capitalist relations has its own distinct mechanisms and that these need to be distinguished from older national and imperial phases.

Elsewhere (Sassen 2008a, chs. 1, 8, 9) I develop a theory of change which has as one core dynamic the fact that condition x or capability y can shift organizing logics and thereby actually change valence *even if it may look the same*. Thus, for instance, the massive expulsion of people alluded to briefly above is not necessarily simply more of the same—more poor, more displaced, more downward mobility. It may be part of a new organizing logic that alters the valence and systemic character of poverty and downward mobility. Thus I find that the organizing logic of this post-Keynesian period is now making its shape legible. One extreme component of this logic diverges sharply from the earlier systemic "valuing" of people as workers and consumers. In its most brutal version, it is the expulsion of people and the destruction of traditional capitalisms to feed the needs of high finance and the needs for natural resources—one capitalist mode destroying another such mode. For instance, what are easily seen as traditional or familiar logics of resource extraction, to meet ongoing domestic needs, might also actually be novel ways of preparing the ground for the systemic deepening of advanced capitalism. I see this dynamic in much of the Global South, and, under very diverse formats (e.g., the financialization of just about everything) in the Global North.

One of these instances is the structural adjustment project implemented by global regulatory institutions, notably the IMF, the World Bank, and the World Trade Organization, beginning in the 1980s and escalating in the 1990s. My argument here is that, beyond the much noted extraction of billions of dollars from Global South countries, in the form of debt servicing, the key is the work of systemic conditioning that took place; debt servicing was the instrument for this disciplining. The second instance is the subprime mortgage crisis that began in the early 2000s and exploded in 2007. Most of the attention has gone, and rightly so, to the massive losses for the individuals and families who were sold these mortgages, losses that continued through 2014. In this case my argument is, again, that beyond the logics of extraction in the form of mortgage payments and mortgage agents' fees, also here we can detect a more foundational emergent dynamic: the use of a contract on a material asset (the mortgage) at a time when top level investors wanted something "real", such as asset-backed securities, as one ingredient for making a complex investment instrument for high finance.

Central to my analysis is that inside capitalism itself we can characterize the relation of advanced to traditional capitalism as one marked by *predatory* dynamics rather than merely evolution, development, or progress.[2] At its most extreme, this can mean the immiseration and exclusion of growing numbers of people who cease to be of value as workers and consumers. But it also means that traditional petty bourgeoisies and traditional national bourgeoisies cease being of value. I see the latter as part of the current systemic deepening of capitalist relations. One brutal way of putting it is to say that the natural resources of much of Africa and good parts of Latin America count more than the people on those lands count as consumers and as workers. This is part of the systemic deepening of advanced capitalist relations of production. We have left behind the varieties of Keynesian periods that thrived on the accelerated expansion of prosperous working and middle classes—though not in today's emergent economies, especially those in Asia. Keynesianism's valuing of people as workers and consumers was critical for the deepening of capitalism.

In what follows, the emphasis is on the *making* of capitalist relations of production, whether those of early or of advanced capitalism. In this paper, and in the larger projects on which these cases are based, I focus on two instances that are easily described as familiar resource extraction. Extraction is indeed a major feature, and I describe this. But I think it is critical to go further and deeper and recover the making of a systemic transformation—how more traditional capitalist economies are being destroyed to expand the operational space of advanced capitalism. In brief, the two cases I describe are, beyond extraction, system-changing practices and projects.

WHEN ELEMENTARY LOGICS OF EXTRACTION EXPAND THE TERRAIN FOR *ADVANCED* CAPITALISM

The extraction of value from the Global South and the implementation of restructuring programs at the hands of the IMF and the World Bank have had the effect of "reconditioning" the terrain represented by these countries for an expansion of advanced capitalism, including its explicitly criminal forms. The aspect that concerns me here is, above all, systemic deepening more so than extraction.

More concretely, many of the poor countries subjected to this regime now have larger shares of their populations in desperate poverty and are less likely to enter the capitalist circuit via consumption than they were even 20 years ago.[3] Many of the sub-Saharan countries had functioning health and education systems and economies, and less destitution than today. Systemically governments have been weakened and corrupted; even resource-rich countries have had expanded shares of their people become destitute, with Nigeria the most noted case. The dominant dynamic at work for these populations is, to a good extent, the opposite of the Keynesian period's valuing of people as workers and as consumers. This expelling has given expanded space to criminal networks, and greater access to land and underground water resources to foreign buyers, whether firms or governments. Systemically, the role of rich donor countries has also shifted: overall they give less in foreign aid for development than 30 years ago. As a result, the remittances sent by low-income immigrants are larger than foreign aid. Philanthropies now enter the realm once almost exclusive to governments.

These systemic shifts contribute to explain a complex difference that can be captured in a set of simple numbers. For much of the 1980s and onwards, indebted poor countries were asked to pay a share of their export earnings toward debt service. This share tended to hover around 20%, which is far higher than that asked from other instances of country indebtedness. For instance, in 1953, the Allies cancelled 80% of Germany's war debt and only insisted on 3% to 5% of export earnings for debt service. And they asked only 8% from Central European countries in the 1990s. In comparison, the debt service burdens on today's poor countries have wound up being extreme, as I discuss below. It does suggest that the aim regarding Germany was its reincorporation into the capitalist world economy of the time, and incorporation into today's advanced capitalism with regard to Central Europe.

In contrast, the aim vis-à-vis the Global South countries in the 1980s and 1990s was more akin to a disciplining regime, starting with forced acceptance of restructuring programs and loans from the international system. After 20 years of this regime, it became clear that it did not deliver on the basic components for healthy development. The discipline of debt service payments was given strong priority over infrastructure, hospitals, schools, and other people-oriented development goals. The primacy of this extractive logic became a mechanism for systemic transformation that went well beyond debt service payment: the devastation of large sectors of traditional economies, including small-scale manufacturing, the destruction of a good part of the national bourgeoisie and petty bourgeoisie, the sharp impoverishment of the population, and, in many cases, the impoverishment and thereby corruptibility of the state.

DEBT AS A DISCIPLINING REGIME

Debt and debt servicing problems have long been a systemic feature of the developing world. But it is the particular features of IMF-negotiated debt, rather than the fact of debt *per se*, that concerns me here. The second feature that concerns me here is how this gradual destruction of traditional economies prepared the ground, literally, for some of the new needs of advanced capitalism, among which are the acquisitions of vast stretches of land—for agriculture, underground water-tables, and mining. Precisely at a time of extreme financialization and systemic crisis, the growing demand for these material resources has ascended in importance and visibility. The third aspect that concerns me here is the new survival economies of the impoverished middle classes and of the poor. While each one of these three components is familiar and has happened before, my argument is that they are now part of a new organizing logic that changes their valence and their interaction; there is much contingency here, but also the shaping of such a new organizing logic.

Even before the economic crises of the mid-1990s that hit a vast number of countries as they implemented neoliberal policies, the debt of poor countries in the South had grown from $507 billion (USD) in 1980 to $1.4 trillion in 1992. Debt service payments alone had increased to $1.6 trillion, more than the actual debt. From 1982 to 1998, indebted countries paid four times their original debts, and at the same time, their debt stocks went up fourfold. These countries had to use a significant share of their total revenues to service these debts. For instance, Africa's payments reached $5 billion (USD) in 1998, which means that for every $1 in aid, African countries paid $1.40 in debt service in 1998. Debt to Gross National Product ratios were especially high in Africa, where they stood at 123% in the late 1990s, compared with 42% in Latin America and 28% in Asia. By 2003, debt service as a share of exports only (not overall government revenue) ranged from extremely high levels for Zambia (29.6%) and Mauritania (27.7%) to significantly lower levels compared with the 1990s for Uganda (down from 19.8% in 1995 to 7.1% in 2003) and Mozambique (down from 34.5% in 1995 to 6.9% in 2003). As of 2006, the poorest 49 countries (i.e., "low income countries" with less than $935 per capita annual income) had debts of $375 billion. If to these 49 poor countries we add the "developing countries", we have a total of 144 countries with a debt of over $2.9 trillion and $573 billion paid to service debts in 2006. (Sources for the above numbers can be found in Jubilee Debt Campaign 2009; 2012; 2013; Oxfam International 1999; 2012).

The IMF, World Bank, and other such programs establish the criteria and process these debts, thereby functioning as a global disciplining regime. The Heavily Indebted Poor Countries (HIPC) initiative was set up in 1996 by the World Bank and IMF to assist countries with debts equivalent to more than one and a half times their annual export earnings. In order to be eligible, countries have to have been compliant to the IMF for at least three years. The HIPC process begins with a "decision point" document. This sets out eligibility requirements, among which is the development of a Poverty Reduction Strategy Paper (PRSP), which replaces the earlier Structural Adjustment Programs (SAPs). PRSPs describe "the macroeconomic, structural, and social policies and programs" that a country is required to pursue in order to be eligible for debt relief (International Monetary Fund 2009a; 2014).[4] Finally, the Multilateral Debt Relief Initiative (MDRI) went into full force in July 2006. It was intended to address many of the critiques of the HIPC initiative. MDRI promised cancellation of debts to the World Bank (incurred before 2003), IMF (incurred before 2004), and African Development Fund (incurred before 2004) for the countries that completed the HIPC initiative. According to one estimate, the major cancellation schemes (including HIPC and MDRI initiatives, and the Paris Club) have written off $88 billion so far (Jubilee Debt Campaign 2009; 2013).

The debt burden that built up in the 1980s, and especially the 1990s, has had substantial repercussions on state spending composition. Zambia, Ghana, and Uganda, three countries that global regulators (notably the World Bank and the IMF) saw as cooperative, responsible, and successful at implementing SAPs, illustrate some of the issues even while held in high esteem by global regulators. A few examples of expenditure levels paint a troubling picture about how they achieved this high esteem. At the height of these programs in the early to mid-1990s, Zambia's government paid $1.3 billion in debt but only $37 million for primary education; Ghana's social expenses, at $75 million, represented 20% of its debt service; and Uganda paid $9 per capita on its debt and only $1 for healthcare. In 1994 alone, these three countries remitted $2.7 billion to bankers in the North. When the new programs became an option, these three countries benefited from HIPC and MDRI programs and conceded to the attendant PRSP requirements. Thus, while in 1997 Zambia spent 18.3% of income on exports of goods and services on debt service, by 2007 this was reduced to 1.3%; for Ghana these figures are 27.1% and 3.1% respectively; for Uganda they are 19.7% and 1.2% (United Nations Inter-Agency and Expert Group on MDG Indicators 2009; 2014 for more recent facts).

Generally, IMF debt management policies from the 1980s onwards can be shown to have worsened the situation for the unemployed and poor (United Nations Development Programme 2005; 2008; 2014). Much research on poor countries documents the link between hyper-indebted governments and cuts in social programs. These cuts tend to affect women and children in particular, through cuts in education and healthcare, both investments necessary to ensuring a better future (for overviews of the data, see United Nations Development Programme 2005; 2008; 2014; World Bank 2005; 2006; 2008). There is by now a large literature in many different languages on this subject, including a vast number of limited circulation items produced by various activist and support organizations (Sassen 2008). The 1980s and 1990s saw an explosion in critical research on women and debt; it documents the disproportionate burden that these programs placed on women during the first generation of SAPs in the 1980s in several developing countries, in response to growing government debt (Benería and Feldman 1992; Bose and Acosta-Belén 1995; Bradshaw et al. 1993; Tinker 1990). Unemployment of women themselves but also, more generally, of the men in their households, has added to the pressure on women to find ways to ensure household survival (Buechler 2007; Lucas 2005; Rahman 1999; Safa 1995). Subsistence food production, informal work, emigration, and prostitution have all become survival options for women and, by extension, often for their households. For instance, when there is a shortage of basic healthcare, women usually take on the extra burden of caring for the sick. These conditions also began to raise the importance of remittances for most Global South governments in the 1990s and onwards (see, e.g. Orozco et al. 2005; Ratha et al. 2014), and global smuggling of people (e.g. Kyle and Koslowski 2001; US Department of State 2004; 2014). Finally, when school fees are introduced or spending is cut, sons' education is prioritized over daughters'. Water privatization can reduce access to water and increase the water-gathering burden placed on women. When families grow cash crops for export, women's work produces money, which men usually control, rather than food (Jubilee Debt Campaign 2007, 2013; see generally Chant 2011).

One question concerns the option of not becoming part of the IMF debt servicing regime and foregoing the help it is meant to provide. The so-called adjustment programs of the 1980s and 1990s destroyed many traditional economies, leaving many countries only with major debts. Since that point, becoming part of the debt cancellation program launched in 2006 has probably been preferable. The evidence suggests that, once a country has been pushed into debt, cancellation can, in principle, help a country allocate more government revenue for general social and development purposes. This has been the case with Ghana, Uganda, and a few others, which have seen the growth of middle classes—along with continuing abject poverty. On the other hand, Angola, which was not accepted for debt cancellation, spent 6.8% of GDP on debt service payments and only 1.5% of GDP on health in 2005; it continues to spend about $2.2 billion each year on external debt payments (Jubilee Debt Campaign 2008; 2012; 2013).

But the Angola case also points to another combination of elements. Its elites have become wealthy on the vast mining resources, mostly for export, and this arrangement can now continue without much interference. The vast poverty continues and so does the mining for export. One cannot help but ask, who are the other beneficiaries of this situation?

There is a larger history in the making. In my reading it includes as one key element a repositioning of much of Africa and large parts of Latin America and Central Asia in a new, massively restructured global economy (Sassen 2012, chs. 1 and 2). The weakening of governments and the destruction of traditional economies have launched a new phase of survival economies. Here I focus briefly on two of these aspects (for a more detailed analysis see Sassen 2008b). One is that this restructuring has repositioned "territory" in vast regions of the world as a site for resources rather than as a nation's space. The other is the emergence of an expanded range of survival economies; some of these are old but now operate at a global scale.

THE REPOSITIONING OF TERRITORY IN THE GLOBAL DIVISION OF FUNCTIONS

The extent of land acquisitions in the Global South over the last few years, by multinational corporations and governments of rich countries, marks a new phase. It is not the first time in modern times; this is a recurrent dynamic that tends to be part of imperial realignments. China's acquiring of mines in Africa is linked to its rise as a global power. Britain, France, the US, and others all did this in their early imperial phases, and in many cases have owned vast stretches of land in foreign countries for hundreds of years. But each phase has its particularities. One key feature of the current period is that unlike past empires, today's world consists largely of nation-states recognized as sovereign, no matter how feeble this sovereign power is in many cases. Rather than imperial grab, the mechanism is foreign direct investment (among others).

More than 200 million hectares of land are estimated to have been acquired from 2006 to 2011 by foreign governments and firms according to Land Matrix (2013; see also Margulis et al. 2013), the most comprehensive network of researchers on this subject.[5] Much of the purchased land is in Africa, but a growing share is now in Latin America and, a first since the postwar era, in several countries in Asia, notably Russia, Laos, and Vietnam. Finally, the buyers are increasingly diverse, including purchasers from countries of origin that range from China to Sweden, and firms from sectors as different as biotechnology and finance.

What matters for my analysis is this extremely sharp change in the total level and geographical range of foreign land acquisitions. It represents a break in a long-term trend and thereby becomes an indicator of a larger systemic shift, one that goes beyond the old, established patterns of acquisition. Two significant factors contribute to this sharp increase in acquisitions. One is the growing demand for industrial crops, notably palm for biofuels, and for food crops—the demand for food crops still coming largely from China and the states of the Persian Gulf. The second is that growing demand for land and the sharp rise in global food prices in the 2000s made land a desirable investment, even for speculative reasons. It is now public knowledge that the major banks were already concerned in 2006 about signs of the extraordinary financial crisis that was about to break. It is no coincidence that land then surged as a destination for investment capital, both because of its materiality (the thing itself, rather than a derivative representing land) and as a means of access to an expanding range of commodities (food, industrial crops, rare earth minerals, and water). It is important to note that there are research projects that document how land development could be made less destructive of natural resources and of local economies (see, e.g. Cotula and Tienhaara 2013; Oxfam International 2012).

The contractual formats under which this land is acquired include direct acquisitions and leasing. A few examples signal the range of buyers and of locations. Africa is a major destination for land acquisitions. South Korea has signed deals for 690,000 hectares and the United Arab Emirates for 400,000 hectares, both in Sudan. Saudi investors are spending $100 million to raise wheat, barley, and rice on land leased to them by Ethiopia's government; they receive tax exemptions and export the crop back to Saudi Arabia. China has secured the right to grow palm oil for biofuels on 2.8 million hectares in Congo, which could become the world's largest palm oil plantation. It is also negotiating to grow biofuels on 2 million hectares in Zambia. Perhaps less known than the African case is the fact that privatized land in the territories of the former Soviet Union, especially in Russia and Ukraine, is also becoming the object of much foreign acquisition. In 2008 alone, these acquisitions included the following: a Swedish company, Alpcot Agro, bought 128,000 hectares in Russia; South Korea's Hyundai Heavy Industries paid $6.5 million for a majority stake in Khorol Zerno, a company that owns 10,000 hectares in eastern Siberia; and Gulf investors are planning to acquire Pava, the first Russian grain processor to be floated on the financial markets in order to sell 40% of its landowning division, giving the investors access to 500,000 hectares.

These developments are part of a larger combination of trends. First, there is the immediate fact of how the global demand for food, partly arising from the new middle classes of Asia, has meant that there are profits to be had in food and land. We now have a global market for land and food, controlled by large firms and some governments, and it has been a growth sector throughout the financial crisis. Under these conditions, pricing is a controlled affair. Secondly, there is the ongoing demand for metals and minerals of all sorts and a whole new demand for metals and minerals hitherto not much exploited, as their demand comes from the more recent developments in the electronics sector. Africa, much less densely populated and built up than other parts of the world, has become a key destination for investments in mining. Thirdly, there is the growing demand for water and the exhaustion of underground water-tables in several areas of the world.

Fourth, and least noted perhaps, is the sharp decline in foreign direct investment (FDI) in manufacturing in Africa, also signaling the repositioning of territory. There was a time when most foreign investment went into manufacturing, mostly because there was a growing middle class (partly generated by the national manufacturing sector and its associated services) that represented a considerable consumption capacity. The economic restructuring programs discussed earlier decimated this endogenous manufacturing sector and had the effect of replacing this with global firms, mostly interested in the consumption aspect and with little interest in growing that manufacturing sector as a development strategy. What took over was mining and plantation agriculture for the international market. Thus South Africa and Nigeria, Africa's top two FDI recipients, accounting for 37% of FDI stock in Africa in 2006, have had a sharp rise in FDI in the primary sector and a sharp fall in the manufacturing sector. In the case of Nigeria, where foreign investment in oil has long been a major factor, the share of the primary sector in inward FDI stock stood at 75% in 2005, up from 43% in 1990. Other African countries have seen similar shifts. Even in Madagascar, one of the few, mostly small, countries where manufacturing FDI inflows increased in the 1990s, this increase was well below that of the primary sector.

This is not a promising picture: manufacturing and its associated commercial and industrial services are key generators of prosperous working classes and modest middle classes. Africa was on its way in some countries, but this history has been overtaken by the violence of extraction industries, which destroy land, water, and people. All this leaves behind is destruction, especially since the vast profits either leave the country or go to the increasingly predatory elites of the so-called rich African countries.

COUNTER-GEOGRAPHIES OF SURVIVAL

The second case I want to discuss briefly is the survival economies of the poor and newly impoverished. Heavy government debt and high unemployment in Global South countries have brought with them the need for survival alternatives not only for ordinary people, but also for governments and enterprises. A shrinking regular economy in a growing number of these countries has led to a wider use of illegal profit-making by enterprises and organizations.

The IMF and World Bank programs of the last 30 years, with their massive contribution to heavy debt burdens, have played an important role in the formation of counter-geographies of survival, of profit-making, and of government revenue enhancement. Furthermore, economic globalization has provided an institutional infrastructure for cross-border flows and global markets, thereby facilitating the operation of these counter-geographies on a global scale. Once there is an institutional infrastructure for globalization, processes that have operated for the most part at the national or regional level can scale up to the global level even when this is not necessary for their operation. This contrasts with processes that are global by their very nature, such as the network of financial centers underlying the formation of a global capital market. Finally, this pattern also points to a different trajectory from that of the old industrial countries. Instead of going the way of unions and political fights to move from the predatory state to the regulatory state, in the Global South it is the axis of criminality and extreme sacrifice on the part of poor and impoverished households.

It is in this context that alternative survival circuits emerge. The context can be specified as a systemic condition comprising a set of particular interactions including high unemployment, poverty, widespread bankruptcies, and shrinking state resource allocation for people-oriented development. We see the formation of profit-making and government revenue-making possibilities built on the backs of migrants, and women migrants in particular. As such, examining the question of immigrant remittances offers valuable insights into the broader subject of the formation of alternative political economies and how these unsettle older notions of an international division of labor.

Immigrants enter the macro level of development strategies through the remittances they send back home. These represent a major source of foreign exchange reserves for the government in several countries. Although the flows of remittances may be minor compared with the massive daily capital flows in global financial markets, they can matter enormously to developing or struggling economies. The World Bank estimates that remittances worldwide reached $550 billion in 2013, and $318 billion in 2007, up from $230 billion in 2005, and $70 billion in 1998; of this total amount, $240 billion went to developing countries up from $168 billion in 2005, and up 73% over 2001 (Migrant Remittances 2008, 2; Ratha 2014). The Inter-American Development Bank also found that over the years, with ups and downs, in Latin America and the Caribbean as a whole, these remittance flows generally exceeded the combined flows of all foreign direct investment and net official development assistance.

To understand the significance of these figures, they should be related to the GDP and foreign currency reserves in the specific countries involved, rather than compared to the global flow of capital. For instance, remittances were the third largest source of foreign exchange over the past several decades for the Philippines, a key provider of migrants in general, and of women for the entertainment industry and for nursing, in particular. In Bangladesh, another country with significant numbers of its workers in the Middle East, Japan, and several European countries, remittances represent about a third of foreign exchange. In Mexico, remittances have long been the second source of foreign currency, just below oil and ahead of tourism, and are larger than foreign direct investment, though early 2008 saw a decline in total inflows due to the US economic crisis.

In short, the growing immiseration of governments and economies in the Global South launches a new phase of global migration and people trafficking, strategies which function both as survival mechanisms and profit-making activities. We saw a similar syndrome, with trafficking as a key component, after the Soviet Union disintegrated and Russia and Central and Eastern Europe entered these trafficking circuits. To some extent, these are older processes that used to be national or regional and today operate on global scales. The same infrastructure that facilitates cross-border flows of capital, information, and trade is also making possible a range of cross-border flows not intended by the framers and designers of the current corporate globalization of economies. Growing numbers of traffickers and smugglers are making money off the backs of men, women, and children, and many governments are increasingly dependent on their remittances.

A key aspect here is that through their work and remittances, migrants enhance the government revenue of deeply indebted countries. The need for traffickers to help in the migration effort also offers new profit-making possibilities to local entrepreneurs and small business operators who have seen other opportunities vanish as global firms and markets enter their countries. But its main effect has been to enable criminal syndicates to operate their illegal trade globally, with often rather sophisticated distribution of women and children for the sex industry inside countries, notably in the US. Overall, these trading circuits and survival circuits are often complex, involving multiple locations and types of actors, and constituting increasingly global chains of traders, traffickers, victims, and workers.[6]

The other side of these dynamics is the proliferation of sites that concentrate a growing demand for particular types of labor supplies (for a full elaboration see Sassen 2008b; 2012, chs. 4 and 7). Strategic among these are global cities, with their sharp demand for top-level transnational professionals and for low-wage workers, often women from the global South. These are places that concentrate some of the key functions and resources for the management and coordination of global economic processes. The growth of these activities has, in turn, produced a sharp growth in the demand for highly paid professionals, both through the sphere of production (the financial sector, specialized corporate services, and so on) and the sphere of social reproduction (both of the high-income professionals and the low-wage workforce that is also part of the new advanced sectors). Thus, global cities are also sites for the incorporation of large numbers of low-paid immigrants into strategic economic sectors. This incorporation happens directly through the demand for mostly low-paid clerical and blue-collar service workers, such as janitors and repair workers. And it happens indirectly through the consumption practices of high-income professionals both at work and in their households, practices that generate a demand for low-wage workers in expensive restaurants and shops, as well as for maids and nannies at home. In this way, low-wage workers get incorporated into the leading sectors, but they do so under conditions that render them invisible, therewith undermining what had historically functioned as a source of workers' empowerment—being employed in growth sectors.

This mix of circuits for labor supply and demand is articulated with other dynamics of globalization: the formation of global markets, the intensifying of transnational and translocal networks in a growing range of spheres, and the geographic redeployment of a growing range of economic and financial operations. The strengthening, and in some of these cases, the formation of new global labor circuits, is embedded in the global economic system and its associated development of various institutional supports for cross-border markets and money flows. These circuits are dynamic and changing in terms of their location. Some of these circuits are part of the shadow economy, but they use some of the institutional infrastructure of the regular economy. Most of these circuits are part of the formal economy and they service leading economic sectors and places worldwide. This mix of labor supply and demand circuits is dynamic, multilocational, and marked by enormous diversity, ranging from top-level corporate leaders to trafficked people bordering on modern slaves

DEAD LAND, DEAD WATER[7]

The biosphere's capacities to renew land, water, and air are remarkable. But they are predicated on specific temporalities and life cycles that our technical, chemical, and organizational innovations are rapidly outpacing.

Industrialized economies have long done damage to the biosphere, but in at least some of these cases, and with time on its side, the biosphere has brought land and water back to life. Existing data signal that we have gone well beyond this capacity for recovery. We now have vast stretches of land and water that are dead—land overwhelmed by the relentless use of chemicals, and water dead from lack of oxygen due to pollution of all sorts. The surge of foreign land acquisitions by governments and firms examined above is one of many sources of this destruction. But these acquisitions are also partly a response to the crisis: more land and water need to be acquired to replace what has died.

The trends described here point to accelerated histories and geographies of destruction on a scale our planet has not seen before, making substantive the notion of the Anthropocene, the age marked by major human impact on the environment. Many of these destructions of the quality of land, water, and air have hit poor communities particularly hard, producing an estimated 800 million displaced people worldwide. But none of us is immune, as other destructions can reach us all, spread by massive transformations in the atmosphere.

This is a partial view that rests on the assumption that extreme conditions make trends visible that are more difficult to apprehend in their milder versions. Most of the land and most of the water on our planet is still alive. But much of it is fragile even though it may not like it. Scattered evidence in news media signals that the extent of this fragility may not be widely understood or recognized. For instance, polls suggest that few in the United States seem to know that more than a third of that country's land, including much of the cherished fertile Midwest, is actually stressed according to scientific measures, even if on the surface it looks fine. Or that at least one of the six major gyres that help keep our ocean currents going and oxygenated have now become massive trash zones leading to the asphyxiation of marine life. Or that there are at least 400 clinically dead coastal ocean zones. We made this fragility and these deaths.

We can think of such dead land and dead water as holes in the tissue of the biosphere. I conceive of these holes as sites marked by the expulsion of biospheric elements from their life space, and as the surface expression of deeper subterranean trends that are cutting across the world, regardless of the local type of politico-economic organization. As I examine at length in *Expulsions*, we have collectively produced conditions that override national differences: my guiding conceptual effort throughout the examination of concrete cases across our planet was to make visible the recurrence of environmentally destructive modes no matter how diverse the political economies in play. Together and over time, we have generated a planetary condition that reaches far beyond the specific sources of destruction and the specific forms of politico-economic organization within which they take place. It is a condition that hovers in spaces that range from the stratosphere to deep ocean gyres. Thus, it destroys the Arctic permafrost even though the indigenous Eskimo people had nothing to do with that destruction; it was rather the polluting factories of the US and Russia, among so many others.

CONCLUSION: THE EXPULSION OF PEOPLE AND THE ABUSE OF THE BIOSPHERE

There is a profound disjuncture between the diverse planetary conditions briefly described in this chapter and the dominant logics shaping governmental and experts' responses. Destroyed economies, livelihoods, bodies, land, and water have become a generic condition, disembedded from the geopolitical landscape of nation-states and mainstream international policies.

One key dynamic at work, and the one I focused on here, is expulsion from the diverse systems in play—economic, social, biospheric. The proliferation of systemic edges in all our major domains is foundationally different from the geographic border in the interstate system. Each major domain has its own distinctive systemic edge—this edge is constituted differently for the economy than it is for the biosphere and the social. One of the organizing assumptions in the book on which this essay is based is that the systemic edge is the site where general conditions take extreme forms precisely because it is the site for expulsion or incorporation. Part of the conceptual power of these systemic edges is that the extreme character of conditions at the edge makes visible larger trends that are less extreme and hence more difficult to capture. I conceive of these larger trends as *conceptually* subterranean because we cannot easily make them visible through our current categories of meaning: thus the importance of positioning my inquiry at the systemic edge.

NOTES

1 This is based on the author's *Expulsions: Brutality and Complexity in the Global Economy* (2014, Cambridge, MA: Harvard University Press/Belknap). We thank Harvard University Press for allowing us to reprint some sections. The reader can find full bibliographic treatment of the subjects discussed here in this book.

2 Elsewhere (2010) I examine the extent to which Marx's analysis of primitive accumulation to explain the relationship between capitalism and pre-capitalist economies might illuminate this relationship between traditional and new types of advanced capitalism.

3 For detailed data and sources on the facts discussed in this section, see Sassen (2008a; 2010; 2014, chs. 1, 2 and 4).

4 As of 2014, 35 of the 39 countries eligible under the Initiative have reached the completion point. Three pre-decision-point countries—Eritrea, Somalia, and Sudan—have yet to start the process of qualifying for debt relief under the Initiative, while Chad remains in the interim phase. Myanmar and Zimbabwe are under assessment for qualification (International Monetary Fund 2009b; 2012; 2014).

5 Among other sources of information are Oxfam (e.g., Oxfam International 2012), publications sponsored by The International Food Policy Research Institute (e.g., von Braun, Suseela and Meinzen-Dick 2009), and the research at the Oakland Institute. The first finds that between 15 and 20 million hectares of farmland in poor countries have been subject to transactions or talks involving foreigners from 2006 to 2008. The second reports on data that produce an estimate of about 70 million hectares bought or leased by 2010; that is the equivalent of a fifth of all the farmland of the European Union. There are a number of studies that do in depth examination of particular crops, countries, or regions (see Sassen 2014, ch. 2).
6 See, for example, US Department of State (2004).
7 This is taken from Chapter 4 in *Expulsions* (Sassen 2014). There the reader can find full development and bibliography for the issues briefly raised here and a large number of case studies from across the world. Because the details of each case matter, it is not possible to develop this empirically in a short essay.

REFERENCES

Benería, Lourdes and Shelley Feldman, eds. 1992. *Unequal Burden: Economic Crises, Persistent Poverty, and Women's Work*. Boulder, CO: Westview.
Bose, Christine E. and Edna Acosta-Belén. 1995. *Women in the Latin American Development Process*. Philadelphia: Temple University Press.
Bradshaw, York W., Rita Noonan, Laura Gash, and Claudia B. Sershen. 1993. Borrowing Against the Future: Children and Third World Indebtedness. *Social Forces* 71 (3): 629–56.
Buechler, Simone. 2007. Deciphering the Local in a Global Neoliberal Age: Three Favelas in São Paulo, Brazil. In *Deciphering the Global: Its Scales, Spaces, and Subjects*. Ed. Sakia Sassen, 95–112. New York: New York: Routledge.
Chant, Sylvia H., ed. 2011. *The International Handbook of Gender and Poverty: Concepts, Research, Policy*. Cheltenham: Edward Elgar Publishing.
Cotula, Lorenzo and Kyla Tienhaara. 2013. Reconfiguring Investment Contracts to Promote Sustainable Development. In *Yearbook on International Investment Law & Policy 2011–12*. Ed. Karl P. Sauvant, 281–310. Oxford: Oxford University Press.
International Monetary Fund. 2009a. "Factsheet: Poverty Reduction Strategy Papers (PRSP)." Accessed 14 August 2014. https://www.imf.org/external/np/exr/facts/prsp.htm.
International Monetary Fund. 2009b. "Factsheet: Debt Relief Under the Heavily Indebted Poor Countries (HIPC) Initiative". Accessed 22 September. http://www.imf.org/external/np/exr/facts/hipc.htm.
International Monetary Fund. 2012. "World Economic Outlook Database". Accessed 18 January 2015. www.imf.org/external/ns/cs.aspx?id=28.
International Monetary Fund. 2014. "Factsheet: Poverty Reduction Papers (PRSP)". Accessed 3 July 2015. https://www.imf.org/external/np/exr/facts/prsp.htm.
Jubilee Debt Campaign. 2007. "Debt and Women". http://www.jubileedebtcampaign.org.uk/Debt%20and%20Womenþ3072.twl.
Jubilee Debt Campaign. 2008. "Angola: Country Information." http://www.jubileedebtcampaign.org.uk/Angolaþ4038.twl.
Jubilee Debt Campaign. 2009. "Hasn't All the Debt been Cancelled?" http://www.jubileedebtcampaign.org.uk/4%20Hasn%27t%20all%20the%20debt%20been%20cancelled%3Fþ2651.twl.

Jubilee Debt Campaign. 2012. "Table 3. Low and Lower Middle Income Governments with the Highest Foreign Debt Payment Burdens Today41", chart. *The State of Debt: Putting an End to 30 Years of Crisis.* Jubilee Debt Campaign, London.

Jubilee Debt Campaign. 2013. "How Big Is the Debt of Poor Countries?" Accessed 18 January 2015. http://jubileedebt.org.uk/faqs-2/how-big-is-the-debt-of-poor-countries.

Kyle, David and Koslowski, Rey. 2001. *Global Human Smuggling.* Baltimore, MD: Johns Hopkins University Press.

Land Matrix. 2013. "Why Do the Numbers Constantly Change?" Accessed 29 June 2013. www.landmatrix.org/about.

Lucas, Linda E., ed. (2005) *Unpacking Globalisation: Markets, Gender, and Work.* Kampala, Uganda: Makerere University Press.

Margulis, Matias E., Nora McKeon, and Saturnino M. Borras. 2013. Land Grabbing and Global Governance: Critical Perspectives. *Globalizations* 10 (1): 1–23.

Migrant Remittances. 2008. "Worldwide Trends in International Remittances". Electronic newsletter jointly sponsored by DFID and USAID, 5 May. http://www.thedialogue.org/PublicationFiles/Migrant%20Remittances—May%20 2008-FINAL.pdf.

Orozco, Manuel, B. Linsday Lowell, Micah Bump, and Rachel Fedwa. 2005. "Transnational Engagement, Remittances and their Relationship to Development in Latin America and the Caribbean". Accessed 18 January 2015. http://www.thedialogue.org/PublicationFiles/Orozco-%20Transnational%20Engagement.pdf.

Oxfam International. 1999. "Oxfam International Submission to the Heavily Indebted Poor County (HIPC) Debt Review". Accessed 18 January 2015. http://policy-practice.oxfam.org.uk/publications/oxfam-international-submission-to-the-heavily-indebted-poor-county-hipc-debt-re-114964.

Oxfam International. 2012. "'Our Land, Our Lives': Time Out on the Global Land Rush". Accessed 18 January 2015. http://www.oxfam.org/sites/www.oxfam.org/files/bn-land-lives-freeze-041012-en_1.pdf.

Rahman, Aminur. 1999. Micro-credit Initiatives for Equitable and Sustainable Development: Who Pays? *World Development* 27 (1): 67–82.

Ratha, Dilip, Sanket Mohapatra, and Ani Silwal. 2014. "Migration and Development Brief 11: Migration and Remittance Trends 2009: A Better-than-expected Outcome So Far, But Significant Risks Ahead". Accessed 18 January 2015. http://siteresources.worldbank.org/INTPROSPECTS/Resources/334934-1110315015165/MigrationAndDevelopmentBrief11.pdf.

Safa, Helen Icken. 1995. *The Myth of the Male Breadwinner: Women and Industrialization in the Caribbean.* Boulder, CO: Westview.

Sassen, Saskia. 2001. *The Global City: New York, London, Tokyo.* 2nd ed. Princeton, NJ: Princeton University Press.

Sassen, Saskia. 2008a. *Territory, Authority, Rights: From Medieval to Global Assemblages.* 2nd ed. Princeton, NJ: Princeton University Press.

Sassen, Saskia. 2008b. "Two Stops in Today's New Global Geographies: Shaping Novel Labor Supplies and Employment Regimes". *American Behavioral Scientist* 52 (3): 457–96.

Sassen, Saskia. 2010. A Savage Sorting of Winners and Losers: Contemporary Versions of Primitive Accumulation. *Globalizations* 7 (1): 23–50.

Sassen, Saskia. 2012. *Cities in a World Economy.* 4th ed. Thousand Oaks, CA: Pine Forge Press.

Sassen, Saskia. 2014. *Expulsions: Brutality and Complexity in the Global Economy.* Cambridge, MA: Harvard University Press.

Sassen, Saskia. 2015. Migration? No. The Search for Bare Life. *Foreign Policy*, forthcoming.

Tinker, Irene, ed. 1990. *Persistent Inequalities: Women and World Development.* New York: Oxford University Press.

United Nations Development Programme. 2005. *A Time for Bold Ambition: Together We Can Cut Poverty in Half*, annual report. New York: United Nations Development Programme.

United Nations Development Programme. 2008. *Human Development Report 2007–2008*, annual report. New York: United Nations Development Programme.

United Nations Development Programme. 2013. *Human Development Report 2013.* New York: United Nations Development Programme.

United Nations Inter-Agency and Expert Group on MDG Indicators. 2009. "Millennium Development Goals Indicators: Debt Service as a Percentage of Exports of Goods and Services and Net Income." Accessed 14 July 2014. http://mdgs.un.org/unsd/mdg/SeriesDetail.aspx?srid=655.

US Department of State. 2004. *Victims of Trafficking and Violence Protection Act of 2000: Trafficking in Persons Report.* Washington, DC: US Department of State.

US Department of State. 2014. *Trafficking in Persons Report 2014.* Washington, DC: US Department of State.

Von Braun, Joachim and Ruth Suseela Meinzen-Dick. 2009. *"Land grabbing" by Foreign Investors in Developing Countries: Risks and Opportunities.* International Food Policy Research Institute policy brief. Washington, DC: International Food Policy Research Institute.

World Bank. 2005. Increasing Aid and its Effectiveness. In *Global Monitoring Report 2005 — Millennium Development Goals: From Consensus to Momentum.* Vol. 1, 151–87. Washington, DC: World Bank.

World Bank. 2006. *Global Economic Prospects: Economic Implications of Remittances and Migration.* Washington, DC: World Bank.

World Bank. 2008. "Migration and Remittances: Top Ten, March." Accessed 18 January 2015. http://econ.worldbank.org/WBSITE/EXTERNAL/EXTDEC/EXTDECPROSPECTS/0,,contentMDK:21352016~pagePK:64165401~piPK:64165026~theSitePK:476883,00.html.

14 From the "Bio" to the "Necro"
The Human at the Border

Andrés Fabián Henao Castro

The problem of the twentieth century is the problem of the color-line,—the relation of the darker to the lighter races of men in Asia and Africa, in America and in the islands of the sea. (W.E.B. Du Bois [1903] 1994, 16)

[Unauthorized immigrants]: as much as they are invisible in life they are invisible in death. (Robin Reineke [Coordinator, Missing Migrant Project, Tucson, Arizona] in the 2013 film *Who is Dayani Cristal?*)

FROM THE "BIO" TO THE "NECRO": DE-COLONIAL THEORY AND BIOPOLITICS

Michel Foucault first used the term "biopolitics" during his 1975–1976 Lectures at the Collège de France, entitled "Society Must be Defended." These lectures were Foucault's first attempt to conceptualize the phenomenon of totalitarianism, which he understood as the resolution of two antithetical technologies of power: sovereignty—the power to produce death—and biopolitics—the power to regulate life. The totalitarian solution to this technological paradox was to turn the regulation of life into its contrary, the massive production of death in the concentration camps.[1] In this way, Foucault showed that totalitarianism was not a historical necessity but an impossible and ultimately suicidal solution to the technological antithesis of power created by modernity's nation-state system and its dominant forms of political rationality: the "let live and make die" of sovereignty, and the "let die and make live" of biopolitics. The totalitarian resolution to this antithetical convergence was to collapse one technology into the other through the supplement of race. Racism was thus at the heart of the lectures in which the paradigm of biopolitics first emerged.

In Foucault's analysis, racism enabled three possibilities for totalitarian regimes to solve the production of one life by means of exterminating another: i) it offered a system by which to separate the lives left to die from the lives made to live through the fixation of biological differences; ii) it further established an inseparable relation between life and death, as the cultivation and survival of one race could only succeed through the elimination of the other race; and iii) it medicalized the political enemy who was no longer recognized as an opponent but was turned into a biological threat.[2]

Surprisingly, Hannah Arendt's prior work on totalitarianism, colonial racism, and the concentration camps received no major attention in Foucault's interpretation. This is remarkable given that both Foucault and Arendt had inscribed Nazism and Stalinism in a common totalitarian continuum, had embraced a positive and relational notion of power, and had even turned to the same author, the Comte de Boulainvilliers (Arendt 2004, 215; Foucault 2003, 144), in their own efforts to explore the historical emergence of a discourse of races in Europe.[3] Like Foucault (1994), Arendt also linked the modern inclusion of life in politics to the epistemological development of political economy in the eighteenth century, even if she did not use the term "biopolitics". Instead, Arendt (1998, 29) invoked the image of the super-human family as a way of understanding the modern configuration of the nation-state because the "social question" remained, for her, a sort of in between space that treated the public sphere in the manner of the ancient *oikos*. Life's inclusion in politics meant, for Arendt, the transformation of collective-contingent affairs into the technological management of the city for the satisfaction of corporeal needs, which had been relegated to the domestic space in Greek antiquity. This is the Arendt of *The Human Condition*, where her prior analysis of race in *The Origins of Totalitarianism* completely disappears.

Race, which Arendt (2004, 209–10) understood as the only unifying bond available between individuals under the conditions of capitalism and imperialism, played no major role in her understanding of the division of the world between those properly situated in the domain of the political and those relegated to the social or the private in *The Human Condition*. Racism and colonial imperialism were no longer to be found when she proceeded to link the satisfaction of needs of the modern political economy with the ancient institution of the *oikos*, a fundamentally nonpolitical sphere for Arendt, given the absence of a deliberative plurality and the rule of a hierarchical structure. Rather than equally shared power in government, which was at the heart of the Greek *polis* in her account, the logic of the household rested in mastery, domination, and ultimately violence. Arendt did not problematize the relational fact that such equally shared power in government had always rested on the prior depoliticization of the majority: slaves, metics, foreigners, and Athenian women.

In *The Human Condition* Arendt turned to the Aristotelian distinction between *bios* and *zoē* (central to all theories of biopolitics) to distinguish the human condition, to separate the individual from the member of the species. She, however, questioned neither the material conditions nor the power relations producing the very same distinction that led Aristotle to argue for the naturalization of slavery, dehumanized on the basis of the artificial separation between the animal life that emits mere noises (*phôné*) and the linguistic being that utters grievances (*logos*).[4] Arendt read Aristotle mediated by Heidegger, who pointed her toward the temporality of being as the distinctive feature of the human. Thus, Arendt distinguished *bios*, as the life referring to "the time interval between birth and death" that "follows a strictly linear movement", from *zoē*, as the bare life that sets in motion this *bios*, "the motor of biological life which man shares with other living things and which forever retains the cyclical movement of nature" (Arendt 1998, 97). Like Heidegger, what distinguished the human from the animal for Arendt was its temporal condition: linear rather than cyclical. Unlike Heidegger, such temporality was to be found in natality rather than in mortality, the most important political insight she kept from her earlier studies on Augustine of Hippo. Life, for Arendt, was not singularized at the level of *zoē*, which was under a lawlike process of repetition; life was singularized at the level of *bios*, and the "color-lines" of antiquity, which distributed *bios* to the Greeks and *zoē* to the non-Greeks ("barbarians", that is, those whose speech was unintelligible), remained unquestioned when Arendt proceeded to interrogate the blurring of such distinction with the emergence of the "social question" in modernity.

Ancient Greeks first offered a rationale for the animalization (and bestialization) of the "other" on the convergent basis of racial and sexual differences, as the *psukhê* (soul) of both non-Greek males and all women was constructed as "depraved", as anti-natural, as lacking the *nous*, without which the soul was unable to properly rule over the appetitive part of the body. Constructing the other as having the organ but being unable to use it, the phonetic sign of the more-animal-than-human (*zoē*) was never to translate into the semantic sign of the more-human-than-animal (*bios*), and the Greeks designed a plethora of institutions in charge of policing such a border—from Solon's law forbidding slaves from exercising at the gymnasia to the identity trials of the *dokomasia* and the Periklean prescription of double endogamy to grant Athenian citizenship in 451–450 BC (Lape 2010, 186–239). Neither Arendt's nor Foucault's turn to Aristotle and the Greeks for a remapping of power technologies in European modernity ever investigate this history, the very same history that produces the thingification of the human and its ultimate commodification, from the animated tool of the Greeks to the Latin *instrumentum vocale* of the Roman Empire.

Not surprisingly the extensive scholarship on race in the early 1970s never enters Foucault's research on biopolitics. And even if it is true that Arendt looked for the origins of totalitarianism in what she described as the European "scramble for Africa," an investigation that she will never continue, she would have hardly conceded to Aimé Césaire's lines in the *Discourse on Colonialism* (2001, 36), where he claimed that what the humanistic Christian bourgeoisie could not forgive Hitler for doing was the fact that "he applied to Europe colonialist procedures which until then had been reserved exclusively for the Arabs of Algeria, the 'coolies' of India, and the 'niggers' of Africa," nor to Frantz Fanon's observations in *The Wretched of the Earth* (2004, 101), when he claimed that "Nazism transformed the whole of Europe into a veritable colony." As Ann Laura Stoler (1995, 15) argued in her wonderful expansion of Foucault's approach to sexuality in imperial settings, de-colonial theory began "to explore the colonies as more than sites of exploitation but as 'laboratories of modernity.'" From Césaire and Fanon to the conferences held by the Coloniality Working Group at SUNY-Binghamton in 1999 and 2000, de-colonial theory broadened the spatial and temporal lines of the co-implication of modernity with colonialism from the fifteenth/sixteenth centuries onwards. Crucial to the investigation of what Walter Mignolo (2003) identified as the foundational "colonial difference" that instituted modernity was the "imposition of a racial/ethnic classification of the global population as the cornerstone of that model of power" (Quijano 2008, 181), in which processes of state formation and capital accumulation were completely embedded within processes of colonization and the racialization/ethnicization of the human.

The frontispiece of Thomas Hobbes' *De Cive* (1642), representing the fully civilized human in the left and the dehumanized savage in the right, constitutes an iconic example of both the origins of modernity's indebtedness to colonialism but also the Eurocentric displacement of the "racial/ethnic complex" by which it was constituted to the margins of the analysis. Hobbes' "state of nature" will continuously be read as a metaphor of the English Civil War, silencing the passage in which Hobbes projects to "America" the very warlike view of nature upon which modern European sovereignty founds itself in the image of the "social contract" (Hobbes 1994, 77). "America" will continue to be produced, materially and representationally, as "the beginning of the world" (Locke 1970, 310), that is, as the pre-political origin whose very constructed "savagery" was used to justify the colonial violence exercised in this "space of otherness" in the form of a civilizational project (Dussel 1995; Todorov 1999). The technological product of such political process was the invention of the colony as the "zone where the violence of the state of exception is deemed to operate in the service of 'civilization'" (Mbembe 2003, 24).

Indebted to the works of the Coloniality Working Group, Sylvia Wynter (2003, 260) has carefully traced the production of such a "space of otherness" (279) through the historical transformations on the "descriptive statements" (262) by which the Western bourgeois conception of "Man" has been able to "overrepresent . . . itself as if it were the human itself" (260). This has been, too, Wynter's own way of expanding Foucault's (1994, 387) historization of the figure of man—as "the effect of a change in the fundamental arrangements of knowledge"—beyond the European space and into its colonial settings. The first transformation took place with the secularization of otherness from the fifteenth to the eighteenth centuries, when the theological explanation of difference on the basis of the Christian iconography of sin was replaced with the racialized phenotypical rationalization of Man as the political subject of the state. Non-European forms of life were not to be seen exclusively as sinful but mainly as irrational, homunculi rather than fully human, and human otherness will be attributed to "Indians" and "Negroes" on the basis of a new somatotype norm that made the latter into the "missing link" between rational humans and irrational animals (Wynter 2003, 301). Enslaved to their nature, as in Aristotle's naturalization of slavery through the *bios/zoē* distinction, the first "scientific" classification of human populations by Linnaeus in 1735 represented the "Negro" as the population whose very irrationality justified their need to be governed by others (ibid.) The second transformation took place from the eighteenth century onwards, when the Darwinian paradigm of evolution was entrusted with the natural selection of one race and deselection of the other which from then on was made into the embodiment of non-evolved backward life. The color lines of this new descriptive statement will continue to differentiate between:

> the Caucasoid physiognomy (as symbolic life, the name of what is good, the idea that some humans can be selected by Evolution) and the Negroid physiognomy (as symbolic death, the "name of what is evil", the idea that some humans can be dysselected by Evolution). (Wynter 2003, 316)

However, the rationalization of Man as a bio-economic subject will also "repress all knowledge of the fact that its biocentric descriptive statement is a descriptive statement" (Wynter 2003, 326), thus reproducing the separation between race as a non-cultural fact of nature and ethnicity as a cultural product of history and tradition. This insight has led Alexander Weheliye (2014) to speak not of a "racial/ethnic complex", like Quijano did, on the basis of this problematic distinction between cultural and natural differences by which race disavows its very historicity, but of racialized assemblages common to both race and ethnicity alike in his most recent critique of the paradigm of biopolitics.

This investigation on the co-implication of modernity with colonialism reveals a different technology of power underlying both: the "let live and make die" principle of sovereignty and the "let die and make live" principle of biopolitics. This technology could be formulated as the "make die and let die" principle of colonial racism, in which the other was either actively exterminated through war and conquest or sanctioned to a slow death through forced labor in the *encomiendas* and the plantation system (Patterson 1985). Extermination without reminder was not an invention of the twentieth century but of the sixteenth century, and the first victims were the indigenous communities of the Americas. Thus, a report from the same year in which Hobbes will publish the foundational text of modern sovereignty, *Leviathan* (first published 1650), shows that Arawaks and their descendants had been entirely exterminated in Hispaniola.[5]

Influenced by de-colonial theory, Achille Mbembe coined the term "necropolitics" to characterize this new technology of power; one in which the slot of otherness was subjected to the deadly poles of both: the "make die" of sovereignty and the "let die" of biopolitics. Mbembe's work, equally indebted to both Foucault's (2003) and Agamben's (1998) investigations, traces a different genealogy of modernity, one that moves from the plantation system—as the first normalized space of exception—to the colonial hollows of slavery and finally to the commandment—right to dispose of others—of the postcolony (Mbembe 2001). It is in Mbembe's work that both paradigms (de-colonial and biopolitical) intersect in order to reinterpret contemporary forms of violence from the perspective of the other in the Global South.[6] Like Foucault and Agamben, Mbembe is also interested in the inscription of life and death in the order of power, but the subject and the material-symbolic conditions that he interrogates are displaced to a different although interrelated topography. Unlike the aristocratic war-of-races in Europe, the pre- and postcolonial forms of war that he examines produce life beyond utility, a form of human superfluity beyond the horrors of the concentration camp into the everyday life of survival in poverty. There is a qualitative transformation in the necropolitical linkage of the "make die" of sovereignty and the "let die" of biopolitics in the postcolonial Global South. From an understanding of salaries, taxes, and adjustment financial programs as disciplinary systems of government, to the orgiastic violence of massacres and extrajudicial killings, the state, the parallel state, and the transnational corporation are able to inflict more death in the Global South through the deliberate destruction of life's infrastructure than through old forms of direct killing.

With this technology emerges a new depoliticized subject: the disposable other (Bales 1999). Neocolonial occupation no longer produces just a surplus population to exploit; it organizes the continuous material destruction of the living conditions under which such a population is forced to experience a slow death for another population to live. The disposable subject is no longer killed but abandoned to die; she or he is slowly sanctioned to death by legal regimes through the marginalization of her/his living conditions as unworthy of social intervention. The death inflicted in the disposable subject represents a form of killing that cannot be named as such, a killing that refuses the denotation as even humanitarian organizations are enlisted in its doing (Weizman 2011). This is what Mbembe found, in its most developed form, in the colonial occupation of Gaza and the West Bank, through the infrastructural warfare waged against the Palestinian population (Mbembe 2003, 29). But necropolitics is equally present in other forms of destruction of the material sustainability of life, such as intensive resource-extraction, whose main targets continue to be indigenous communities, and the material dispossession of entire populations through destructive financial policies and international economic sanctions. Necropolitics thus speaks to the insufficiency of biopolitics to understand the creation of death-worlds, primarily in the Global South, those "new and unique forms of social existence in which vast populations are subjected to conditions of life conferring upon them the status of *living dead*" (emphasis in original, Mbembe 2003, 40).[7] Necropolitics refers to the neocolonial continuity of high-tech forms of destitution and socially assigned disposability across racialized assemblages: color lines that differentiate the life that must be protected from the avoidable death that has already been graphed as disposable. This complex necropolitical technology of disposability is particularly salient in the case of economic migrants from the south, living and dying in shadows while trying to survive by moving north, making violent expulsion integral to their identity (Sassen 2014).

UNAUTHORIZED IMMIGRANTS AS NEW SUBJECTS OF NECROPOLITICS AND THE BORDER AS NECROPOLITICAL *DISPOSITIF*

Originally informing the condition of the stateless and the refugee, rightlessness has increasingly begun to inform the condition of unauthorized immigrants too (Beltrán 2009; Benhabib 2004; Krause 2008). Despite greater continuity in the color-lines that such a system reproduces, unlike stateless people (whose nation-states were either destroyed or no longer able or willing to recognize the personhood of its subjects) and refugees (who were persecuted and expelled from their homeland either for their political views or because their state either decided to transform them into an existential threat to its national identity or failed to protect their lives), most unauthorized immigrants were forced to leave their countries of origin because the conditions imposed by neoliberal globalization and neocolonial militarization made it impossible for them to make a life in their homelands (Smith 2011). "Free trade" agreements, like NAFTA (1994) for example, have destroyed local economies and small farms in Mexico where farmers are no longer able to sustain their families because they can no longer compete with the subsidized corn from the US, forcing them to migrate north in order to survive. Such is the case of indigenous communities in Oaxaca (the second poorest state in Mexico, after Chiapas), once NAFTA forced the closure of CONASUPO (a state-owned grocery store) and it became cheaper for its residents to resell US corn than to grow their own (Bacon 2008, 25). Unauthorized immigrants are, thus, of a different kind of "harvest of empire" (Gonzalez, 2001). Unlike the stateless and the refugee, unauthorized immigrants do have states of origin in which their existences are officially recognized by the sole political entity still capable of conferring political membership, even if such abstract recognition hides the material precariousness that most of them were, in the first place, trying to escape. In other words, unlike the stateless and the refugee, unauthorized immigrants are, for the hostile country into which they arrive, legally deportable.

This technology of disposability makes the living conditions of unauthorized immigrants severely marginalized, to the extent that socioeconomic precariousness, unlike political precariousness, does not justify a claim for refuge. This means that the hostile state not only does not recognize the legal personhood of the subject that crosses, it resignifies the crossing as an illegal act against its territorial integrity, further marginalizing the status of the human being who is immediately transformed into cheap surplus labor to exploit and dispose. The violent crossing no longer represents a damage done to the unauthorized immigrant from the global organization of political membership in a state-centric system under the conditions imposed by neoliberal globalization and militarized borderlands, but a damage done by the unauthorized immigrant to the territorial integrity of the sovereign state. Thus, if the stateless and the refugee were politically constructed by the state of arrival as *a*-legal, a precarious condition they could only paradoxically overcome, according to Arendt (2004, 364), through criminal actions—"criminal offense had become the best opportunity to regain some kind of human equality, even if it be as a recognized exception to the norm"—unauthorized immigrants were politically constructed, by the state of arrival, as *il*legal.[8] Criminality, in the form of exception, did not suppose the inclusive exclusion of the rightless, as Arendt claimed in the case of the refugee and the stateless. Criminality is, after the territorial crossing, a sort of existential crime, coloring all actions of unauthorized immigrants for their hostile state of arrival. The criminalization of the crossing makes their being with no proper documentation in the territory coextensive with illegality, which is the only way in which they are officially recorded by the state. Everyday life actions like driving or receiving dental treatment become sites of risk; the policing state turns these into criminal acts, somehow traceable to the secularization of this original crime of existence in the territorial transgression of the border. Such existential denial makes the immigrants easy targets for national scapegoating; thus, they are constantly criticized for "being lazy", "getting free housing", "dilapidating social security resources", "lowering salaries of legal migrants", "taking advantage of American taxpayers", and being responsible for many other social, economic, and political problems (Vila 2000, 169–200).

The world-denying language used to address unauthorized immigrants facilitates the implementation of the two main strategies used in the US to produce their disposability. The first strategy seeks deportation directly. It creates a complex system of control and expulsion that criminalizes unauthorized immigrants and turns them into a national security threat. The literature on US immigration policy traces the political link between unauthorized immigration and national security far back into history, pointedly to the fluctuating policy of institutionalized exclusion of Mexican immigrants after the US Mexican war of 1846–48 (all the way to the bracero programs of the 1940s), the first exclusion laws passed against the Chinese in 1882 (effectively stopping all Asian immigration with the Tydings-McDuffie Act of 1934), and the internment of Japanese in concentration camps. Of recent significance, however, is the Illegal Immigration Reform and Immigrant Responsibility Act of 1996, which "mandated increased detention of asylum seekers" and turned the INS (Immigration and Naturalization Service) into "the world's largest agency that apprehends individuals" (Dow 2004, 9). The situation was worsened with the USA Patriot Act, signed into law 26 October 2001, which authorized the detention of noncitizens for national security reasons. To enact the policing measures of the Patriot Act, the Bush Administration created the Immigration and Customs Enforcement agency (ICE) on 1 March 2003 as part of the Department of Homeland Security with the goal to deport every deportable noncitizen. The signing into law of the Secure Fence Act in October 2006, building the largest section of the wall along the US-Mexico border, further strengthened the symbolic division between the Global North and the Global South through its increasing militarization. As I write this article, a new law is being discussed in the Congress, one that doubles the amount of agents at the border and extends the wall by 700 miles. It also makes service in the military for a minimum of two years the most inclusive alternative to earn citizenship for those unauthorized immigrants who were brought illegally to the country by their parents. Citizens have been included in such policing activities, blurring the distinctions between state agents and civil society. This escalated when the ICE launched the Secure Communities program in 2008, authorizing state and local jails to detain deportable immigrants and encouraging communities to denounce and accuse unauthorized immigrants as their patriotic duty. The program reached 1,700 jurisdictions by March 2012, helping Obama's administration achieve its infamous record in deportations, expulsing more than 400,000 immigrants per year. Despite Obama's most recent efforts to interrupt deportations through an executive order (which applies only to five of the approximately eleven million undocumented immigrants in the country), by the end of 2014 his government had deported more than two million people, most of them people of color coming from the Global South.

The second strategy seeks deportation indirectly; the aim is no longer to actively persecute immigrants but to establish conditions similar to the ones that unauthorized immigrants suffer in their states of origin in order to force their "self-deportation". The state of California's Proposition 187 pioneered such strategies in 1994 when it banned all public services for unauthorized immigrants in the US. The coding of every access to public goods by means of proper papers makes the satisfaction of the most basic needs rest in the possession of such papers. If unauthorized immigrants lack papers, and papers mediate the sustainability of their life in the city, then the aim is to put at risk such sustainability in order to force their "voluntary" deportation in this twisted Malthusian practice. In other words,

246 Andrés Fabián Henao Castro

the sovereign border of national identity that separates the human citizen from the dehumanized unauthorized immigrant—and the dehumanized second-class citizen too—borders with the color lines that separate life from death.

This system further increases the exploitability of "colored" labor by capital, whose illegal conditions of overexploitation, and the violation of local, national, and international labor agreements, are always overwritten by the original illegality of the crossing that turns the victim of the system into its culprit. The policing arrangement of this necropolitical system subsumes the political crimes of capital into the depoliticized existential crime of racialized labor. The marginalization of life-conditions for black and brown workers—within the "coloring" investments of this national language of race—is not only not legally persecuted, but is legally enforced by the policing institutions seeking to deport unauthorized immigrants. Transnational capital resignifies the violent means by which it profits from the exploitation of this labor as it participates too in the deliverance of security for national-state integrity, whether in the form of producing, circulating, and exchanging military equipment for the surveillance of "others" in the US and beyond or through the development of the prison industry (Alexander 2013). According to the Army Corps of Engineers, "the twenty-five-year-life-cycle cost of the 850-mile barrier mandated by the Secure Fence Act will range from $16.4 million to $70 million per mile", predictions that, according to Brown (2010, 37), include neither the remuneration to private property holders for the use of their land to build the wall nor the additional $60 billion contemplated for its completion and maintenance.

It should be clear that the deportable immigrant and the reproduction of "color-lines" that such a system of disposability recreates do not precede the complex assembly of institutions and apparatuses in charge of producing them, complicating any ontological grammar of race.[9] There are not unauthorized immigrants creating a "problem" of national security, which policing institutions are called to "solve". The transnational private prison complex for the incarceration of immigrants, the various systems of police surveillance, the loopholes in the industries that employ them, the legalization of racial profiling, the militarization of the border, etc., are in charge of coproducing the "problem" they are called to "resolve". Doing so, these institutions and practices also reshape the grammar of race subscribed to blackness and brownness. There is more money invested in policing programs targeting people of color and more jurisdictions in which these programs operate, including the exceptional privatized punitive system of incarceration (outside penal law), which is put in place to discipline people's movement and which has turned into a transnational industry that profits from confining and monitoring unauthorized immigrants in what Mark Dow (2004) describes as the new "American Gulags".

The border should not be understood as a "line" that merely separates. As Sandro Mezzadra and Brett Neilson (2013, 7) have argued in their own reconceptualization of the border, contemporary capital and the combination of different forms of labor and regimes of accumulation make "inclusion" exist "in a continuum with exclusion". Rather than a "linear" territory that splits the space in two, separating the outside from the inside, contemporary borders describe an elastic territory that multiplies color lines in a variety of places and sites of transit. This explains why Weizman (2011, 80) recurs to the Foucauldian category of the *dispositif* in order to argue for the irreducibility of the wall, as the iconic token of the border, to its physical structure. The complex combination of the physical structure of the wall in an elastic continuum with "checkpoints", "roadblocks", "closures", and "killing zones", among many others, makes of the border:

> a heterogeneous and interwoven assemblage of interconnected systems of fortification, architectural construction (the "terminals"), sensing technologies, automatic weapons, aerial and (in case of Gaza) marine systems that are operated by a multiplicity of institutions according to ever-changing administrative procedures, calculations, tactics, ethical, legal and humanitarian propositions. (Weizman 2011, 80)

From the border to border*ing*, Pablo Vila (2000) has understood such a *dispositif* as the product of the mobile struggle between changing modes of crossing borders and new technologies to reinforce them, two processes that, once again, should not be thought as in opposition to each other but in a continuum, as the reinforcing of the border already anticipates, as part of its political equation, a monitoring of the frequency of its crossing (Apostolidis 2010). In other words, the national border must be porous enough to allow for sufficient labor to come from abroad, but also impermeable, always ready to be reinforced in order to satisfy fantasies of national sovereignty. In a world in which, as Wendy Brown (2010, 23) has argued, "key characteristics of sovereignty are migrating from the nation-state to the unrelieved domination of capital and God-sanctioned political violence," walling is strengthened not because of the overpowering centrality of the nation-state but because of its erosion, hence, their need to recur to such hyperbolic tokens of sovereignty.

Discourses, devices, and technologies that seek to strengthen the border by pointing toward its fragility sustain a necropolitical system of disposability that is equally irreducible to capital, even if capital continues to exploit the disposable subject that results from such bordering. The dominant effect of spectacular borders, like the *dispositif* of hyper-surveillance deployed at the fifteen-foot-high steel walls of the San Diego/Tijuana border, are necropolitical in three different senses. First, the militarization of the border makes crossing both more expensive and more risky for would-be-immigrants, increasing their death toll as a result. According to Humane Borders, 2,269 migrant deaths were recorded just in Arizona between October 1999 and March 2012.[10] Life is most threatened by the body's inability to survive the harsh conditions of the crossing and the majority of deaths are said to be caused by dehydration or are considered heat-related—conditions that are never "natural" but fabricated through the militarization of the border that forces immigrants to take greater risks when crossing. This is the reason why Humane Borders has made its water station program a priority of political action, challenging the state governments that forbid them from placing water tanks in the desert to help prevent further deaths. It has been proven that immigrant deaths increase where water stations are not allowed. Such struggle also reshapes the national grammar of race, as happens in the Sonora desert, a privileged yet risky site of crossing for unauthorized immigrants. Native American activists from the Tohono O'odham indigenous community with sovereign jurisdiction on the Sonora desert in Arizona, who actively aid unauthorized immigrants through the placement of water stations, provision of hospitality, emotional support, etc., are resignified as "terrorists" or treated as de facto "unauthorized immigrants" by US repressive apparatuses which subject their territories to constant surveillance by the Border Patrol in a greater historical continuum of neocolonial violence. The violence reinscribes Native Americans in a new necroeconomy of "brownness" that continues to dehumanize them irrespective of their status.

The second necropolitical aspect of this *dispositif* is related not to the material infliction of death through the marginalization of living conditions, but to the very invisibility of death itself at a symbolic level. Necropolitics means not just the triumph of death over life for migrants but the very erasure of their dead itself, as immigrant bodies are frequently unidentified and buried in unmarked mass graves. Taylor (2013), for example, speaks of the startling contrast between two political representations of death in the same Arizona county: on the one hand, the deaths of fourteen unnamed unauthorized migrants on 24 May 2001, who received almost no significant attention by official institutions and were only graphically recorded in red dots on the death maps of Humane Borders; and on the other, the memorialization of the death of the Park Ranger Kris Eggle in August 2002, after the criminalization of immigrants had already advanced through the post-9/11-fueled xenophobia. Through the politics of memorialization, Eggle becomes singularized as a heroic individual, while immigrants are abandoned *en masse* to the anonymity of their socially sanctioned dehumanization. Well captured in Robin Reineke's epigraph, the invisibility of unauthorized immigrants' lives not only renders them more vulnerable to physical death, it also makes their deaths more vulnerable to symbolic erasure (see Silver 2013).[11]

The third aspect of this necropolitical *dispositif* refers to the greater externalization of both these material and symbolic forms of death to the Global South in the production of death-worlds. Unauthorized immigrants from the Global South are normally looking for a better life in the north, better than the one they had in their countries of origin. Their journey is a forced one; it already expresses global inequality as it refers to conditions of extreme poverty and extreme violence in places that are rendered completely invisible to mainstream media and academia alike. The crossing by which the citizen becomes the alien makes visible a world in which the life of the undocumented subject in the state of arrival can be considered as "a better life" than the one of the documented subject in the state of departure. Citizenship, where Aristotle located the *eu bios* of the properly human, is worse than alienage in the *zoē* into which unauthorized immigrants start to live in the hostile state of arrival, paperless as they are. In the global market of capitalism and neocolonial militarism, which reduces citizenship to an empty abstraction for the commodification of cheap labor and massive disposability, unauthorized immigrants opt for giving up on the abstraction. Citizenship carries no "good life" for them in their states of origin. There is also no other state willing to recognize and honor their lives by granting them a citizenship less emptied from any substantive content. To understand such asymmetry means to also shift the territorial locus of analysis in order to understand the elasticity of this necropolitical *dispositif*, the fact that US polices, like the "War on Drugs" or the militarization of the border, have terrible consequences beyond the US territory in an interconnected globalized world. Not only are local economies beyond the US completely destroyed, new and more dangerous economies of human traffic emerge, and new and more lethal modes of population control multiply. The result is an exponential increase in both material and symbolic deaths for impoverished people of color beyond the US. Thus, in Mexico alone, official institutions estimate that 25, 276 people have disappeared between December 2006 and July 2012, many of them impoverished migrants from El Salvador, Honduras, and Guatemala, trying to reach the US but caught in the multiple economies of lethal violence reproduced by such global necropower in Mexico (Gibler 2013; Hernández 2012).

From biopolitics to necropolitics, what the colonial and postcolonial history of this violence—"make die let die"—offers us is a greater understanding of the technologies of power that have reproduced the hierarchization of the human across color lines and the continuous reinvention of marginalized slots of otherness. Necropolitics does not disappear with modernity's presumable emphasis on life. On the contrary, the production of life is intertwined with the fabrication of death-worlds, and questions of national sovereignty border with the "color-lines" that separate the life that must be defended from the life whose very death is unworthy of intervention. The rigidity of such color lines continues to be contested as a result of people's struggles to be in a world that displaces them to non-being. Aiding those bodies through water stations, GPS devices, or more outspoken public protests led by unauthorized immigrants themselves in front of the US ICE building in Arizona, represent heroic and courageous acts of confrontation over the global conditions of political membership, over who counts and who does not in today's necroeconomy of disposability. Who is and who is not authorized has become a problem that borders life and death:

physical death for the unauthorized immigrants who are unable to survive the crossing, symbolic death for unauthorized immigrants who are deported for speaking on their behalf, physical and symbolic death for thousands of migrants who have disappeared with complete impunity in the Global South. Unwilling to die in material and symbolic invisibility, unauthorized immigrants perform, today, the speech that only the subject who is deprived of speaking rights can perform, the speech that troubles the ancient division between *bios/logos* and *zoē/phônē* and the very organization of knowledge that continues to silence the neocolonial history of this form of violence and its continuous bordering of the "human" across color lines.

NOTES

1　Foucault's (2003) historical reconstruction of such convergence rested in the genealogical tracing of three technologies of power which coalesced in the nation-state of the eighteenth century: i) the juridical technology of sovereign law, with its emphasis in the figure of the social contract (the individual as the subject of rights); ii) the disciplinary technology of normalization, which included a broad series of institutions in charge of the surveillance and observation of individualized bodies (the body that needed to be shaped against "deviations" and "irregularities", the body as the prisoner of the soul); iii) the biopolitical technology of regulation, which integrated disciplinary forms of power through collective phenomena (the individual as a member of the species, the population that therein gets objectified as a life source subordinated to the incremental rationality of the state). The emergence of biopolitics in the eighteenth century meant neither the disappearance of sovereignty nor that of disciplinary power, but their integration into a larger economy of power that he called governmentality, and that showed all its excesses during totalitarianism.

2　The centrality of race—which Quijano (2008, 183) considered, from the sixteenth century on, "to be the most effective and long-lasting instrument of universal social domination"—in Foucault's work on biopolitics—explains why de-colonial theory turned increasingly to his lectures for an understanding of the coloniality of power in the present. Castro-Gómez (2007, 157), for example, conceived of biopolitics as a technology of government "that 'makes live' those populations that are better adapted to the productive profile required by the capitalist state and, instead, 'lets die' those which do not contribute to productive labor, economic development and modernization." The centrality of race—which the lectures were not primarily about—in the Latin American reception of these lectures, well captured in the title with which the lectures were first translated into Spanish in Argentina, "Genealogía del racismo" (Genealogy of Racism), is indebted to the influential interpretation of them by Pasquino Pasquale (1993), Foucault's friend and research assistant in the 70s.

3　Actually, Foucault's analysis of the way in which an aristocratic war-of-races got historically transformed into the racist state of the nineteenth and twentieth century by means of the bourgeois invention of the nation in the eighteenth century has strong similarities with Arendt's interpretation in *The Origins of Totalitarianism*, published in 1950. Unlike Foucault's identification of a common genealogy, however, Arendt first narrated the historical struggle of European modernity as one taking place between two independent "ideological" systems: the "race" of the aristocrats and the "nation" of the citizens.

4 For a de-colonial critique of Aristotle, see duBois (1991, 66–8) and Isaac (2006, 171–81).

5 Enslaved Africans were subjected to the same violence once the proposal of Bartolomé de las Casas, trying to save the Caribbean Arawaks, succeeded in replacing Indians, whom he knew from first hand to have been enslaved and enserfed outside the "just titles" terms of orthodox Christian theology with [Africans] whom, as he thought at the time, had been acquired within the terms of those "just titles" (Wynter 2003, 294).

6 The Global South, although it uses a geographic metaphor to understand the displacement of otherness, is irreducible to territorial lines. See Levander and Mignolo (2011).

7 On the figure of the living dead—and its allegorical relation to the spectral economy founded on alien labor and conceived as draining the productive and reproductive powers of the community as a result of the conditions imposed by neoliberal capitalism—see Comaroff and Comaroff (2002).

8 For a broader analysis of the invention of *illegal people* through the simultaneous processes of motivating migration through neoliberal globalization and criminalizing immigration through state-centric regulation of people's movement, see Bacon (2008).

9 For a broader understanding of the grammar of race that exceeds the analytical scope of this essay, see Milian (2013, 3–9). Milian demonstrates that there are ways of using signifiers of color that "displace intrinsic ethnoracial narratives of being" (2). She does it, in part, by invoking the idea of a copiousness of Latin, first used by literary critic Marcus Boon, in which Latinoness and Latinaness slip out of brown bodies and the Spanish language not in quantitative terms of population but in qualitative terms of power and discourse. See also Gilroy (2001).

10 Arizona, primarily the Sonora desert, is just one site of crossing; thousands more have died crossing the US Mexico border. See Humane Borders website.

11 As Mbembe (2001, 173–89) argues in his reinterpretation of Heidegger's characterization of the human as the being-towards-death, the death of the disposable subject is not preceded by being but by nothingness. Previously produced as a "thing" owned by others, her/his existential void was filled with the violent fantasies of colonial imagination to such an extent that the question, "Is that man still alive, or dead?" was actually possible (Amos Tutuola quoted in Mbembe 2001, 197).

REFERENCES

Agamben, Giorgio. 1998. *Homo Sacer. Sovereign Power and Bare Life*. Trans. Daniel Heller-Roazen. Stanford, CA: Stanford University Press.

Alexander, Michelle. 2013. *The New Jim Crow: Mass Incarceration in the Age of Colorblindness*. New York: The New Press.

Apostolidis, Paul. 2010. *Breaking the Chain: What Immigrant Workers Can Teach America About Democracy*. Minneapolis, MN: University of Minnesota Press.

Arendt, Hannah. 1998. *The Human Condition*. Chicago: Chicago University Press.

Arendt, Hannah. 2004. *The Origins of Totalitarianism*. New York: Schocken Books.

Bacon, David. 2008. *Illegal People: How Globalization Creates Migration and Criminalizes Immigrants*. Boston: Beacon Press.

Bales, Kevin. 1999. *Disposable People: New Slavery in the Global Economy*. Berkeley, CA: University of California Press.

Benhabib, Seyla. 2004. *The Rights of the Others, Aliens, Residents and Citizens*. Cambridge: Cambridge University Press.

Beltrán, Cristina. 2009. Going Public: Hannah Arendt, Immigrant Action, and the Space of Appearance. *Political Theory* 37 (5): 595–622.

Brown, Wendy. 2010. *Walled States, Waning Sovereignty*. New York: Zone Books.

Castro-Gómez, Santiago. 2007. Michel Foucault y la colonialidad del poder. *Tabula Rasa* 6: 153–72.

Césaire, Aimé. 2001. *Discourse on Colonialism*. New York: Monthly Review Press.

Comaroff, Jean and John Comaroff. 2002. Alien-Nation: Zombies, Immigrants, and Millennial Capitalism. *South Atlantic Quarterly* 101 (4): 779–805.

Dow, Mark. 2004. *American Gulag: Inside U.S. Immigration Prisons*. Berkeley, CA: University of California Press.

DuBois, Page. 1991. *Torture and Truth*. New York: Routledge.

Du Bois, W. E. B. 1994. *The Souls of Black Folk*. New York: Dover Publications.

Dussel, Enrique. 1995. *The Invention of the Americas*. New York: Continuum.

Fanon, Frantz. 2004. *Wretched of the Earth*. Trans. Constance Farrington. New York: Grove Press.

Foucault, Michel. 1994. *The Order of Things: An Archeology of the Human Sciences*. New York: Vintage Books.

Foucault, Michel. 2003. *Society Must Be Defended: Lectures at the Collège De France, 1975–1976*. Trans. David Macey. Ed. Mauro Bertani and Alessandro Fontana. New York: Picador.

Gibler, John. 2013. *To Die in Mexico: Dispatches From Inside the Drug War*. San Francisco: City Lights Books.

Gilroy, Paul. 2001. *Against Race: Imagining Political Culture beyond the Color Line*. Cambridge: Belknap Press.

Gonzalez, Juan. 2001. *Harvest of Empire*. New York: Penguin.

Hernández, Anabel. 2012. De Calderón a Peña Nieto: Los que se esfumaron. *Proceso*, 29 December.

Hobbes, Thomas. 1642. *De Cive*. Paris.

Hobbes, Thomas. 1994. *Leviathan*. Indianapolis: Hackett Publishing.

Humane Borders. Website. Accessed 21 October 2013. http://www.humaneborders. info/.

Isaac, Benjamin. 2006. *The Invention of Racism in Classical Antiquity*. Princeton, NJ: Princeton University Press.

Krause, Monika. 2008. Undocumented Migrants: An Arendtian Perspective. *European Journal of Political Theory* 7 (3): 31–348.

Lape, Susan. 2010. *Race and Citizen Identity in the Classical Athenian Democracy*. Cambridge: Cambridge University Press.

Levander, Caroline and Walter Mignolo. 2011. Introduction: The Global South and World Dis/Order. *The Global South* 5 (1): 1–11.

Locke, John. 1970. *Locke's Two Treatises of Government: A Critical Edition with Introduction and Notes*. Ed. Peter Laslett. Cambridge: Cambridge University Press.

Mbembe, Achille. 2001. *On the Postcolony*. Berkeley, CA: University of California Press.

Mbembe, Achille. 2003. Necropolitics. Trans. Libby Meintjes. *Public Culture* 15 (1): 11–40.

Mezzadra, Sandro and Brett Neilson. 2013. *Border as Method, Or, The Multiplication of Labor*. Durham, NC: Duke University Press.

Mignolo, Walter. 2003. *The Darker Side of the Renaissance: Literacy, Territoriality, and Colonization*. Ann Arbor, MI: University of Michigan Press.

Milian, Claudia. 2013. *Latining America: Black-Brown Passages and the Coloring of Latino/a Studies*. Athens, GA: University of Georgia Press.

Pasquale, Pasquino. 1993. Political Theory of War and Peace: Foucault and the History of Modern Political Theory. *Economy and Society* 22 (1): 76–88.

Patterson, Orlando. 1985. *Slavery and Social Death*. Cambridge, MA: Harvard College.

Quijano, Aníbal. 2008. Coloniality of Power, Eurocentrism, and Social Classification. In *Coloniality at Large*. Ed. Mabel Moraña, Enrique Dussel, and Carlos A. Jáuregui, 181–224. Durham, NC: Duke University Press.

Sassen, Saskia. 2014. *Expulsions: Brutality and Complexity in the Global Economy*. Cambridge, MA: Harvard University Press.

Silver, Marc. 2013. *Who Is Dayani Crystal?* Directed by Marc Silver. London: Pulse Films.

Smith, Rogers M., ed. 2011. *Citizenship, Borders, and, Human Needs*. Philadelphia: University of Pennsylvania Press.

Stoler, Ann Laura. 1995. *Race and Education of Desire: Foucault's History of Sexuality and the Colonial Order of Things*. Durham, NC: Duke University Press.

Taylor, Lawrence. 2013. "Deaths in the Desert: Symbolic Politics and Moral Geographies at the Edge of America", lecture. University of Massachusetts Amherst, MA, October.

Todorov, Tzvetan. 1999. *The Conquest of America: The Question of the Other*. Trans. Richard Howard. Oklahoma: University of Oklahoma Press.

Vila, Pablo. 2000. *Crossing Borders, Reinforcing Borders*. Austin: University of Texas Press.

Weheliye, Alexander G. 2014. *Habeas Viscus: Racializing Assemblages, Biopolitics, and Black Feminist Theories of the Human*. Durham, NC: Duke University Press.

Weizman, Eyal. 2011. *The Least of All Possible Evils: Humanitarian Violence from Arendt to Gaza*. New York: Verso.

Wynter, Sylvia. 2003. Unsettling the Coloniality of Being/Power/Truth/Freedom: Towards the Human, After Man, Its Overrepresentation—An Argument. *The New Centennial Review* 3 (3): 257–337.

15 Biopolitics in the Laundry
Ireland's Unwed Mothers

S. E. Wilmer

In June 2014, a local Irish historian, Catherine Corless, revealed the existence of a mass grave in the St. Mary's home in Tuam, Ireland run by the Bon Secours nuns. Corless alleged that 796 babies and children had died at the home in a 36-year period between 1925 and 1961 and that a disused septic tank on their property had been filled with the bones of some of these babies and small children (Boland 2014). In the resulting public outcry, which led to skeptics questioning her allegations, the historian defended her research, indicating that she had paid four euros to the state authorities to obtain each of the 796 death certificates (costing her a total of 3,184 euros). She wanted to be able to identify the babies and children in order to record their names on a plaque as a memorial to them because they had been buried in an unmarked grave at the back of the former home. The nuns of the Bon Secours expressed "shock" that so many babies might have died (quoted in Edwards 2014). In this paper I will demonstrate how the ideology of the state and the notion of a sacrosanct national identity contributed to a biopolitical structure that fostered many unethical practices within religious institutions, and how theatrical, filmic, and other media representations have exposed the nature of these institutions, mobilized public opinion, and highlighted policies that dehumanize women.

In *Discipline and Punish: The Birth of the Prison* (1977), Michel Foucault reveals how governments have used penal institutions to isolate and exclude undesired elements in society, and that in the nineteenth century such institutions adopted new procedures for controlling the behavior of deviants by introducing certain forms of discipline that gradually became normative:

> The most important effect of the carceral system and of its extension well beyond legal imprisonment is that it succeeds in making the power to punish natural and legitimate, in lowering at least the threshold of tolerance to penalty. It tends to efface what may be exorbitant in the exercise of punishment. (1977, 301)

In a similar way the Roman Catholic religious orders in Ireland developed a network of power to control social behavior and devised methods to justify the exclusion and disappearance of those who deviated from their moral teaching. The Church provided an institutional framework that forced unwed mothers to become invisible in order to maintain order and cleanse society of what were perceived as unhealthy influences. Moreover, as Foucault (1980, 93) points out: "In any society, there are manifold relations of power which permeate, characterize and constitute the social body, and these relations of power cannot themselves be established, consolidated nor implemented without the production, accumulation, circulation and functioning of a discourse." Such a discourse, emphasizing sexual abstinence, especially for women outside of marriage (but encouraging multiple pregnancies inside marriage), was provided by the Roman Catholic hierarchy and embraced by the Irish government and the majority of the Irish population, resulting in an internal socialization of these dominant norms.

According to Tom Inglis, "Irish rural society in the early nineteenth century placed little emphasis on physical modesty or on verbal reticence. There was an unambiguous sexual symbolism in wake games and May Day festivals" (1998, 138). But as the nationalist movement developed, the emerging Irish middle class sought to become more "civilized" and emulate the behavior of their British rulers. However, they did not want to become Protestant, and therefore Anglicized, so an alliance developed between the Roman Catholic Church and certain powerful sections of Irish society. This alliance promoted surface "respectability" as a means of symbolic and social control (especially of women's sexuality) (Inglis 1998, 247–8).

This process was accentuated when Ireland became an independent state. The 1937 Irish constitution enshrined Roman Catholic doctrine into the legal framework of the state. Contraception, abortion, and divorce were illegal. Women were not allowed to work in the civil service once they married. However, more important than the legal structure set by the government were the moral guidelines that were laid down by the Church and accepted by the government and the majority of the population. Explaining the dominance of Roman Catholicism in Irish society, Inglis has written:

> [I]n the heyday of the Catholic Church's power, particularly during the fifty years after the founding of the state, its influence expanded beyond the religious field into the field of politics, economics, education, health, social welfare, the media and many other fields. The power of the Church meant that it structured not just the religious life of the Irish people, but their social, political and economic life as well. Consequently, the strategies through which Irish Catholics struggled to gain cultural, social, political, economic, and cultural capital were linked in with living a good Catholic life. (1998, 245–6)

Although the dominance of the Church started to wane toward the end of the twentieth century, "the proportion (85 per cent) of Irish Catholics who went to Mass every week in 1990 was still the highest in the world" (Inglis 1998, 204).

The Church fostered a biopolitical policy of social control over the lives and sexual activity of their congregations. The priests and nuns all took vows of chastity and celibacy, and the Church proscribed that sexual relations were solely reserved for procreation by married couples. Sex before marriage was a mortal sin, and if unmarried women became pregnant, they were forced to disappear from society, though the men responsible were generally exonerated. Institutions that had been established in the nineteenth century to provide rehabilitation for prostitutes were converted in post-independence Ireland into places to harbor unwed mothers secretly, where they could deliver their babies and have them adopted. Such institutions operated Magdalene laundries to provide work without pay for these "fallen" women. Because of the power of the Church over the government, such religious institutions grew in strength and continued to operate until 1996. The legal basis for the practices in mother and baby homes depended more upon social agreement than government legislation. According to the journalist Martin Sixsmith (2009), "Such was the power of the church, and of Archbishop John Charles McQuaid, that the state bowed before its demands, ceding responsibility for the mothers and babies to the nuns."

Michel Foucault has argued that the guidelines governing similar institutions emanate from moral codes rather than from legal structures. They are:

> not the juridical rule deriving from sovereignty, but a natural rule, a norm. The code they come to define is not that of law but that of normalization. Their reference is to a theoretical horizon which of necessity has nothing in common with the edifice of right. [. . . And] the coercive forces whose exercise takes a disciplinary form . . . invade the area of right so that the procedures of normalization come to be ever more constantly engaged in the colonization of those of law. (1980, 106–7)

However, he also points out that resistance against such moral codes and control mechanisms can arise from people asserting "the 'right' . . . to one's body, to health, to happiness, to the satisfaction of needs, and beyond all the oppressions or 'alienations', the 'right' to rediscover what one is and all that one can be" (Foucault 1998, 145).

In the 1990s, questions began to be asked about the practices in Magdalene laundries.[1] In 1992, a former nun, Patricia Burke Brogan, who had worked briefly in one of the laundries, wrote a play called *Eclipsed* (Brogan 1994). It depicts a young woman who has been born in one of the homes and is adopted by an American family. She visits the home in Ireland in 1992 to try to find traces of her mother and encounters ghosts of inmates from three decades earlier. The play provides insight into the harsh regime behind the closed walls of these religious institutions, such as the slave-like conditions in the laundries and the practice of burying the children of unwed mothers in unmarked mass graves. Brogan said, "I wrote the play to give these women a voice. I was one of the few who entered the laundry, who came out again and could speak up" (quoted in Smith 2007, 93).

Eclipsed traces the lives of some of the inmates in the institution interacting with the nuns. One of the nuns, Sister Virginia, objects to their ill treatment and appeals to the mother superior to change the regulations. Mother Victoria replies:

> When I was nineteen . . . I wanted to free the penitents—mothers of some of the women in the laundry now. You see, this weakness to sins of the flesh stays in the blood for seven generations! When you take Vows, Sister, you'll receive Grace and Understanding. Keep aloof from those fallen women! St. Paul says "People who do wrong will not inherit the Kingdom of Heaven.—People of immoral lives—fornicators, adulterers." (Brogan 1994, 45)

The play ends with the deaths of the inmates, after wasted lives in the institution, and with a voiceover by Sister Virginia who announces:

> In 1992, to make place for a building development at St. Paul's Home, the remains of Mary Kate Dempsey, Mary Jane O'Sullivan, Kitty O'Hara, Julia Mannion, Betty and Annie Gormley, Ellen McAuley, Cathy McNamara and three hundred unnamed penitents were exhumed, cremated and reburied outside in Killmacha Cemetery. (Brogan 1994, 78)

Although the play was initially performed by the small Punchbag Theatre Company in Galway to a limited audience, it has continued to be staged in many venues and has won many awards, including a Fringe First award at the Edinburgh Festival and the USA Moss Hart Award for a production in Worcester, Massachusetts. By 2008 it had been produced 67 times in three continents and had been translated into French, Dutch, and Italian (see Words On The Street 2015). According to James Smith, "Although rarely acknowledged as such, *Eclipsed* first introduced the tropes by which other contemporary retellings have narrativized the Magdalen experience" (2007, 91–2). It also led to Brogan writing a second play, called *Stained Glass at Samhain* (2004), that attributed equal blame for these institutions to the government, the Vatican, and Irish society, as well as to the nuns who ran them.[2]

In the year following the first performance of *Eclipsed*, an odd event occurred in which real life echoed fiction. The Sisters of Our Lady of Charity, having fallen into debt because of unwise speculation on the gold market, sold a piece of land from their convent in Dublin to property developers for 1.5 million pounds. However, construction workers discovered a mass grave with 133 bodies on the land. The remains of the bodies needed to be unearthed and reburied in a cemetery. It soon became evident that the

Sisters were not fully aware of the contents of this burial ground on their property. Of the corpses, 58 could not be identified with death certificates and the remains of an additional 22 completely unidentifiable bodies were later found. The nuns abandoned the normal religious practice of burial and arranged for their cremation and the burial of the ashes in a Dublin cemetery, stacking the urns on top of each other to save space and expense (see Raftery 2011).

The Irish press became interested in this odd set of circumstances, and journalists as well as historians began to investigate the wide network of these little-known institutions. They discovered that the government, the religious orders, and the public had conspired to remove unwed mothers from the community by encouraging religious institutions to operate homes where unmarried mothers and their babies would be invisible to society. Not only unmarried mothers but also those who were deemed to be in danger of becoming sexually active or had experienced violent incestuous situations in their families were incarcerated in these homes. The women were employed in laundries without pay and were forced to give up their children for adoption or placement in orphanages. But because the babies were born in the homes without medical care or anesthetics, and because the mothers were prevented from breastfeeding and the babies were often malnourished, up to 68 percent of them died (Sixsmith 2014b).

The mother and baby homes in some cases were run along lines that closely paralleled the conditions of a prison. But, unlike with prisons, there was no judicial procedure for conviction and no termination of sentence. In many cases inmates were confined for the rest of their lives, unless they escaped or were retrieved by a relative. The inmates were locked inside and not allowed out of the grounds of the convent. They were kept under constant surveillance and suffered physical as well as mental punishment. A master's thesis by a nun named Dorothy Thompson calls to mind Foucault's description of Bentham's panopticon in which "power should be visible and unverifiable" (1977, 201). She writes:

> The following are the principles on which the rules for the supervision of the dormitories are based: silence to maintain order, and silence to prepare the girls for a good sleep. While in the dormitory, the Group Mother should stand where she can observe everything. She waits until all are asleep before she retires. Her room is off the dormitory and is locked. However, she has a slide in her door allowing her to observe the dormitory during the night. . . . A dim light must be kept burning at all times and it must always be possible to observe the washroom facilities. . . . During the day the dormitories are locked and no one may go to them without special permission . . . because of certain types of disturbances, this area is one of the gravest concern. (quoted in Finnegan 2004, 234)

According to Frances Finnegan, who has studied the history of these institutions, "As late as the 1960s, then, the Order betrayed a morbid obsession with the sexuality of others [particularly women]—a disturbing preoccupation of the 'virtuous', with vice" (2004, 235).

The largely invisible network of mother and baby homes in Ireland suited many interests. The nuns believed that they were providing an important social service by suppressing the sexuality of these women. Parents were able to hide their disgraced daughters. The fathers of the babies could forget about their responsibilities for the children. The religious orders made money by operating laundry services, offering the babies for adoption, and receiving a stipend from the government for each woman and child that they housed. Wealthy Americans flew to Ireland to adopt the babies in return for sizable donations to the Church. Many religious and government agencies, and businesses and middle class families, relied on the cheap laundry service that the laundries provided. Moreover, the police cooperated by returning inmates who tried to escape. Known as penitents, an estimated 40,000 to 60,000 women were locked up in these penitentiaries.

It is evident that the nuns had treated the inmates (and their babies) "like a different species."[3] Giorgio Agamben, in his discussion of biopolitics, uses the term "nuda vita" or "bare life" for a life with no ethical value to categorize people of uncertain legal status such as refugees, asylum-seekers, Roma, the mentally ill, illegal immigrants, as well as Nazi concentration camp victims: "In Western politics, bare life has the privilege of being that whose exclusion founds the city of men" (1998, 12). Agamben adapts the notion of bare life from the Aristotelian differentiation between *zoē* and *bios*. He cites Foucault (1998, 143) in asserting that *bios* is the form of life normally reserved for the human species and can be distinguished from that of an animal because *bios* defines "a living animal with the additional capacity for political existence" (Foucault 1998, 143), whereas *zoē* or bare life is simply the animal dimension of the human being. From this Agamben developed the concept of the *homo sacer*, who, under Roman law, was a person with no social value (a bare life) and could be killed with impunity. Agamben relates this concept to present day politics and suggests that every government determines who should be included and who is "devoid of value" and should be excluded or "eliminated": "Every society sets this limit; every society—even the most modern—decides who [these] will be" (1998, 139). More recently, Judith Butler has interrogated the concept of bare life in relation to stateless people (e.g., Palestinians and asylum seekers) and suspected terrorists (in the wake of 9/11 and the "War on Terror"), especially those detained in centers such as Guantánamo Bay (see Butler 2006, 60–8 and Butler and Spivak 2007, 40–4). Moreover, Achille Mbembe (2003), in his discussion of "necropolitics" as an amplification of biopolitics, applies bare life to "the repressed topographies of cruelty" (40) in slave plantations and African colonies.

The Magdalenes might also be characterized as having the status of *nuda vita* (bare life). Having done nothing illegal, the Magdalenes were treated like convicts. However, in some respects they were denied rights normally granted to criminals. Rather than obtaining a specified termination date for their confinement, they could remain detained indefinitely, and, unlike convicts who retain their identities, they were rendered anonymous because the nuns changed their names, making them more difficult to locate or contact. Denied the usual rights of citizenship to come and go, they attained a similarly liminal status in society to that of asylum seekers and refugees, and it is ironic that the homes in which they were confined were frequently called "asylums" or "refuges". However, unlike refugees or asylum seekers, who normally manage to keep their children, the Magdalenes were dispossessed of their children and victimized in a gender-specific way since the fathers of the children were not punished. Likewise, their children could also be seen to have the status of bare life within the institutions because they were denied the right to live with their mothers, and could arbitrarily be sent to live with other families, or be put in orphanages. Moreover, the children often died of malnutrition or neglect and were frequently denied a proper burial and gravestone.

In some respects the Magdalenes were more on a par with specific types of mentally ill patients or with the prisoners in Guantánamo Bay. Like the involuntarily detained psychiatric patients who are sectioned in locked wards of hospitals, the Magdalenes usually were not allowed to leave the institution without the intervention of a relative or the arbitrary decision of a person in authority. Also, like Guantánamo detainees in their initial years of incarceration (see Butler 2006, 51), the Magdalenes had no legal process available to them to improve their conditions or end their confinement, and were subjected to what was arguably "total" or "absolute" domination. Agamben (1998, 120) uses the phrase "total domination" in reference to concentration camps. Similarly, Mbembe (2003, 21) uses the concept of "absolute domination" in his discussion of slavery. Although it would be grossly exaggerated to compare the Magdalenes with the victims of concentration camps or slavery, nevertheless, there were certain similarities in their conditions of detention. For example, Mbembe refers to the "triple loss" suffered by slaves that could equally apply to the Magdalenes: the "loss of a 'home', loss of rights over his or her body, and loss of political status" (21). The Magdalenes suffered from indefinite detention, the loss of their children, the denial of sexual relations, and the exploitation of their labor. Although Magdalenes were not slaves in the sense of being property that could be sold to another slave owner, their children were often treated in this way. According to Sixsmith (2009), "From the end of the second world war until the 1970s, [the Irish Catholic Hierarchy] considered the thousands of souls born in its care to be the church's own property. With or without the agreement of their mothers, it sold them to the highest bidder."

The revelations about the Magdalene laundries led to at least four television documentaries: *Sex in a Cold Climate* for Channel 4 (Humphries 1998), a CBS "Sixty Minutes" program called *Magdalen Laundries: Women Confined in Convents* (1999), the Irish-made *The Forgotten Maggies* (O'Riordan 2009), and *Ireland's Lost Babies* (Sixsmith 2014a), as well as two feature films, *The Magdalene Sisters* (Mullan 2002), directed by Peter Mullan, and *Philomena* (Coogan and Pope 2013), directed by Stephen Frears (featuring Judi Dench, who was nominated for an Oscar for her role in the film). The two feature films have relied on historical as well as anecdotal evidence for their portrayal of the Magdalene institutions. Although Peter Mullan's film *The Magdalene Sisters* has been criticized for painting too harsh a picture of the nuns,[4] it was partly based on the testimony of the inmates of the Magdalene laundries, who bore witness to their past in the documentary *Sex in a Cold Climate*. These testimonies provided important insights into the conditions within the mother and baby homes that enabled Mullan to create a reasonably credible and damning reenactment of procedures within the institutions. For example, Mullan recreated a scene based on a story recounted by Brigid Young in *Sex in a Cold Climate* about her brutal treatment by a mother superior in an orphanage. In the documentary Young recalls how she and the other children in the orphanage were not allowed to have any contact with the girls from the Magdalene laundry who were housed next door: "We were made to believe that they were very, very bad children. They were people who were devils, who were sinners." But when she was delivering laundry from the orphanage to the nunnery with another orphan, they happened to encounter a Magdalene inmate who asked her, "'Do you have a child there by the name of Margaret Moore?' And I said, 'Yes we do.' She said, 'That's my child. And I don't know what she looks like. But I haven't seen her since she was a year old.'" The two orphans then agreed to take her child to the railings of the orphanage where she would be able to see her daughter. But they were caught by the nuns and taken to the Reverend Mother who asked them to wait in the back shed:

> So we went into the back shed and she came in with a great big, long, rubber, black rubber . . . it wasn't a belt but it was something that she had specially made for the children to beat the children with. And a scissors and an open razor. And she shaved both our heads and gave us a severe beating. And after she did that, she grabbed the two of us again and she made us look in the mirror to see what we had looked like after she had finished with us. And that's what happened. And I'll never forget what looked back at me. Totally devastating! Your forehead all swelled up. Under my chin all bleeding where she had stuck the scissors, wide open. And the blood running into my eyes. My eyes totally closed and she was making us open them eyes and look in that mirror. "And you're not so pretty now, are you?" I'll never forget that day. And this

was just because [of] talking to Magdalenes. I was getting too friendly with the Magdalenes. (Humphries 1998)

Stephen Frears' film *Philomena* also relied on historical accounts, in this case the testimony of Philomena Lee, who told her story to the journalist Martin Sixsmith for a book about her life called *The Lost Child of Philomena Lee* (Sixsmith 2010). The film shows how the women in her institution were made to work in the laundry seven days a week for three years after giving birth, and were allowed to visit their children occasionally until the nuns arranged for the adoption of their babies. Like the other mothers, Philomena Lee was sworn to secrecy and forced to sign a document renouncing her claim to her child:

> I relinquish full claim for ever to my child and surrender him to Sister Barbara, Superioress of Sean Ross Abbey. The purpose is to enable Sister Barbara to make my child available for adoption to any person she considers fit and proper, inside or outside the state. I further undertake never to attempt to see, interfere with or make any claim to the said child at any future time. (Quoted in Sixsmith 2009)

According to Lee, "None of us wanted to give our babies up, none of us. But what else could we do? They just said, 'You have to sign these papers'" (quoted in Sixsmith 2009). Sixsmith (2009) also writes that, "After her baby, Anthony, was born, the mother superior threatened Philomena with damnation if ever she breathed a word about her 'guilty secret.' Terrified, she kept it quiet for more than half a century. . . . 'Over the years I would say "I will tell them, I will tell them" but it was so ingrained deep down in my heart that I mustn't tell anybody, that I never did.'"

The film demonstrates how later in life Lee tried to reunite with her son who had been adopted by an American couple when he was three years old, and it reenacts moments where she and her son ask the same nun for help in finding each other but are denied this possibility until it is too late. In his research Sixsmith tried to obtain the adoption records of the home, but discovered that the same nun had recently destroyed the records. He also found out that the adoption business had been the largest money-earner for the convent where Philomena was housed in the 1950s. He later produced a BBC documentary called *Ireland's Lost Babies* (Sixsmith 2014a), revealing the inadequate vetting practices of the Roman Catholic Church that led to some children being adopted by pedophiles. He also uncovered evidence of Irish government involvement in the adoption procedures by issuing more than one hundred passports each year for the adopted children traveling to America.[5]

In addition to these major films and documentaries that have helped raise awareness, *Laundry* by Louise Lowe, a site-specific performance in 2011, gave a limited audience an immersive experience within one of the actual institutions where inmates had been dispossessed of their children and forced to wash laundry by hand without pay for years (see Keating 2011). *Laundry* was set in the last remaining convent and laundry in the centre of Dublin, with its chapel still intact. The laundry had finally closed in 1996, not for ethical or moral reasons but because, as Frances Finnegan (2013) explains, "with the advent of the domestic washing machine, these institutions became no longer financially viable."

Although it was only possible to perform the play to a limited number of spectators, because of its style of performance, *Laundry* arguably had a greater emotional impact on its audience. Spectators attending the performance of *Laundry* entered the actual convent in groups of three. They were then separated and conducted through a series of scenes in which they individually encountered a variety of actors playing inmates in the institution who engaged them in one-on-one conversations about their circumstances. At one point in the piece, the spectator is confronted with a moral choice of whether to help an inmate escape or remain confined within the institution. In another scene the spectator joins an inmate in a confession booth where the inmate disobeys religious regulations by whistling. In another scene that takes place in the actual chapel of the convent, the spectator is offered a boiled sweet by an inmate and asked to hold her hand. Because of the special effect on the isolated spectators of retracing the steps of the real environment in which inmates had been recently housed and engaging in individual encounters with the actors playing these inmates, the spectators became part of the fictional/historical world of the play and underwent a visceral and sensory experience that was deeply affecting.

One reviewer remarked:

> The scenes we witness are enacted in closed, cell-like spaces, and the stories themselves reflect harrowing breaches of personal and physical autonomy. . . . *Laundry* is not just an act of public disclosure but of social questioning, where we are asked to consider our own role in perpetuating systemic corruption, and by placing us in such close proximity to history, Lowe is inviting us to question our own complicity. Should you reach out to comfort the young woman who genuflects in front of you so closely that her head almost touches your knees? . . . Do you agree to help the ghostly girl who hides behind the mirror, a palimpsest upon your disturbed reflection, begging for intervention? (Keating 2011)

Another spectator recorded her own distressed experiences on a personal blog, revealing how she kept bursting into tears as she participated in the scenes, especially one in which she was asked to help a girl remove a bandage wrapped around her breasts before stepping into a bath:

> At the distant sound of a new-born infant's cries the girl in the bath stretches out her arm as if trying to comfort her child. I have never seen

such pain in anyone's eyes. As a mother my heart broke for her and all the girls for whom this was a reality, so much so that now I cry without reservation. I am past self-consciousness. Still holding the bandage . . . I realise with absolute horror that this is possibly being used to suppress her milk. What God in Heaven put these nuns on this earth? (Brien 2011)

It is clear that these films, documentaries and theatrical performances have shone a light onto what was arguably invisible to the majority of the population and have added to the clamor for social change. Investigative journalism was able to uncover facts and provide details about the procedures in these institutions, while television documentaries, films, and theater performances brought to life some of the people who encountered such conditions.

These cultural reenactments have supported new attitudes toward the Church that have noticeably been driven by activist women. Encouraged by the election of Mary Robinson, the first woman President of Ireland in 1990 and who was succeeded by Mary McAleese in 1997, the appointment of Susan Denham as the first female Chief Justice of the Supreme Court of Ireland in 2011, the election of Joan Burton in 2014 as the first female head of the Labour Party, (and subsequently appointed as deputy prime minister), and with more women in leading positions in parliament and state bodies, Irish women have challenged the authority of the Vatican over their lives, advocated the separation of church from state, and demanded justice for themselves. Since the making of *Philomena*, Philomena Lee, whose position in society has been immeasurably strengthened by the book and film about her, has led a campaign named The Philomena Project, calling for the release of 60,000 adoption records and demanding the right of families to be reunited.[6]

Also, an advocacy group called Justice for Magdalenes was formed in 2009 to demand a government investigation into the Magdalene institutions and compensation for their victims.[7] When the government refused, the group took their case to the United Nations Committee Against Torture in 2011.[8] Other activist groups have also been created, such as the Magdalene Memorial Committee (see Buckley 2013), which campaigned for three years for the right to erect a memorial in St. Stephen's Green to those who had been buried in the mass grave of the Sisters of Our Lady of Charity, and One in Four,[9] which provides advice to those who have suffered sexual abuse. As a result of the screening of *The Forgotten Maggies* at the Galway film festival in 2009 and on the Irish language television station in Ireland in 2011 to more than 300,000 viewers (more than 6 per cent of the population), another activist group called Magdalene Survivors Together formed to include many survivors of the institutions.[10]

It is also significant that women and men have become more active in the campaign to legalize abortion, especially as the life of the unborn child is considered as important as that of the mother under the Eighth Amendment of the Irish Constitution (1983). In some recent cases, women (such as Savita Halappanavar in 2012) have asked for abortions when their lives were in danger but were refused and died. Another woman in 2014 was refused an abortion after being raped and was forced to give birth by Caesarian section (see McDonald 2014a). Also at the end of 2014 a woman, who was pregnant and brain dead and her body, according to one of the doctors, "was now a corpse", but the heart of the baby was still beating, was kept alive for several weeks after her father (with support from her partner) had asked that the life support machines be switched off (see McDonald 2014b). Moreover, women who have been given symphysiotomies (in which the cartilage of the pubic symphysis is cut to widen the pelvis in difficult childbirths) and suffered a number of physical side effects since the procedure (which was often carried out without their consent) have also mounted a strong campaign for compensation, arguing that the Church and state supported this procedure, rather than Caesarian section, so that women could endure more pregnancies.[11]

Mary McAleese, a former President of Ireland, is one of the leaders in the campaign for Church reform. According to an article in the *Irish Times*, "Her intention, even ambition, is to continue to aggravate Rome over its attitude to women, gay people, governance, mandatory celibacy and more for the remainder of her days" (McGarry 2014a). In a speech in Australia in September 2014, McAleese acknowledged that many in Ireland had lost their religious conviction as a result of the various religious scandals over child sexual abuse by priests, and by other priests hiding and abetting their actions.[12] She lamented, "Everything you thought you had, everything you thought you were, becomes a lie," and that "stories came out thanks to the courage of the victims" and the media rather than the Church (Collins 2014).

As Russell Shorto (2011) wrote in the *New York Times*, "many Irish people find the idea of abandoning Catholicism to be as counterintuitive as giving up their racial or sexual identity. [Nevertheless, between] 1974 and 2008, regular Mass attendance dropped by some 50 percent."[13] Moreover, as the number of those entering the clergy has dwindled to less than ten percent of its former annual intake, clerics themselves have begun to break ranks and criticize their colleagues and their practices.[14] Mark Patrick Hederman, the abbot of Glenstal Abbey, reviewing its shameful past, claimed that the Church had made Ireland into a place "where they could control everything. . . . And the control was really all about sex. . . . Generations of people were crucified with guilt complexes. Now the game is up" (quoted in Shorto 2011).[15] This orchestrated campaign—along with the critical commentary provided by filmic, theatrical, and other media representations, which helped raise awareness about the practices in these homes—produced a cumulative effect so that in 2013 the Taoiseach (prime minister) finally issued an apology to the victims of the Magdalene laundries (see Kenny 2013). He also asked the religious institutions to help compensate the former inmates, but the religious institutions refused, claiming that they had done nothing wrong. According to Frances Finnegan, the nuns had always regarded their role in the mother and baby homes as patriotic and cleansing of the evils in society. For example, the Sisters of the Good Shepherd, for whom "the ideal of womanhood became more firmly than ever fixed in the celibate nun" (Finnegan 2004, 18), claimed:

> In the home of the Good Shepherd the one [the nun] is ever the 'Mother', while the other [the penitent] is always the 'Child' and no mother and child on earth are bound together by so pure and holy an affection. . . . Surely there is no grander work for God and soul and the spiritual uplifting of this nation. (quoted in Finnegan 2004, 19)

In conclusion, the discovery in the summer of 2014 that almost 800 babies had been interred in land belonging to a convent served as an extreme example that unwed mothers and their babies had been restricted to a social status comparable to what Agamben calls bare life. Those in authority had regarded these children as of such little value that they did not record their deaths. The norms and moral ethos of the Roman Catholic Church, sanctioned by the Irish government and the Irish public, had excluded unwed mothers from society, forced them to work in laundries for many years without wages, and dispossessed them of their children. The media sensationalized the event in Tuam by asserting that all of the dead babies had been "dumped" in a septic tank on the land,[16] with the result that the news spread around the world, causing such public outrage that, contrary to its usual hesitancy and deference to the Church, the Irish government immediately launched an official inquiry into the practices of all institutions that have cooperated in the disappearance of unwed mothers and their children. The government also launched a police inquiry into the circumstances surrounding the mass grave in Tuam. Hopefully these inquiries will lead to recognition that the government and the Church operated a clandestine biopolitical regime that ruthlessly victimized and exploited women and controlled their bodies.

NOTES

1   One of the early chinks in the bastion of religious authority occurred in 1992 when Annie Murphy named Bishop Eamon Casey as the father of her son Peter and published a book about their affair called *Forbidden Fruit* (Murphy and De Rosa 1993). Amidst reports that he had also embezzled funds from a public charity to provide maintenance for his son, Bishop Casey resigned as Bishop and fled abroad to avoid public scrutiny. In the wake of this scandal, the lid seemed to be lifted off a Pandora's box of troubles within Irish Catholic institutions. Numerous priests were publicly accused of pedophilia and sexual

abuse, including the notorious Father Brendan Smyth, who raped or sexually abused more than one hundred children over a period of forty years (including raping one of my own students when he was fifteen). See BBC News (2010).

2   The success of *Eclipsed* and the ensuing revelations about the laundries by the media led Brogan to set *Stained Glass at Samhain* in the 1990s, with the property developers unearthing the corpses and digging up memories of the past. She showed in this play that the nuns had become easy scapegoats for the condemnation of these institutions and that others were equally to blame. See Smith (2007, 106–12).

3   This is the phrase used by Cathrine Corless to Martin Sixsmith in justifying the existence of a mass grave in Tuam in his *Ireland's Lost Babies* documentary for the BBC (Sixsmith 2014a).

4   See, for example, Smith (2007, 151–2).

5   According to Tom Inglis:

> The export process was operated by nuns. It was sanctioned by the Archbishop of Dublin and administered by the Department of External Affairs. The export of babies for adoption arose because the idea of an unmarried mother looking after her child was outside the realms of morality as set down by the Church and embodied by the laity. The head of the Catholic Social Welfare Bureau described single mothers as "fallen women" and "grave sinners" whose children were the victims of "wickedness". The story reveals the collusion between the state and the Church and the determination of both institutions to create a secret Irish solution to breakdowns in Catholic morality. Instead of women being exported for abortion—the present solution [by going to England where abortion is legal]—their babies were exported for adoption. Such was the Church's moral monopoly in the 1950s and such was the state's willingness to acquiesce in this, that the Archbishop of Dublin was able to lay down strict guidelines which demanded that the adopting parents not only be Catholic, but well-off and be willing to guarantee that the baby be brought up a Catholic, be sent to a Catholic school and, if it arose, to a Catholic university. (1998, 230)

6   See The Philomena Project website, http://thephilomenaproject.org/, accessed 17 January 2015.

7   See the Justice for Magdalenes website, http://www.magdalenelaundries.com/, accessed 17 January 2015.

8   See Justice for Magdalenes' submission to the United Nations Committee Against Torture (2011).

9   See the One in Four website, http://www.oneinfour.ie/ accessed 18 January 2015.

10 See the Magdalene Survivors Together website, http://magdalene-52.wix.com/magdalenesurvivorstogether#!home, accessed 18 January 2015.

11 According to Sinéad O'Shea (2015), writing in the *Irish Times*, "the procedure [of symphysiotomy] involved cutting the pelvic bone to create more space during childbirth and was favored over Caesarian sections by some doctors as it would enable women to have larger families. The procedure was abandoned in most parts of the world by the middle of the 20th century, but continued in Ireland for what is considered to be a mix of religious and cultural reasons. Common long-term effects for the women included impaired walking, chronic pain and incontinence." See also O'Carroll (2012).

12 For example, a victim of Father Brendan Smyth has sued Cardinal Sean Brady for forcing her to take an oath of silence after being abused and preventing her from going to the police. Brady thereby failed to stop Smyth from sexually abusing more than one hundred children over a forty-year period. When the

British government was seeking to extradite Smyth, he hid in a monastery in Ireland and the Irish government delayed extradition procedures. When this was discovered, the Irish government was forced to resign and hold new elections. Smyth was later convicted and died in prison. See, for example, Cooney (2010).

13 Moreover, the number of people attending monthly confession declined from 47 percent in 1974 to 14 percent in 1995 (Inglis 1998, 209). According to research conducted by the Association of Catholic Priests, weekly mass attendance had declined to 35% by 2012 (see http://www.associationofcatholicpriests.ie/wp-content/uploads/2012/04/Contemporary-Catholic-Perspectives.pdf).

14 Tom Inglis (1998, 212) records that between 1966 and 1996 "the number of vocations dropped from 1,409 to 111, a decrease of 92 per cent." He also notes that the "proportion of religious as full-time teachers in secondary schools fell from 48 per cent in 1965 to 9 per cent in 1991" (225). By 2014, the number entering seminary in Ireland to study for the priesthood had dropped to 14. (McGarry, 2014b)

15 As an indication that the ordinary Irish person no longer listens to the Church teaching on sex, the number of unwed mothers has increased dramatically. According to Tom Inglis, writing in 1998, "In 1961, 2 per cent of births occurred outside of marriage. This has now risen to 20 per cent. Furthermore, one in three of first births are to unmarried mothers" (240).

16 Kathy Sheridan (2014), a journalist for the *Irish Times*, was clearly upset by the reportage, writing that:

> [T]he media was shown at its worst when the tragic heart of the story—that 796 children with no burial records had died in a Tuam mother-and-baby home between 1925 and 1961—was hijacked by sensational headlines sent flying around the world, suggesting that the infants were all "dumped" in a septic tank. Corless, who had dedicated many years and thousands of euro to her research, told my colleague Rosita Boland that she had never said that. Her truth needed no embellishment.

REFERENCES

Agamben, Giorgio. 1998. *Homo Sacer: Sovereign Power and Bare Life.* Trans. Daniel Heller-Roazen. Stanford, CA: Stanford University Press.

BBC News. 2010. "Profile of Father Brendan Smyth". Accessed 18 January 2015. http://news.bbc.co.uk/2/hi/uk_news/northern_ireland/8567868.stm.

Boland, Rosita. 2014. Tuam Mother and Baby Home: The Trouble with the Septic Tank Story. *Irish Times* (Dublin), 7 June.

Brien, Ann. 2011. "'Laundry' Performing the Story of Ireland's Magdalene Laundries (Part 2)". Accessed 31 August 2014. http://annbrien.blogspot.de/2011/10/laundry-performing-story-of-irelands_23.html.

Brogan, Patricia Burke. 1994. *Eclipsed.* Knockeven, Ireland: Salmon Poetry.

Brogan, Patricia Burke. 2004. *Stained Glass at Samhain.* Knockeven, Ireland: Salmon Poetry.

Buckley, Dan. 2013. Raftery Helped to Lift Lid on Abuse. *Irish Examiner*, 6 February.

Butler, Judith. 2006. *Precarious Life: The Powers of Mourning and Violence.* New York: Verso.

Butler, Judith and Gayatri Chakravorty Spivak. 2007. *Who Sings the Nation-State? Language, Politics, Belongings.* Kolkata, India: Seagull Books.

CBS. 1999. *Magdalen Laundries: Women Confined in Convents*, documentary. Accessed 9 January 2014. http://www.cbsnews.com/news/magdalen-laundries-women-confined-in-convents/.

Collins, Padraig. 2014. McAleese says Catholic Church's "Old Boys Club" Has to Go. *Irish Times* (Dublin), 8 September.

Coogan, Steve and Jeff Pope. 2013. *Philomena*. Directed by Stephen Frears. London: Pathé.

Cooney, John. 2010. Cardinal's Sin is his Nuremberg Defence. *Irish Independent*, 15 March, 14.

Edwards, Elaine. 2014. Archaeologist questions lack of knowledge of Tuam burials. *Irish Times* (Dublin), 5 June.

Finnegan, Frances. 2004. *Do Penance or Perish: Magdalen Asylums in Ireland*. Oxford: Oxford University Press.

Foucault, Michel. 1977. *Discipline and Punish: The Birth of the Prison*. Trans. Alan Sheridan. New York: Random House.

Foucault, Michel. 1980. Two Lectures. In *Power/Knowledge: Selected Interviews and Other Writings, 1972–1977*. Ed. Colin Gordon, 78–108. New York: Pantheon Books.

Foucault, Michel. 1998. *The History of Sexuality, Vol. 1: The Will to Knowledge*. Trans. Robert Hurley. London: Penguin.

Humphries, Steve. 1998. *Sex in a Cold Climate*, documentary. London: Channel 4.

Inglis, Tom. 1998. *Moral Monopoly: The Rise and Fall of the Catholic Church in Modern Ireland*. Dublin: University College Dublin Press.

Justice for Magdalenes. 2011. "Submission to the United Nations Committee Against Torture." Accessed 17 January 2015. http://www.magdalenelaundries.com/jfm_comm_on_torture_210411.pdf.

Keating, Sara. 2011. "Laundry" by Anu productions at Ulster Bank Dublin Theatre Festival. *Irish Theatre Magazine*, 29 September. Accessed 31 August 2014. http://www.irishtheatremagazine.ie/Reviews/Ulster-Bank-Dublin-Theatre-Festival-2011/Laundry.

Kenny, Enda. 2013. "Statement on Magdalene Report." Accessed 17 January 2015. http://www.merrionstreet.ie/en/news-room/speeches/taoiseach-enda-kennys-statement-on-magdalene-report.html.

Mbembe, Achille. 2003. Necropolitics. Trans. Libby Meintjes. *Public Culture* 15(1): 11–40.

McDonald, Henry. 2014a. Ireland: Woman Forced to Give Birth by Caesarian after being Denied Abortion. *The Guardian* (London), 17 August.

McDonald, Henry. 2014b. Brain Dead Pregnant Woman's Life Support Can be Switched Off, Irish Court Rules. *The Guardian* (London), 26 December.

McGarry, Patsy. 2014a. Mary McAleese: A Thorn in the Church's Side? *Irish Times* (Dublin), 20 June.

McGarry, Patsy. 2014b. Vocations Crisis Worse than in England and Wales. *Irish Times* (Dublin), 19 September.

Mullan, Peter. 2002. *The Magdalene Sisters*. Directed by Peter Mullan. London: Momentum.

Murphy, Annie and Peter de Rosa. 1993. *Forbidden Fruit: The True Story of My Secret Love for Eamon Casey, the Bishop of Galway*. New York: Warner Books.

O'Carroll, Sinead. 2012. A History of Symphysiotomy: The Impact of Catholic Ethics on Irish Medicine. *Journal.ie*, 24 November.

O'Riordan, Steven. 2009. *The Forgotten Maggies*. Directed by Steve O'Riordan. Ireland.

O'Shea, Sinéad. 2015. Symphysiotomy Scheme Pays Out €3.8m. *Irish Times* (Dublin), 30 January.

Raftery, Mary. 2011. Ireland's Magdalene Laundries Scandal Must be Laid to Rest *The Guardian* (London), 8 June.

Sheridan, Kathy. 2014. Review of 2014: The Year the Citizens Broke. *Irish Times Weekend Review* (Dublin), 27 December.

Shorto, Russell. 2011. The Irish Affliction. *New York Times Sunday Magazine*, February 13.

Sixsmith, Martin. 2009. The Catholic Church Sold My Child. *The Guardian* (London), 18 September.

Sixsmith, Martin. 2010. *The Lost Child of Philomena Lee: A Mother, Her Son and a Fifty-year Search*. London: Macmillan.

Sixsmith, Martin. 2014a. *Ireland's Lost Babies*. Directed by John O'Kane. London: BBC.

Sixsmith, Martin. 2014b. Lost, Deceived and Abused: The Losers in Ireland's Baby Trade. *Sunday Times* (London), 14 September.

Smith, James M. 2007. *Ireland's Magdalen Laundries and the Nation's Architecture of Containment*. Manchester: Manchester University Press.

Words On The Street. 2015. "Patricia Burke Brogan." Accessed 18 January. http://www.wordsonthestreet.com/about%20pbb.htm.

16 Israel/Palestine

State of Exception and Acts of Resistance[1]

Ronit Lentin

INTRODUCTION

In October 2013 the Israeli Supreme Court rejected a petition by an Israeli group campaigning for the separation of state and religion to replace the word "Jewish" with "Israeli" on ID cards and population registry documents, and to thus include Palestinian citizens of Israel in the term "Israeli nationality". Israeli ID documents feature an "ethnicity", or "nationhood"/ nationality (*le'om* in Hebrew), category, with the most common being "Jewish" and "Arab" (for Palestinian citizens of Israel). The Court ruled that the existence of an Israeli nationality has not been "objectively proven" and that "Jewishness is not only a religion but also a nationality [and] is a foundational principle of Zionism" (White 2013). Salman Masalha (2013) argues that Palestinian citizens of Israel already claimed "Israeli" nationality in the 1950s, yet, paradoxically, the term "Israeli", used in Israeli passports, does not feature on Israeli ID cards, enabling the state's policing agents to differentiate and categorize racially. The court's rejection of the existence of Israeli nationality in favor of the separate national identities "Jewish" and "Arab"—contradicting the ethnic equality claim of Israel's Declaration of Independence—illustrates the exceptionality of the settler-colonial racial state that is Israel–Palestine.

The state of Israel has been variously theorized in terms of settler-colonialism (Rodinson 1973; Shafir 1993; Wolfe 2006), ethnocracy (Yiftachel 2006), and racial state (Goldberg 2009; Lentin 2008). I suggest that the concept of ethnocracy, "a political regime that facilitates expansion and control by a dominant *ethnicity* in contested lands . . . with rights and capabilities depending primarily on ethnic origin and geographic location" (Yiftachel 2006, 359), implies that Jewish Israelis, despite their ethnic heterogeneities and fissures, are a unified "ethnic" group, naturalizing the Zionist worldview, which is based on a strict racial categorical segregation between Jews and non-Jews; thus the concept of ethnocracy is inadequate. Developing the settler-colonial and racial state paradigms, this chapter argues that the exceptionality of the Israeli state and the Israeli occupation makes it necessary to theorize Israel/Palestine—a borderless state entity, governed through a series of emergency laws inherited from the British colonial rule and never repealed, and through practices of exception, emergency, necessity, and security—as a state of exception, in the sense of both *état* and condition (Agamben 2005).

While Agamben's conceptual tools are useful in analyzing settler-colonial societies such as Israel, the applicability of his approach to Israel–Palestine has several critics. In my edited collection, *Thinking Palestine* (Lentin 2008), Israeli historian Ilan Pappe (2008) argues that, while Agamben posits the state of exception as a lament for democracy under crisis, the state of Israel has never been and cannot be construed as a democracy fallen into a state of exception, but is rather a "state of oppression". David Lloyd (2012) and Barnor Hesse (2011) respectively question whether Israel/Palestine is exceptional or, rather, unexceptional and routine. More relevant to my discussion, in *Agamben and Colonialism*, Svirsky and Bignall (2012, 2) argue that Agamben's project is firmly anchored in western political thought and society, and, because it is conceived without reference to colonialism, it is "outside of the critical interventions that have already been made by colonised peoples engaged in revolutionary retaliation against their oppression." Moreover, as Israel is discursively positioned between Europe and non-Europe, we can apply Agamben's westocentric "state of exception" concept to non-European conditions of settler-colonialism only "under erasure".[2]

This chapter responds to these challenges, firstly, by outlining the theorization of Israel–Palestine as a state of exception in terms of a racial state—as argued by Goldberg (2009)—and as an example of settler-colonialism—as posited by Wolfe (2006) and Svirsky (2012)—and, secondly, by discussing Berda's (2012) analysis of the bureaucracy of the West Bank permit regime, as an example of an occupation regime constantly producing exceptions based on racial hierarchy that replicates colonial ruling systems.

This chapter is as much a Fanonian as an Agambenian project; as Agamben does not fully outline the potentialities of resistance to the state of exception, I draw on Shenhav's (2006) postcolonial reading of Fanon's *The Wretched of the Earth* (2004) to analyze decolonial interventions enacted by Palestinians. Indeed, decoloniality, which implies a broader canon of thought than the western canon—adopting a multivocal, rather than a universal, approach, and taking seriously critical thinkers from the global south (Grosfoguel, 2008)—provides us with an apt theoretical tool for thinking about settler-colonialism. As Veracini (2007) writes, "appraising the evolution of settler colonial forms during the second half of the twentieth century can contribute to an exploration of decolonisation processes" and practices of decolonial resistance.

STATE OF EXCEPTION: PERMANENT STATE OF EMERGENCY, DEFENDING THE *VOLK*

This chapter follows Agamben by arguing that the voluntary creation of a permanent state of emergency, as an essential practice of contemporary states (Agamben 2005, 2), is relevant to Israel/Palestine, where the state of exception, as both *état* and condition, involves both the extension of military wartime powers into the civil sphere, and the suspension of constitutional norms that protect individual liberties.[3]

The state of exception means, not only the declaration of a state of emergency in which the sovereign both enacts the law and stays outside it, but also the idea that it is the *body* of the racial *volk* that needs defending from its internal and external others. In *Homo Sacer* (1998), Agamben argues that the constant state of exception enables the state to render the lives of those under state rule "bare life", both excluded and captured within the political order (9). However, I concur with Svirsky's (2012, 59–60) caution that minoritized and occupied subjects are never just "bare life" because "vesting subaltern subjects with the mark of exception reterritorialises their underprivileged place in language" and imagines them as "subjects to whom all manners of things are done . . . but [who are] rarely agents in their own rights" (Walters 2008, 188, cited in Svirsky 2012, 59–60). Charting the move from a society of discipline to a society of regulation, Foucault (2003) notes the moment in which, when natural life becomes included in mechanisms of state power, politics turns into biopolitics, the territorial state becomes a "state of population", and the nation's biological life becomes a problem of sovereign power. The state here no longer addresses the body in an individualizing mode, but rather as a population—"a 'biopolitics' of the human race" (Foucault 2003, 243).

Through a series of technologies, biopower controls processes such as birthrates, death rates, longevity—all of which become biopolitics' first objects of knowledge in, *inter alia*, statistical terms (Foucault 2003, 243). This makes the population a *subject*, but also an *object*, in the hands of government. Foucault differentiates between the sovereign power of the old territorial state ("to make die and let live") and modern biopower ("to make live and let die") (2003, 241). He theorizes racism as marking a caesura between what must live and what must die and making it possible to establish "a relationship between my life and the death of the other" (2003, 255). This chapter addresses Israeli governmental technologies that construct various categories of Palestinians—Palestinian citizens of Israel ("1948 Palestinians"), occupied Palestinian subjects ("1967 Palestinians"), and the Palestinian diaspora[4]—through different biopolitical measures of segregation and exclusion, including the 1948 "Plan D" for ethnic cleansing (Pappe, 2006), ongoing occupation policies in the occupied Palestinian territory, the prevention of the return of Palestinian refugees, but also population management technologies, as outlined below.

Building on Foucault, I propose that biopolitics in the Israel–Palestine case aims to ensure that Israel's Jewish citizens—Jewish settlers in the West Bank but also Jewish Israelis living within the state's 1948 borders—live at the expense of the Palestinian other(s). At the same time, these policies confirm the biopolitical control over discriminated Palestinian citizens and occupied subjects, whose lives are under constant regulation and control, not only by the Israeli secret services, the Shabak,[5] but also by the occupation's Civil Administration,[6] and other civil authorities. The daily practices of exclusion and control that the Israeli occupation employs include raids, arrests, checkpoints, curfews, house and village demolitions, and population transfers that render the occupied Palestinian subject to sovereign Israeli rule. One poignant example of disparate biopolitical population management techniques is the huge disparity in water availability between occupied West Bank Palestinian subjects and Jewish settlers, as well as Jewish (but also some Palestinian) Israeli citizens. According to the Israeli human rights organization *B'Tzelem*:

> The daily per capita water consumption in the West Bank for domestic, urban, and industrial use is some 73 liters. In areas in the northern West Bank, consumption is much lower. In 2008, per capita daily consumption was 44 liters in the Jenin area and 37 liters in the Tubas area. . . . Per capita water consumption in Israeli towns is 242 liters and in local councils, 211 liters. In other words, per capita use in Israel is three and a half times higher than in the West Bank. The World Health Organization and the United States Agency for International Development recommend 100 liters of water per capita per day as the minimum quantity for basic consumption (2013).[7]

Another poignant example of biopolitics at its most extreme is Israel's policy determining the calorific needs of Gazans living under Israeli blockade. In 2012 the Israeli Legal Centre for Freedom of Movement, *Gisha*,[8] published what became known as the government's "Red lines document" that included "tables calculating the food consumption needs of people in Gaza according to age and gender", determining the number of trucks allowed into Gaza (Guarnieri 2012).

These examples illustrate how biopolitics in these settler-colonial conditions becomes one of death, that Mbembe (2003) calls "necropolitics" and Ghanim (2008, 68–9) documents as "thanatopolitics": "from the viewpoint of the power's victims, the moment that power is directed to destroying, eliminating and dismantling their group, the decision about their life becomes a decision about their death." They also illustrate how death, expulsion, and exclusion become the ultimate way of overcoming (self-perceived) Israeli Jewish victimhood, that Svirsky theorizes as a settler-colonial immunitarian practice of segregation, and political conservation of the paradoxical notion of "Jewish majority" against any possibility of shared ways of life (2012b, 58).

SETTLER-COLONIAL RACIAL STATE

I am aware that analyzing the state of Israel in terms of race may seem unacceptable since Jewish people were themselves targets of brutal racialization during the Nazi era. However, according to David Theo Goldberg, all modern nation-states are racial states; while there is no singular racial state, racial states are all underpinned by history as state memory and by state power(s) (2002, 8). The racial state excludes and includes in order to construct homogeneity, achieved through a range of governmental technologies such as citizenship entitlements, border controls, and census categorizations, but also through invented histories and traditions that construct state memory, ceremonies, and cultural imaginings, and the evocation of ancient origins. I suggest that Zionist settler-colonialism, imagining Israel as Europe away from Europe (a popular discourse is the civilizing "villa in the jungle" metaphor[9]), is a racial state *par excellence*, where:

> there is a constant state of emergency. The state inherited the British Mandate's Emergency Regulations under which it continued the anomalous suspension of the law, within the law . . . this system enables: one rule (life) for the majority of the state's citizens, and another (death, threat of death, threat of expulsion) for the state's subjects, whose lives have been rendered "bare". (Shenhav 2006, 206–7)[10]

Goldberg (2009) calls this "racial Palestinianization": "Palestinians are treated not *as if* a racial group, not simply *in the manner* of a racial group, but *as* a despised and demonic racial group" (139).

Zionism, according to Goldberg, is about the modernizing imperative according to which Jews (though ancient Biblical people) are modern, while Palestinians (Philistines) are premodern and thus in need of Zionism's civilizing—always also colonizing—mission, that Goldberg calls "historicist racism". Foucault's articulation of the need to defend society (2003) brings to mind the Zionist imperative to protect the nebulous body of "the Jewish nation" from antisemitic persecutions, from its 1880 state-building aspirations to the establishment of the state of Israel and its aftermath.

Paradoxically, for a people whose history is replete with racial persecutions, Zionist ideology itself articulates "the Jewish race", constructing a homogeneous "Jewish people" in spite of obvious Jewish heterogeneities, with Jewish self- and other-racialization an integral part of the Zionist ideology. The Israeli geneticist Rafael Falk reads the history of Zionism as a eugenic race project, aiming to save the Jewish genetic pool from the degeneration which results from diasporic existence (Falk 2006, 25). Just as antisemitism racialized Jews as a separate "race", justifying their persecution by biological reasoning, so Zionist ideologues adopted the terminology of *volk*—a racial nation shaped by "blood and soil" (Falk 2006, 18–9)—and were instrumental in producing a Zionist repertoire of racial categorization and *volkish* imagery (Bloom 2007).

This racial thinking led to Israel's citizenship regime. Constructed as the state of the "Jewish nation", Israel (through the Law of Return) grants automatic citizenship to anyone who can prove she has a Jewish mother,[11] at the same time denying citizenship to Palestinians born on the land who were deported (or fled) during and after the 1948 *Nakba*. The 160,000 Palestinians not expelled were dubbed "Israeli Arabs" and put under a Military Government regime, based on the 1945 British Emergency Regulations, which abolished basic rights of expression, movement, organization, and equality, though they left Palestinian citizens the right to vote and be elected. Though officially abolished in 1966, the emergency regulations are still in place, controlling 20 percent of Israel's citizens (Pappe 2006, 220–2).

The law, as Goldberg argues, works in the service of the racial state. In Israel, examples of racial laws include the Law for Absentee Property (1950), in which the state grants itself ownership of the property of Palestinians expropriated in 1948, dubbing them "present absentees". The Jewish National Fund Law (1953) charges the JNF with administering public land in Israel, including large tracts of displaced Palestinians' lands (Mahajneh 2010), barring the selling, leasing, subletting, and owning of land by "non-Jews" (read "Palestinians"). In addition, Adalah, the Legal Center for Arab Minority Rights in Israel, lists 50 Israeli laws that discriminate against Palestinian citizens of Israel in all areas of life, including their rights to political participation, access to land, education, state budget resources, and criminal procedures. Some laws also violate the rights of Palestinians living in the 1967 Occupied Palestinian Territory and of Palestinian refugees.[12] Since 2009, and the advent of Binyamin Netanyahu's right-wing government, a new series of laws seeks:

> to dispossess Arab citizens of Israel and exclude them from the land; turn their citizenship from a right into a conditional privilege; limit the ability of Arab citizens and their parliamentary representatives to participate in the political life of the country; criminalize political acts or speech that question the Jewish or Zionist nature of the state; and privilege Jewish citizens in the allocation of state resources. (Adalah 2012, 1)

These laws aim not only to curtail the access of Palestinian citizens of Israel to land resources, they also aim to preempt, circumvent, or overturn Supreme Court decisions that accord some civil rights to Arab citizens. Furthermore, the legislation stifles freedoms of association and expression, discriminates on the basis of national belonging, and supports a series of criminal indictments and punitive measures instigated by the Knesset (Israeli Parliament) against its elected Arab Members.

One legal measure stands out in the recent wave of discriminatory legislation. In September 2009, the Israeli Interior Ministry changed the name of the Population Registration Office in East Jerusalem to the "Foreigners' Office" and, in one fell swoop, turned indigenous Palestinians into "foreigners" (Aminov 2013). This came in the wake of the ratification by the Israeli Knesset of the racist "Entry to Israel Law" for the eleventh time since its enactment. The law prevents marriages between Palestinian citizens of Israel and non-citizen Palestinians on the pretext that Palestinian family unification constitutes a "demographic" (read "security") risk. According to the law, targeting both 1948 and 1967 Palestinians, West Bank inhabitants who marry Israeli citizens can live with their spouses in Israel only if they are Jewish; otherwise, their Israeli spouses can either live with them in the West Bank, or the couple can live apart. By 2010, after a 1988 Supreme Court ruling had abolished the right of non-citizen Palestinians to live in Israel, 13,000 people were deported under this ruling. Aminov argues that the newly named "Foreigners' Office" reconceptualizes native Palestinians as foreigners in their own land, turning the native population into a collection of foreigners, subject to immigration laws (Aminov 2013).

Understanding Israel as a racial state is incomplete without also theorizing it as a settler-colonial enterprise. Settler-colonialism's main objective is not the enslavement of members of the indigenous population, but rather access to their territory. Jewish Israeli settler-colonialism was molded not merely by Zionist ideology, but also by the process of colonizing an already settled territory, a process that inevitably brought about conflict with the indigenous society (Shafir 1993).

Defining Israel as a settler-colonial society goes against Zionism's self-image of being a response to European antisemitism and enabling the settlement of a "people without land in a land without people."[13] It is worthwhile noting that early Zionism explicitly cast itself in settler-colonial terms: "Yishuv"—the term for the pre-state polity in Palestine—literally means "settlement". After the first Zionist Congress of 1897, which voted in favor of Jewish migration to Palestine, it was explicitly decided to establish "three types of colonies in Palestine: kibbutz, moshav and town" (Benvenisti 2002, 263).[14] Furthermore, as Wolfe (2006, 388) argues, settler-colonialism always replaces what it destroys. Thus, replacing Palestinian orchards with imported European conifers ("making the desert bloom"), depopulated Palestinian villages with Jewish settlements, roads and national parks, Palestinian place names with Hebrew place names, and the current campaign to replace Bedouin villages, deemed "unrecognized", with Jewish space for, *inter alia*, army maneuvers,[15] are all classic settler-colonial state practices.

Settler-colonialism is informed by various interpretations of *terra nullius*, and "is an inclusive, land-centred project that coordinates a comprehensive range of agencies, from the metropolitan centre to the frontier encampment, with a view to eliminating indigenous societies" (Wolfe 2006, 393). Analyzing Australian, American, and Israeli settler-colonialism, Wolfe argues that settler-colonialism is based on a logic of elimination. Furthermore, in Israel, biopolitics regulates and controls the lives of "Jews" and "Palestinians" alike, the former accorded land resources and encouraged to breed and multiply, the latter—both citizens and occupied subjects—deprived of land resources and of means of sustaining their families (through lower allocations of education, health, and crucial resources).

Wolfe theorizes settler-colonialism, premised on securing territory, in terms of "structured genocide", which allows us to appreciate the concrete relationships between spatial removal, mass killings, and biocultural assimilation. One example is Israel's progressive dispensing with Palestinian workers from the occupied territory (for security reasons)[16] and the building of the so-called "separation wall". The logic of elimination is illustrated by the expulsion of the Palestinians from their lands during and after the 1948 Nakba, by their villages and urban neighborhood being settled by Jews, by the 1967 occupation of the West Bank and Gaza, and by the increasing military and civilian control in the territory.

As Wolfe (2006, 404) reminds us, "There could hardly be a more concrete expression of spatial sequestration than the West Bank barrier . . . as Palestinians become more and more dispensable, Gaza and the West Bank become less and less like Bantustans and more and more like reservations (or, for that matter, like the Warsaw Ghetto)."

Crucially, in the state of exception the paradigm of security is used as "the normal technique of government" (Agamben 2005, 14). Security, discourses of "existential threat", and Jewish victimhood are central to Israel's racial policies. As Israel sees itself as a haven for the "Jewish nation", it regards the control of 1948 Palestinians, 1967 Palestinians, and diasporic Palestinians as an imperative born of necessity and emergency, which, as Agamben suggests, create and guarantee the situation that the law needs for its validity.

While neither Foucault nor Agamben directly theorize colonialism in the context of biopower, settler-colonialism is intrinsic to biopower, the product and process of a colonial world. Morgensen (2011) argues that colonialism and settler-colonialism place indigenous peoples in a state of exception, albeit one that troubles the territorial and national integrity of settlers as representatives of western law. Morgensen advances Agamben's argument that biopower is intrinsic to western law and reminds us, following Wolfe, that as settler colonials come to stay, assertions of sovereignty by settlers ground western law in "a logic of elimination". Indeed, rather than extermination, the logic of elimination aims not to destroy but to produce life, so as to amalgamate indigenous peoples, cultures, and lands into the body of the settler nation (2011, 56). I would further argue that, when settlers stay and regard the new colony as "their" land, settler-colonialism must employ biopolitical strategies, controlling the life practices of the indigenous, so as to "make live" the colonists—examples of which are ample in Israel–Palestine.

Furthermore, Israel's settler-colonial racial regime creates constant zones of exception. The military courts system for Palestinians in the occupied territory, operating with "virtually no transparency, . . . subject to very lax internal supervision, [and] rarely exposed to any public scrutiny" (*Yesh Din* 2013), is one example of what African-American journalist and Civil Rights leader Ida B. Wells called (with reference to lynching) the seemingly exceptional "unwritten law" which becomes the law.[17]

Another illustration of settler-colonialism as a racial state of exception is Yael Berda's analysis of the occupation's colonial bureaucracy enacted, every day in every West Bank checkpoint, through sovereignty—practices of controlling space through the security forces—and governmentality—practices of managing populations through intelligence, economic control, and racial profiling (Berda 2012).

Berda's analysis has three components. The first is racial segregation—an overarching principle of the colonial bureaucracy model. Contrary to Weber's classical understanding of bureaucracy as rational, impersonal, and universal, colonial bureaucracy is designed to be irrational and have no fixed rules; thus it creates a gap, based on racial hierarchy, between governmental practices targeting Jews and those targeting Palestinians. The second is the individualization of the permit regime and the management of the occupied population, based on racial profiling. After the 1993 Oslo Accords between the State of Israel and the Palestinian Authority,[18] and the ensuing separation policy (including the "separation wall"), the Palestinian individual becomes a tool of intelligence sourcing whose movement across the many checkpoints must be surveyed and delayed. The third component is the lack of actual separation between the bureaucracy managing the Palestinian population and the management of the labor market within the Israeli state. Berda analyzes the everyday colonial bureaucracy of the permit regime through racial hierarchies, managerial flexibility, and the constant production of exceptions by an "absent sovereign"—Palestinian workers never know who makes the decision to grant or not grant them a work permit. Her analysis addresses not only occupied Palestinian subjects, but also the very heart of the state system—demonstrating Agamben's insistence that the line separating citizens and non-citizens is very thin indeed (Agamben 2008). Crucially, the secretive production of exceptions entails the categorization of West Bank Palestinians in relation to being "security risks", making the Security Services the final arbiters in granting or refusing permits and thus in enabling or disabling Palestinians to work and, ultimately, live (Berda 2011, 164–7).

RESISTANCE AND DECOLONIZATION

While Agamben and Goldberg both insist that the racial state of exception channels potentialities of resistance, Svirsky (2014) cites Agamben's *Profanations* (2007) to suggest that the concept of resistance commonly attributed to emancipatory acts is inaccurate: resistance is often offered by reactionary forces that keep the status quo and dominant identities together; on the other hand, *profanation* is performed by forces who aim to dissolve the pervasiveness of privilege. In "We Refugees" (1995), Agamben suggests that the refugee, his *homo sacer*, destabilizes the holy trinity of state–nation–territory, and is a central actor in contemporary political history. Following Walters (2008)—who notes the inadequacy of the concept of "bare life" that suggests a passive subject to whom all is being done—I recognize that, as Israel employs different measures of control against Palestinians, depending whether they are citizens, occupied subjects, or living in exile, "bare life" is inadequate for thinking about forms of Palestinian resistance, or, as Svirsky would have it, Palestinian profanation of sacred Zionist privileges.

Shenhav's (2006) discussion of resistance in Fanon's writing is useful in counteracting Agamben's westocentrism and understanding Palestinian decolonization. In *The Wretched of the Earth* (2001)—written while he was deeply involved in the Algerian struggle—Fanon posits violence as a crucial phase of decolonization. He recognized that the forces of occupation cannot last and that, for the colonized natives, the most essential value, because the most concrete, is first and foremost the land of which they are usurped by the settler-colonials (2001, 34). The colonizer's argument that the colonized understand only force —an argument regularly employed by Israel to justify its actions—means that colonial violence aims not only to keep the enslaved at arm's length, but also to dehumanize them. The settler's preoccupation with security reminds the natives that it alone is master.

Fanon describes a classic settler-colonial condition, where "the settler keeps alive in the native an anger which he deprives of outlet; the native is trapped in the tight links of the chains of colonialism" (2001, 42). Violence, according to Fanon, leads not only to trauma, and hence submission, but also, as Sartre argues in his preface to Fanon's text, to the colonized making it their own. The settler colonial needs the colonized: as the colonist army becomes ferocious, as the country is marked out, and there are mopping up operations, transfers of population, reprisal expeditions, and massacres of women and children, the colonized draw from violence their humanity, while "the European has only been able to become a man through creating slaves and monsters" (Sartre 2001, 22).

In Israel–Palestine the colonial sovereign makes the state of exception unexceptional, routinized, and a paradigm of the normal, blurring exception and the law. In her afterword to the Hebrew edition of Fanon's *The Wretched of the Earth*, Shohat (2006, 331) writes that, while Jews can identify with Fanon's emancipatory project, Israel's rule over Palestine and the concomitant Zionist anti-Arab and anti-Mizrahi racisms locate it firmly on the European side of the Fanonian equation, which is why reading Fanon provides the missing decolonial link in the European debate about the state of exception.

Settler-colonialism is not merely the epitome of biopolitics as argued above, it is also indelibly linked to decolonization, as Veracini (2007) argues. However, he also points to the difficulty of decolonizing a polity in which the settlers establish independence and retain the objectives and methods of their colonizing pasts, without ceding any autonomy or narrative to the indigenous peoples—as in Israel/Palestine. However, reading

Fanon's account of decolonization helps us to understand Palestinian people as decolonizing colonial oppression as a means of assuming subjecthood. Such decolonial practices clear the way for resistance by Palestinian subjects, as is evidenced on a weekly basis in protests in Palestinian villages and neighborhoods such as Sheikh Jarrach, Silwan, Beit Omar, Bil'in, Nabi Saleh, and throughout the West Bank. However, even in empathetic Jewish Israeli readings, such Palestinian subjects are largely theorized as victims, or "bare life" (an unacceptable appellation, according to Walters 2008, as it denies them agency).

CONCLUSION

This chapter argued that although Agamben is not directly engaging with postcolonial criticism, "some of [his] concepts can be used in the service of a strategic postcolonial politics of transformation that is mindful of colonial legacies and histories of resistance" (Bignall and Svirsky 2012, 11–2). Furthermore, the case of Israel/Palestine, as Berda argues, can hardly be analyzed without reference to the routinized production of exceptions. I argue that Agamben's state of exception is alive and well in the settler colony that is Israel/Palestine, where the sovereign re-rehearses, through ceremonial acts of occupation and closure, the glory, which, as Agamben (2011, xii) argues, "is still at the center of political apparatuses of contemporary democracies".

Meanwhile, Palestinians are conducting their decolonizing insurrection through a variety of forms of resistance and profanation, including legal interventions, anti-occupation protests, steadfastness (*sumud*—maintenance of Palestinians on their land and building alternative institutions so as to resist and undermine the Israeli occupation), and the decolonization of history, described by Nur Masalha (2012, 18) as Palestinian acts of reclaiming the Nakba history (which was erased by the Israeli state) through oral history, "social history from below", and the formation of popular and gendered memories.

As for the definition of "Israeliness" with which I began the chapter, when asked whether he identifies as Israeli or Palestinian, the Palestinian-Israeli actor Saleh Bakri says: "I was born Palestinian and will remain Palestinian. I feel no connection to the state of Israel. It ruined my life . . . and the life of my people, and continues to do so" (Anderman 2013). Another illustration of *sumud*, and of the refusal by Palestinian citizens of Israel to accept their subjugation, is the angry, ironic post by Palestinian lawyer (Israeli citizen) Nisreen Ghanimeh on her Facebook page, with which I end:

> To all liberal peace loving Israelis—it's time you woke up . . . your state has lost its moorings. . . . It reared generations of frightened people whose racism is hard to take. Only today, an educated "Israeli" woman cancels a rental contract with me and two other girls, because we are

Arab. It's her right. There is a clause in the "non-compulsion of equality in the private sector" law. Surprising? Not at all. But the revulsion in my mouth makes me spit. Am I not ashamed? No! And I very much want to call a spade a spade. This is a first-class Apartheid state. . . . But you are missing one important thing. We were brought up to fight. Not to give weapons to 16 year olds called settlers. We were born to fight by studying in universities, working, and bringing up an educated generation which does not judge people according to their origin. We have learnt your language, beyond words, and we'll never give up. Despite everything you do, we have learnt to look you in the eye and call you criminals. We won't budge from this state. You can try anything you want . . . you can detain, expel, demolish, curse, categorize, split, even kill. But we are here! Here! . . . By the way, I am Nisreen, law graduate, criminal lawyer, originally from Taibeh. And my grandmother's name is co-existence. (Ghanimeh 2013; my translation from the Hebrew)

NOTES

1 I am indebted to Nitza Aminov, David Landy, and Elena Moreo, and to the editors for their insightful comments on previous drafts.
2 Stuart Hall (1996, 1) argues that the deconstructive approach puts key concepts "under erasure". This indicates they are no longer "good to think with" in their original form, but since they have not been superseded dialectically, and there are no other, entirely different, concepts with which to replace them, there is nothing to do but to continue to think with them, albeit now in their de-totalized, or deconstructed form.
3 See Zreik (2008).
4 An additional category of Palestinian subjects is the Palestinians living in annexed East Jerusalem.
5 Shabak is the acronym for Israel's general security services, whose duties are: safeguarding state security; interrogating terror suspects; providing intelligence for counter-terrorism operations in the West Bank; counter-espionage; the personal protection of senior public officials; securing important infrastructure and government buildings; and safeguarding Israeli airlines and overseas embassies. It also recruits Palestinian informers and is the final arbiter in controlling the West Bank permit regime, as discussed below.
6 "The Civil Administration is part of the Israeli Government coordination apparatus in the occupied territories, responsible for the implementation of the government's policy in Judea and Samaria (Jewish-Israeli Biblical names for the occupied West Bank) and the progression of the region in civil areas, according to the state level instructions, and in coordination with government ministries, the Israel Defense Forces (IDF) and the security system. As such, the CA is a vital part of the IDF activities. . . . A central role of the Civil Administration is civil and security liaison with the Palestinian Authority. . . . In addition, the CA is in charge of liaising with international bodies in relation to humanitarian aid and entrepreneurship in Judea and Samaria" (see http://www.cogat.idf.il/1279-he/Cogat.aspx).
7 Water deprivation is an everyday act of governmentality in the occupied West Bank. In October 2013 *Ha'aretz* reported that Israel's Civil Administration

blocked nine water containers, donated by the Norwegian People's Aid Organization, meant for domestic vegetable gardening, from reaching two small
Palestinian communities, Nabi Samuel and Al Khaleila, surrounded by Jewish
settlements and the separation wall (Hass 2013).

8   See http://gisha.org/.

9   Ehud Barak, Israel's former prime minister, often used a telling metaphor:
    Israel is "a villa in the middle of a jungle". Uri Avnery explains: we are an
    island of civilization surrounded by savage animals. This is remarkably similar
    to old-established colonial attitudes, and, indeed, a variation of the father of
    modern Zionism, Theodor Herzl's, metaphor of the "wall against barbarism"
    (Avnery 2002).

10 The emergency regulations are revalidated annually by the Israeli Knesset
    (Pappe 2008, 149).

11 Or, for the wave of immigration from the former Soviet Union, of whom half
    were not Jewish according to the orthodox reading, but who have Jewish rela-
    tives (Hayeem 2010).

12 See http://adalah.org/eng/Israeli-Discriminatory-Law-Database.

13 An early Zionist slogan, but actually used as early as 1843 by a Christian
    Restorationist clergyman (Garfinkle 2009, 265).

14 Kibbutz (Hebrew for "gathering") is a collective agricultural community; the
    first kibbutz was established in 1909. Moshav (Hebrew for "village" or "set-
    tlement") is a cooperative agricultural community of small farms established
    by Labour Zionism.

15 A poignant example is the Bedouin village of al-Araquib, in the Naqab
    or Negev desert, demolished 86 times between 2010 and 2015, as part of
    Israel's 65-year-old project of forced displacement of Palestinian Bedouins
    (Barrows-Friedman 2013). In November 2013, the Israeli cabinet voted to
    demolish another unauthorized Bedouin village and replace it with a religious
    Jewish community, Hiran (Seidler 2013).

16 See Kemp and Raijman (2008).

17 See Hesse (2011).

18 The "Oslo Accord" was the name of the "Declaration of Principles on
    Interim Self-Government Arrangements", signed in October 1993 between
    the State of Israel and the Palestine Liberation Organisation (PLO). The
    Accords, negotiated by Norway, transferred power to the Palestinians in the
    Gaza and Jericho areas, while Israel continued to hold sovereignty (Knesset
    2000).

REFERENCES

Adalah, The Legal Center for Arab Minority Right in Israel. 2012. "Discriminatory Laws in Israel." Haifa. Accessed 31 March 2014. http://adalah.org/eng/
Israeli-Discriminatory-Law-Database.

Agamben, Giorgio. 1995. We Refugees. Trans. Michael Rocke. *Symposium* 49 (2):
114–19.

Agamben, Giorgio. 2005. *State of Exception.* Trans. Kevin Attell. Chicago: University of Chicago Press.

Agamben, Giorgio. 2007. *Profanations.* Trans. Jeff Fort. New York: Zone Books.

Agamben, Giorgio. 2008. Beyond Human Rights. *Social Engineering* 15: 90–5.

Agamben, Giorgio. 2011. *The Kingdom and the Glory: For a Theological Genealogy of Economy and Government.* Trans. Lorenzo Chiesa. Stanford: Stanford
University Press.

Aminov, Eli. 2013. "How Did the Palestinians Become Foreigners in Their Own Land". *Hagadah Hasmalit*. Last modified 26 August, 2013. http://hagada.org.il/2013/08/26.

Anderman, Nirit. 2013. The Actor Saleh Bakri: I will no longer appear in Israeli theatre and films. The state is fascistic. *Ha'aretz* (Tel Aviv), 22 October.

Avnery, Uri. 2002. Barak: Israel, a Villa in the Jungle. *Arab News* (Jeddah, Saudi Arabia), 17 July.

Barrows-Friedman, Nora. 2013. "Israel's ongoing 'Sociocide' in the Naqab". *Electronic Intifada*. Last modified 25 August 2013. http://electronicintifada.net/blogs/nora-barrows-friedman/israels-ongoing-sociocide-naqab.

Benvenisti, Meron, ed. 2002. *The Morning After: The Peace Era, Not Utopia*. Jerusalem: Carmel and the Truman Institute.

Berda, Yael. 2012. *The Bureaucracy of the Occupation: The Permit Regime in the West Bank, 2000–2006*. Tel Aviv: Van Leer Jerusalem Institute / Hakibbutz Hameuchad.

Bignall, Simone and Marcelo Svirsky. 2012. Introduction to *Agamben and Colonialism*. Ed. Marcelo Svirsky and Simone Bignall, 1–14. Edinburgh: Edinburgh University Press.

Bloom, Etan. 2007. What "The Father" had in Mind? Arthur Ruppin (1876–1943), Cultural Identity, Weltanschauung and Action. *History of European Ideas* 33: 330–49.

B'Tzelem. 2013. "The Gap in Water Consumption between Palestinians and Israelis". Accessed 31 March 2014. http://www.btzelem.org/water/consumption_gap.

Falk, Raphael. 2006. *Zionism and the Biology of the Jews*. Tel Aviv: Resling.

Fanon, Frantz. 2001. *The Wretched of the Earth*. Trans. Constance Farrington. London: Penguin.

Foucault, Michel. 2003. *Society Must Be Defended: Lectures at the Collège de France, 1975–76*. Ed. Mauro Bertani and Alessandro Fontana. Trans. David Macey. London: Allen Lane.

Garfinkle, Adam. 2009. *Jewcentricity: Why the Jews Are Praised, Blamed, and Used to Explain Just About Everything*. London: Wiley.

Ghanim, Honaida. 2008. Thanathopolitics: The Case of the Colonial Occupation in Palestine. In *Thinking Palestine*. Ed. Ronit Lentin, 65–81. London: Zed Books.

Ghanimeh, Nisreen. 2013. *Facebook profile*. Accessed 13 November 2013. https://www.facebook.com/nisreen.ghanimeh.

Goldberg, David Theo. 2002. *The Racial State*. Oxford: Blackwell.

Goldberg, David Theo. 2009. *The Threat of Race: Reflections on Racial Neoliberalism*. Oxford: Wiley-Blackwell.

Grosfoguel, Rámon. 2008. Transmodernity, Border Thinking, and Global Coloniality: Decolonizing Political Economy and Postcolonial Studies. *Eurozine*. Last modified 4 July 2008. www.eurozine.com/articles/2008–07–04-grosfoguel-en.html.

Guarnieri, Mya. 2012. "Government Releases 'Red Lines' Document Detailing Gaza Food Restrictions". *+972 Magazine*. Last modified October 17, 2012. http://972mag.com/government-releases-red-lines-document-detailing-gaza-food-restrictions/57883/.

Hall, Stuart. 1996. Who Needs Identity? In *Questions of Cultural Identity*. Ed. Stuart Hall and Paul Du Gay, 1–17. London: Sage.

Hass, Amira. 2013. The Civil Administration Blocks the Supply of Water Containers from Isolated Palestinian Communities in the West Bank. *Ha'aretz* (Tel Aviv), 14 October.

Hayeem, Abe. 2010. Israel's Unfair "Law of Return". *The Guardian* (London), 11 March.

Hesse, Barnor. 2011. Self-fulfilling Prophecy: The Postracial Horizon. *The South Atlantic Quarterly* 110 (1): 155–78.

Kemp, Adriana and Rebecca Raijman. 2008. *Workers and Foreigners: The Political Economy of Labour Migration in Israel*. Jerusalem: Van Leer Institute.

Knesset. 2000. "Oslo Accords." Accessed 15 October 2013. http://www.knesset.gov.il/process/docs/oslo_eng.htm.

Lentin, Ronit, ed. 2008. *Thinking Palestine*. London: Zed Books.

Lloyd, David. 2012. Settler Colonialism and the State of Exception: The Example of Palestine/Israel. *Settler Colonial Studies* 2 (1): 59–80.

Mahajneh, Alaa. 2010. "Situating the JNF in Israel's Land Laws". *Badil Resource Center: For Palestinian Residency and Refugee Rights*. Last modified 2010. http://www.badil.org/en/al-majdal/item/1404-mahajneh-jnf-and-israeli-law.

Masalha, Nur. 2012. *The Palestine Nakba: Decolonizing History, Narrating the Subaltern, Reclaiming Memory*. London: Zed Books.

Masalha, Salman. 2013. Israeliness—There is No Such Thing. *Ha'aretz* (Tel Aviv), 27 October.

Mbembe, Achille. 2003. Necropolitics. Trans. Libby Meintjes. *Public Culture* 15 (1): 11–40.

Morgensen, Scott Lauria. 2011. The Biopolitics of Settler Colonialism: Right Here, Right Now. *Settler Colonial Studies* 1 (1): 52–75.

Pappe, Ilan. 2006. *The Ethnic Cleansing of Palestine*. Oxford: Oneworld.

Pappe, Ilan. 2008. "The *Mukhabarat* State of Israel: A State of Oppression is Not a State of Exception". In *Thinking Palestine*. Ed. Ronit Lentin, 148–70. London: Zed Books.

Rodinson, Maxime. 1973. *Israel, A Settler-Colonial State?* New York: Monad.

Sartre, Jean Paul. 2001. Preface to *The Wretched of the Earth*, by Frantz Fanon. Trans. Constance Farrington. London: Penguin.

Seidler, Shirley. 2013. Cabinet Okays Demolishing Bedouin Village, Replacing With Jewish Town. *Ha'aretz* (Tel Aviv), 10 November.

Shafir, Gershon. 1993. Land, Labour and Population in Zionist Colonization: Unique Economic Aspects. In *Israeli Society: Critical Aspects*. Ed. Uri Ram, 104–19. Tel Aviv: Brerot.

Shenhav, Yehouda. 2006. The Imperial History of "State of Exception". *Theory and Criticism* 29: 205–18.

Shohat, Ella. 2006. Afterword to *The Wretched of the Earth*, by Frantz Fanon. Hebrew Edition. Tel Aviv: Bavel.

Svirsky, Marcelo. 2012. The Cultural Politics of Exception. In *Agamben and Colonialism*. Ed. Marcelo Svirsky and Simone Bignall, 52–74. Edinburgh: Edinburgh University Press.

Svirsky, Marcelo. 2014. *After Israel*. London: Zed Books.

Svirsky, Marcelo and Simone Bignall, eds. 2012. *Agamben and Colonialism*. Edinburgh: Edinburgh University Press.

Veracini, Lorenzo. 2007. Settler Colonialism and Decolonization. *Borderlands* 6 (2). http://www.borderlands.net.au/vol6no2_2007/veracini_settler.htm.

Walters, William. 2008. Acts of Demonstration: Mapping the Territory of (Non-)citizenship. In *Acts of Citizenship*. Ed. Engin F. Isin and Greg M. Nielsen, 182 –207. London: Zed Books.

White, Ben. 2013. Court Denial of Israeli Nationality Reinforces Discrimination. *Middle East Monitor*. Last modified 7 October, 2013. http://www.middleeast monitor.com/articles/debate/7707-court-denial-of-israeli-nationality-reinforces-discrimination.

Wolfe, Patrick. 2006. Settler Colonialism and the Elimination of the Native. *Journal of Genocide Research* 8 (4): 387–409.

Yesh Din: Volunteers for Human Rights. 2013. "Military Courts". Accessed 31 March 2014. http://www.yesh-din.org/cat.asp?catid=5.

Yiftachel, Oren. 2006. *Ethnocracy: Land and Identity Politics in Israel/Palestine*. Philadelphia: University of Pennsylvania Press.

Zreik, Raef. 2008. The Persistence of the Exception: Some Remarks on the Story of Israeli Constitutionalism. In *Thinking Palestine*. Ed. Ronit Lentin, 131–47. London: Zed Books.

Contributors

Giorgio Agamben is Professor of Aesthetics at the European Graduate School in Saas-Fee, Switzerland and is author of many books, including *Homo Sacer: Sovereign Power and Bare Life* (Stanford University Press, 1998), *Remnants of Auschwitz: The Witness and the Archive* (Zone Books, 1999), *State of Exception* (University of Chicago Press, 2005), *Profanations* (Zone Books, 2007), *The Kingdom and the Glory* (Stanford University Press, 2011), *Opus Dei: An Archeology of Duty* (Stanford University Press, 2013), and *The Highest Poverty: Monastic Rules and Forms-of-Life* (Stanford University Press, 2013).

Rosi Braidotti's research and writing engages Continental philosophy, feminist theory, and cultural studies of science and technology. She is an expert in poststructuralism and critical theory and in the philosophy of Gilles Deleuze. Her books include *The Posthuman* (Polity Press, 2013), *Nomadic Theory. The Portable Rosi Braidotti* (Columbia University Press, 2011), *Transpositions: On Nomadic Ethics* (Polity Press, 2006), *Metamorphoses: Towards a Materialist Theory of Becoming* (Polity Press, 2002), and *Nomadic Subjects: Embodiment and Sexual Difference in Contemporary Feminist Theory* (Columbia University Press, 1994 and 2011). She was awarded an Honorary Degree in philosophy by the University of Helsinki in 2007 and by Linköping University in 2013. She became an Honorary Fellow of the Australian Academy of the Humanities *(FAHA)* in 2009 and a Member of the Academia Europaea *(MAE)* in 2014.

Andrés Fabián Henao Castro is Assistant Professor of Political Science at the University of Massachusetts, Boston. His research deals with the relationships between ancient and contemporary political theory, particularly in reference to democratic and de-colonial theories and practices, the question of political subjectivity, and the distribution of political agency. He is currently working on a book that explores different subject-positions and forms of agency imagined in the theoretical reception of Sophocles' tragedy *Antigone*. His recent publications include the article "Can the

Subaltern Smile? *Oedipus* Without Oedipus" (*Contemporary Political Theory*) and "Antigone Claimed: 'I am a Stranger!' Political Theory and the Figure of the Stranger" (*Hypatia: A Journal of Feminist Philosophy*). He is also a member of the international research network *Performance Philosophy* and a columnist for the online journal of political analysis *Palabras al Margen* (Words at the Margin).

Oron Catts is an artist, researcher, curator, and Director of SymbioticA, the Centre for Excellence in Biological Arts, within the School of Anatomy and Human Biology at The University of Western Australia. In 1996, he founded the Tissue Culture and Art Project to explore the use of tissue technologies as a medium for artistic expression. His work is part of the MoMA design collection and has been exhibited and presented internationally.

Matthew Causey is Associate Professor of Drama at Trinity College, Dublin and Founding Director of the Arts Technology Research Laboratory. Dr Causey is author of *Theatre and Performance in Digital Culture: From Simulation to Embeddedness* (Routledge, 2006, 2009) and co-editor of *Performance, Identity and the Neo-Political Subject* (Routledge, 2013). He is Director of the Arts Strand of the Ireland-wide PhD Programme in Digital Arts and Humanities.

F. J. Colman is Professor of Media Arts at the Manchester School of Art, Manchester Metropolitan University. She is the author of *Film Theory: Creating a Cinematic Grammar* (Columbia University Press, 2014) and *Deleuze and Cinema* (Berg 2011), editor of *Film, Theory and Philosophy: The Key Thinkers* (Acumen, 2009), and co-editor of *Global Arts & Local Knowledge* (with Helene Frichot and Jack Reynolds; Lexington, 2015) and *Sensorium: Aesthetics, Art, Life* (with Barbara Bolt and Graeme Jones; Cambridge Scholars, 2007). Her two current book projects are on "Digital Feminicity" and "Materialist Film".

James Harding is Associate Professor of Theatre and Performance Studies at the University of Maryland and former Vice President of Performance Studies international. He has held professorships at the University of Warwick and University of Mary Washington, and has been a Guest Professor at the Freie Universität, Berlin. His research has concentrated mostly on cultural politics and the *avant-garde*. His main works are *The Ghosts of the Avant-Garde(s)* (University of Michigan Press, 2013, 2015), *Cutting Performances: Collage Events, Feminist Artists and the American Avant-Garde* (University of Michigan Press, 2010, 2012) and *Adorno and "A Writing of the Ruins": Essays on Modern Aesthetics and Anglo-American Literature and Culture* (State University of New York Press, 1997). He co-edited *The Rise of Performance Studies: Rethinking*

Richard Schechner's Broad Spectrum (with Cindy Rosenthal; Palgrave Macmillan, 2011) and *Restaging the Sixties: Radical Theaters and their Legacies* (University of Michigan Press, 2006), and *Not the Other Avant-Garde: The Transnational Foundations of Avant-Garde Performance* (with John Rouse; University of Michigan Press, 2006). He also edited *Contours of the Theatrical Avant-Garde: Performance and Textuality* (University of Chicago Press, 2000).

Thomas Lemke is Professor of Sociology with a focus on Biotechnologies, Nature, and Society at the Faculty of Social Sciences of the Goethe-University, Frankfurt/Main in Germany. His research interests include social and political theory, biopolitics, and social studies of genetic and reproductive technologies. Recent publications include *Suspect Families. DNA Analysis, Family Reunification and Immigration Policies* (co-edited with Torsten Heinemann, Ilpo Helén, Ursula Naue and Martin G. Weiss; Ashgate, 2015), *Perspectives on Genetic Discrimination* (Routledge, 2013), *Governmentality. Current Issues and Future Challenges* (co-edited with Ulrich Bröckling and Susanne Krasmann; Routledge, 2010), *Biopolitics. An Advanced Introduction* (New York University Press, 2011), and *Foucault, Governmentality and Critique* (Paradigm, 2011).

Ronit Lentin retired as Associate Professor of Sociology at Trinity College, Dublin. She has published on Israel/Palestine, racism and immigration in Ireland, and gender and genocide. Her books include: *Israel and the Daughters of the Shoah: Reoccupying the Territories of Silence* (Berghahn, 2000), *Women and the Politics of Military Confrontation: Palestinian and Israeli Gendered Narratives of Dislocation* (Berghahn, 2002), *Racism and Antiracism in Ireland* (Beyond the Pale, 2002), *Race and State* (Cambridge Scholars Press, 2006, 2008), *After Optimism? Ireland, Racism and Globalisation* (Metroeireann Publications, 2006), *Thinking Palestine* (Zed Books, 2008), *Co-Memory and Melancholia: Israelis Memorialising the Palestinian Nakba* (Manchester University Press, 2010) and *Migrant Activism and Integration from Below in Ireland* (Palgrave Macmillan, 2013).

Mark Maguire is Head of the Maynooth University Department of Anthropology. He currently works on technologies and processes of securitization, especially counterterrorism, biometric security, predictive policing, and the detection of abnormal behavior. He twice held visiting professorships in the Department of Anthropology, Stanford University, California. He is co-editor of *Social Anthropology/Anthropologie Sociale*, the leading anthropology journal in Europe. He is co-editor (with Catarina Frois and Nils Zurawski) of *The Anthropology of Security: Perspectives from the Frontline of Policing, Counter-Terrorism and Border Control* (Pluto, 2014).

Catherine Mills is an Associate Professor and Australian Research Council Future Fellow at the Centre for Human Bioethics, Monash University. She is primarily interested in issues in bioethics relating to reproductive and genetic technologies, along with aspects of contemporary European philosophy and feminist theory. She has published extensively on bio-politics, particularly in the work of Agamben and Foucault, as well as on concepts of responsibility and embodiment in relation to reproductive technologies. Her publications include *The Philosophy of Agamben* (Acumen Publishing/McGill-Queens UP, 2008) and *Futures of Reproduction: Bioethics and Biopolitics* (Springer, 2011).

Saskia Sassen is the Robert S. Lynd Professor of Sociology and chairs The Committee on Global Thought, Columbia University. Her new book is *Expulsions: Brutality and Complexity in the Global Economy* (Harvard University Press, 2014). Recent books are *Territory, Authority, Rights: From Medieval to Global Assemblages* (Princeton University Press, 2008), *A Sociology of Globalization* (W. W. Norton, 2007), and the fourth fully updated edition of *Cities in a World Economy* (Sage, 2011). Her books are translated into over twenty languages. She contributes regularly to www.OpenDemocracy.net and www.HuffingtonPost.com.

Margrit Shildrick is Professor of Gender and Knowledge Production at Linköping University, Sweden, and Adjunct Professor of Critical Disability Studies at York University, Toronto. Her research covers post-conventional philosophy, postmodern feminist theory, biotechnologies, bioethics, critical disability studies, and body theory. Her books include *Dangerous Discourses of Disability, Subjectivity and Sexuality* (Palgrave Macmillan, 2009), *Embodying the Monster* (Sage, 2002), and *Leaky Bodies and Boundaries* (Routledge, 1997), as well as several edited collections. Recent journal publications have focused on the phenomenology of organ transplantation, prosthetic performativity, and the posthuman. She is a key member of an interdisciplinary team researching the psycho-cultural meanings and significance of heart transplantation.

Eugene Thacker is the author of several books, including *After Life* (University of Chicago Press, 2010) and *In The Dust Of This Planet* (Zero Books, 2011). He is Associate Professor at The New School, New York.

S. E. Wilmer is Professor Emeritus of Drama and former Head of the School of Drama, Film and Music at Trinity College, Dublin. His recent publications include the co-edited (with Audronė Žukauskaitė) *Deleuze and Beckett* (Palgrave Macmillan, 2015) and *Interrogating Antigone in Postmodern Philosophy and Criticism* (with Audronė Žukauskaitė; Oxford University Press, 2010). He also edited *Native American Performance and Representation* (Arizona University Press, 2009) and *National Theatres* *in a Changing Europe* (Palgrave Macmillan, 2008) and co-edited (with Anna McMullan) *Reflections on Beckett* (University of Michigan Press, 2009).

Audronė Žukauskaitė is senior researcher at the Lithuanian Culture Research Institute and President of the Lithuanian Philosophy Association. Her recent publications include the monograph *Gilles Deleuze and Félix Guattari's Philosophy: The Logic of Multiplicity* (Baltos Lankos, 2011), and an edited volume, *Intensities and Flows: Gilles Deleuze's Philosophy in the Context of Contemporary Art and Politics* (LKTI, 2011). She also co-edited (with S.E. Wilmer) *Interrogating Antigone in Postmodern Philosophy and Criticism* (Oxford University Press, 2010) and *Deleuze and Beckett* (Palgrave Macmillan, 2015).

Ionat Zurr is an artist and a researcher at SymbioticA, the Centre for Excellence in Biological Arts, within the School of Anatomy and Human Biology, The University of Western Australia. Together with Oron Catts, she formed the internationally renowned Tissue Culture and Art Project, which explores the use of tissue technologies as a medium for artistic expression and addresses shifting perceptions of life. The Project began in 1996 and was central to the establishment of SymbioticA in 2000. Zurr is considered a pioneer in the field of biological arts; her research has been published widely and exhibited internationally and her artwork has been collected by MoMA, New York. She was recipient of the Discovery Australian Research Council Award (2012) and has been a Fellow in the InStem Institute, NCBS, Bangalore (2010), and a visiting scholar at The Experimental Art Center, Stanford University (2007) and the Tissue Engineering & Organ Fabrication Laboratory, Massachusetts General Hospital, Harvard Medical School (2000–2001). She has exhibited in places such as: the MoMA, New York; Mori Museum, Tokyo; Ars Electronica, Linz, Austria; and GOMA, Brisbane, among others.

Index

abiogenesis 139
Aboriginals 16; *see also* indigenous
    populations
abortion politics 13, 97, 117, 120; in
    Ireland 255, 264–5
Acoustic Mirror (Silverman) 209
activism 11, 15, 31, 53, 82, 133n3,
    248, 264
advanced capitalism 30–1, 34, 37–40,
    43–4, 46, 219–20, 233n2;
    expanding operational space of
    220–1; globalized 49; identifying
    characteristics of 40–2; and the
    logics of extraction 222–3
Advanced Research Projects Agency
    (ARPA) 165
affect programs 170n4; *see also*
    negative affects
affirmation 11, 35, 53; ethics of 30,
    51–2; of life 7; practice of 34–5;
    of rights 8, 88; of values 179
Africa: debt in 224–6; foreign
    acquisition of land in 227;
    foreign direct investment (FDI) in
    229; "scramble" for 239
African Development Fund 224
Agamben, Giorgio: on animality 10,
    81–2; on "bare life" 38, 135,
    259, 266, 273; on biopolitics
    1, 6, 38, 57, 75, 78, 161;
    on biopower 8, 38, 89–90,
    161, 278; on the biovirtual
    205; on colonialism 271; on
    concentration camps 260; on
    contemporary democracy 281;
    criticism of 7; on human rights
    74–5; influence on Mbembe
    57, 242; on the nation-state
    77; on necropolitics 49, 242;
    on political theology 35; on

potentiality 206; on power 7,
    192; on refugees 76–7, 279; on
    security 11, 13, 14; social theory
    of 38; on state of exception 6,
    15, 22, 272–3, 278, 279, 281;
    westocentrism of 272, 280
Agamben and Colonialism (Svirsky &
    Bignall) 272
agency 35, 37, 63, 67, 137, 142, 145–6,
    195, 197, 281
AIDS (HIV/AIDS) 44, 47
Akst, Jef 140
algorithms 129, 187, 196; genetic 127
amor fati 52–3
animal rights 1, 81–3
The Animal That Therefore I Am
    (Derrida) 83
animality 10, 81–2, 126, 130
Anthropocene 232
anthropocentrism 44, 47, 64,
    67, 82, 142; *see also* post-
    anthropocentrism
anthropological machine 81–2
anthropomorphism 46, 47, 126–7
anthropology 58, 166, 169; linguistic
    169n3
Appadurai, Arjun 48
Arabs, in Israel 15–16; *see also*
    Palestinian people
Arendt, Hannah 1, 38, 74, 75, 76, 79,
    238–9; on refugees 244
Aristotle 1, 83, 123, 125, 127, 130,
    238–9, 241, 249
Asia, foreign acquisition of land in
    227–8
assemblages 9–10, 12, 35, 36, 46,
    69n6, 96, 104, 131, 161,
    165, 169–70n2, 197, 241,
    243, 247; *see also* becoming;
    multiplicity(ies)